D0933913

ANIMAL MINDS

DONALD R. GRIFFIN

ANIMAL MINDS

Beyond Cognition to Consciousness

THE UNIVERSITY OF CHICAGO PRESS
CHICAGO AND LONDON

Donald R. Griffin has been a professor at Cornell, Harvard, and Rockefeller Universities and is now an associate of the Museum of Comparative Zoology at Harvard. His many books include *The Question of Animal Awareness* (1976), *Animal Thinking* (1984), *Listening in the Dark* (1958), *Echoes of Bats and Men* (1959), *Animal Structure and Function* (1962), and *Bird Migration* (1964).

The University of Chicago Press, Chicago 60637
The University of Chicago Press, Ltd., London

Printed in the United States of America

10 09 08 07 06 05 04 03 02 01   1 2 3 4 5

ISBN: 0-226-30865-0 (cloth)

Library of Congress Cataloging-in-Publication Data

Griffin, Donald R. (Donald Redfield), 1915–
Animal minds : beyond cognition to consciousness / Donald R. Griffin.
p. cm.
Includes bibliographical references (p.) and index.
ISBN 0-226-30865-0 (cloth : alk. paper)
1. Cognition in animals. 2. Animal behavior. 3. Animal psychology. I. Title.
QL785.G715 2001
591.5—dc21   00-010006

♾ The paper used in this publication meets the minimum requirements of the American National Standard for Information Sciences—Permanence of Paper for Printed Library Materials, ANSI Z39.48-1992.

To the Memory of

Jocelyn Crane

My Love and Inspiration

CONTENTS

PREFACE

Are animals consciously aware of anything, or are they all "zombies" incapable of conscious thoughts or emotional feelings? (I use the words *aware* and *conscious* in their ordinary meanings, as virtual synonyms, but chapter 1 includes a discussion of definitions.) This is a highly significant scientific question with philosophical and ethical implications. The difference between being totally unconscious and being aware of anything at all is basic, and regardless of the content of an animal's consciousness, the step from a nonconscious to a conscious state is crucial. Obviously, when one is unconscious the content of consciousness is zero.

Some are convinced that only human beings are capable of consciousness. But many others are equally certain that some animals must experience at least simple feelings and thoughts and that their behavior is sometimes guided by fear of dangers, a desire to get something good to eat, or a belief that food can be obtained in a certain place by a particular activity. A major philosophical issue is the problem of other minds, that is, how we can come to understand minds other than our own, including those of other species. Finally, our ethical judgments about how we should treat members of other species are strongly influenced by what we believe about their consciousness.

Very few of us would voluntarily turn ourselves irreversibly into unconscious robots with no mental experiences at all, as discussed at length by Siewert (1998), for what we perceive and feel, and what we think about, are important aspects of our lives. It is self-evident that we are aware of at least some of what goes on around us and that we think about our situation and about the probable results of various actions that we might take. This sort of conscious subjective mental experience is significant and useful because it often helps us select appropriate behavior. Even the simplest and most basic experiences, such as fear and affection, are often very intense, and this intensity may well make them as important to animals as ours are to us. And they are certainly important to our understanding of animals, for we can only appreciate other species fully when we know what, if anything, they think and feel. Thus the challenging scientific question is, how widespread among animals is conscious awareness?

Can scientific investigation of animal mentality tell us whether animals are conscious? The short answer is "not yet," because it is very difficult to gather convincing evidence about whatever conscious experiences may occur in animals. This difficulty has persuaded many of the scientists who study animal behavior—comparative psychologists, ethologists, and behavioral ecologists—that all discussion of animal consciousness is controversial, objectionable, and even immoral. One of my books about animal mentality (Griffin 1984) has been called "the Satanic Verses of Animal Behavior" (Davis, quoted by Thomas and Lorden, 1993). And some scientists attempt to "purge all biological discussions of mentalistic interpretations" (G. Williams 1997).

What has caused this yawning gulf between the commonsense opinion that some animals must experience at least simple thoughts and emotional feelings and the prevailing views of the scientists most directly concerned with animal behavior? Can this gulf be bridged, and if so, how? Have scientists proved conclusively that animals are never conscious, perhaps by means of evidence so complex and technical (like quantum mechanics) that ordinary people cannot understand it? No, almost all biologists and psychologists who study animal behavior avoid any such sweeping claim, and they often grant that some animals are probably conscious at times. But they hasten to argue that there is no way to tell whether they are or not, and that for this reason the subject cannot be investigated scientifically. Many scientists therefore have come to believe that to be concerned with consciousness of any kind, especially nonhuman consciousness, is sentimental and uncritical thinking to be avoided by serious scientists.

This restriction of inquiry has come to be needless and counterproductive, because twentieth-century investigations showed that many animals behave in ways that strongly indicate that they are aware of their situation and how their behavior can affect it. This revised and expanded edition of *Animal Minds* will explore what is known about animal mentality, including cognition, in the sense of information processing in the nervous system, and subjective, conscious awareness. Although the available evidence does not prove conclusively that any particular animal is conscious, it is quite sufficient to open our eyes to an appreciative view of animals in which we attempt to understand what life is like for them.

Curiosity about what life is like for members of other species is not at all new, but during most of the twentieth century such curiosity was strongly discouraged, even repressed, among scientists concerned with animal behavior, as discussed in chapter 2. This has become a sort of taboo that is seriously impeding investigation of important scientific problems. For, once scientists recognize the opportunities for study that will be explored in this book, they can gather objective data that will reduce our ignorance about what it is like to be particular animals under varying conditions. Consciousness is back in style, and there is no longer any reason for scientists to pussyfoot around the subject.

During the 1990s there has been a true renaissance in scientific investigation of consciousness, with an astonishing number of new journals, books, and scholarly journals discussing the subject from a variety of perspectives, for example, Baars (1988, 1997), Crick (1994), Flanagan (1992), Hameroff et al. (1996, 1998), Tomasello and Call (1997), Griffin (1998, in press), Macphail (1998), Rose (1998), Shettleworth (1998), Hendrichs et al. (1999), Page (1999), Heyes and Huber (2000), and Taylor (1999). But almost all of the emphasis has been on human consciousness, along with exciting new studies of brain mechanisms related to consciousness in both human subjects and monkeys. Very little attention has been paid to other animals equipped with central nervous systems made up of very similar neurons, synapses, and glial cells. The extent and nature of consciousness in the enormous variety of animals with which we share this planet is a fundamental and important scientific question that is only beginning to be addressed with the enthusiasm it deserves, as discussed by Allen and Bekoff (1997), Page (1999), Hendrichs et al. (1999), and Hauser (2000).

The customary claim of scientists that no objective, verifiable data about nonhuman consciousness can ever be obtained does not justify neglecting this important subject. I will argue that this view is mistaken and that it has severely restricted important scientific investigation, for there are in fact several kinds of evidence bearing on the question of animal consciousness, as will be discussed in later chapters. One type of evidence is especially relevant, and yet it has been almost completely neglected by scientists. This is animal communication. To appreciate its relevance, we need only ask ourselves how we judge whether our human companions are aware of anything or what the content of their conscious experiences may be. Our chief source of evidence comes from human communication. People often tell us about their conscious experiences, and in addition to their words we make use of nonverbal communication—gestures, tone of voice, and the whole range of "body language."

Communicative behavior is not a human monopoly. Many animals also communicate, in their own ways. When dogs growl, or when injured animals squeal in obvious pain, we routinely and reasonably infer that they are experiencing at least simple emotional feelings. Furthermore, ethologists have learned that some animals express not only basic emotions such as fear or aggressiveness but sometimes also specific semantic information about important matters such as approaching dangers. Therefore it is possible to employ with animals the same basic approach by which we gather evidence about what our human companions are feeling and thinking.

Just as human speech and nonverbal communication tell us most of what we know about the thoughts and feelings of other people, the interpretation and analysis of animal communication can provide us, figuratively speaking, with a window on animal minds. This approach to the question of animal awareness will be discussed in detail in chapters 9–12. Charles Darwin and many other students of animal behavior have interpreted animal communication as expression of emotions, but ethologists have recently discovered that some animals express both feelings and simple thoughts. The communicative signals by which they do so provide us with a promising source of objective data about their mental experiences. We can record and analyze these signals and learn from responses of recipients a great deal about the message conveyed. Experimental playback of recorded signals is often very helpful in identifying the specific information they convey. Such evidence is never totally complete or accurate, but it does provide objective data on which scientists can build.

Many scientists are convinced that without human language no animal can report about any conscious experiences it may have, and therefore we can never gather any convincing objective and verifiable evidence about nonhuman consciousness. To be sure, much of our nonverbal communication is involuntary, and it sometimes reveals feelings and thoughts we would prefer to keep to ourselves. But the question of volition, whether animals communicate intentionally, is a separate one from the question whether their communication expresses conscious thoughts or feelings. Animals may or may not intend to communicate to others, but in either case their messages may well reflect their feelings and thoughts. It is possible that *all* animal communication is comparable to our blushing when embarrassed. But the accumulating evidence of versatility in animal behavior, and especially of animal communication, that will be reviewed in this book renders this view less and less plausible.

The widely held opinion that animal communication never expresses any conscious experience implies that human communication, nonverbal as well as verbal, has some qualitative attribute that is completely lacking in other species. But there is no direct evidence for this assumption, and the difficulty of determining whether an animal is conscious cuts both ways. Animal awareness is as difficult to disprove as it is to prove.

Given the difficulties encountered by scientists searching for conclusive evidence about the content of an animal's conscious experiences, it is a reasonable and conservative position to remain agnostic and admit that we simply do not know. But there is a strong tendency to move from agnostic caution to dogmatically negative assertions or at least implications that animals are not conscious at all.

In a sharp break with the traditional conviction that the mental experiences of animals cannot be studied scientifically, some of us have begun to try. This is now timely because ethologists and comparative psychologists have learned so much about the versatility of animal behavior that we can now investigate animal cognition on the basis of extensive evidence that was not available in the nineteenth century to Darwin, Romanes, or Lloyd Morgan. The difficulties of obtaining convincing evidence remain formidable, partly because the subject has been neglected by the scientists best equipped to investigate it, as discussed by Beer (1992). It is therefore necessary to begin with a variety of intuitions and speculative suggestions that must be tried out in any way that appears promising, in order to learn which, if any, may develop into solid and effective scientific procedures, as discussed by Silverman (1983).

This is "pre-science," which is where scientific advances often begin. Scientists who accept only well-proven methods that can be confidently expected to yield unambiguous data will necessarily be disappointed, because this is unknown territory waiting to be explored, not well-mapped terrain where we know the main features and seek only to pin down specific details. It is not yet possible to formulate cookbook specifications stipulating just how to prove conclusively whether a particular animal is conscious. But I hope that those ready to face the challenges of exploring the significant unknown will be stimulated by the prospect of learning what life is like to our evolutionary kin.

As discussed in chapter 8, cognitive neuroscience has achieved remarkable understanding of the basic neural mechanisms of learning and memory, which are certainly important cognitive processes. One significant discovery is that explicit or declarative and conscious learning is based on cellular and molecular processes very similar to those that govern implicit or unconscious learning. Recent evidence indicates that *both* kinds of learning and memory occur in many animals. As Milner, Squire, and Kandel (1998, 451–54) summarize these discoveries: "Several characteristics have been useful in extending the notion of declarative memory to mice, rats and monkeys. . . . These include its flexibility and the ability to use it inferentially in novel situations. . . . Even though there are differences in detail, the anatomical and functional organization of the medial temporal-lobe system is similar in humans, nonhuman primates, and simpler mammals such as rats and mice. . . . Moreover even the mouse requires this memory system for the storage of memory about places and objects, and this type of memory has many of the characteristics of human declarative memory, affording, for instance, the flexible use of relational information about multiple distal cues. . . . An animal does not need a large brain or even many thousands of nerve cells for perfectly good long-term storage of a variety of different memories."

In an attempt to explore these important questions I have advocated, in the first edition of this book and elsewhere (Griffin, 1976, 1978, 1984, 1985a, 1985b, 1998), that the scientific investigation of animal behavior be expanded to include cognitive ethology—the comparative investigation of mental phenomena, including both unconscious and conscious mental states. It is obvious that complex adaptive behavior requires information processing within the nervous system. Just what information is processed, and how, is one important part of cognitive ethology. But a more fundamental and significant question is the degree to which this is accompanied or influenced by conscious awareness on the animal's part. Several recent books and reviews have concentrated on animal cognition with the strong bias that information processing, but not subjective awareness, is appropriate for scientific analysis, as will be discussed in chapter 2. Partly because of the widespread neglect of animal consciousness, this book will emphasize evidence bearing on its occurrence and content. The conclusions and interpretations of scientists are sharply divided, and in this revised edition I will try to review and interpret in a balanced fashion the full spectrum of these thoughtful contributions and strongly felt views. The intensity of many scientists' discomfort at the mention of animal consciousness is itself an interesting phenomenon that deserves further investigation.

Like traditional ethology, the cognitive approach is grounded on the recognition that the rich spectrum of attributes that have been adapted in biological evolution to enable animals to survive and reproduce include not only structures, functions, and behavior but also cognition. As we learn more about animal consciousness, it too may join the list. A barrier to scientific inquiry about these questions has been the firm conviction that consciousness is a uniquely human attribute, and for that reason it is futile to study it in animals. But this assumption has recently come to seem less and less plausible as more and more aspects of animal behavior have been found to depend on cognitive processes.

I have been pleasantly surprised to find how enlightening it is to review the extensive published literature on animal behavior from this cognitive perspective. Even two of the most severe critics of this approach have recognized that "The development of cognitive ethology has helped emphasize that animals routinely engage in behavior more complex than most ethologists or psychologists would have thought plausible" (Yoerg and Kamil 1991, 278). Thus we can at least hope to learn just how versatile animals actually are by considering what we know of their behavior from their perspectives. Furthermore, I expect that we will find it more and more plausible that they are sometimes consciously aware of their situation and of the likely results of their activities.

Since the first edition of *Animal Minds* was published in 1992, animal cognition has been extensively and sometimes heatedly discussed by scientists, even though animal consciousness has been neglected. But much of the new and relevant empirical evidence has been obtained by ingenious investigations of the versatility of animal behavior that also provide at least suggestive evidence of animal consciousness. Indeed so many pertinent publications are appearing every week that it has become impossible to review them all in this book. Chapter 1 will discuss the reasons why it seems probable that animals are sometimes conscious, and in chapter 2 I will do my best to present a balanced review of the views of those who believe either that no animals are ever conscious or that the subject is wholly inappropriate for scientific investigation, and to explore the weaknesses and limitations of these opinions. Chapters 3 through 12 will then consider several types of evidence that suggest conscious thinking and subjective emotional feelings. The two final chapters will discuss in greater depth some of the general issues and implications that arise from a consideration of animal mentality.

I am especially grateful to several colleagues who have allowed me to draw on their recent work, including material not yet published or in press. These include G. M. Burghardt, J. A. Fisher, M. D. Hauser, H. E. Hodgdon, A. C. Kamil, P. Marler, T. Natsoulas, I. M. Pepperberg, K. Pryor, D. G. Reid, C. A. Ristau, H. Ryden, P. Stander, and E. A. Wasserman. The symposia arranged and edited by C. A. Ristau (1991), M. Bekoff and D. Jamieson (1990), and Cartmill (in press) have been most stimulating and have led to many improvements. Gayle B. Speck has contributed endless corrections, clarifications, and encouragement for which I am especially grateful. Christie Henry and her colleagues at the University of Chicago Press have patiently and thoughtfully disciplined my writing. Finally, the detailed criticisms by Robert M. Seyfarth and G. M. Burghardt have led to many clarifications and improvements, and while not agreeing with all their recommendations, I am glad to acknowledge the beneficial results of their thoughtful efforts.

CHAPTER ONE

In Favor of Animal Consciousness

A hungry chimpanzee walking through his native rain forest comes upon a large *Panda oleosa* nut lying on the ground under one of the widely scattered Panda trees. He knows that these nuts are much too hard to open with his hands or teeth and that although he can use pieces of wood or relatively soft rocks to batter open the more abundant *Coula edulis* nuts, these tough Panda nuts can only be cracked by pounding them with a very hard piece of rock. Very few stones are available in the rain forest, but he walks 80 meters straight to another tree where several days ago he had cracked open a Panda nut with a large chunk of granite. He carries this rock back to the nut he has just found, places it in a crotch between two buttress roots, and cracks it open with a few well-aimed blows. (The loud noises of chimpanzees cracking nuts with rocks had led early European explorers to suspect that some unknown native tribe was forging metal tools in the depths of the rain forest.)

In a city park in Japan, a hungry green-backed heron picks up a twig, breaks it into small pieces, and carries one of these to the edge of a pond, where she drops it into the water. At first it drifts away, but she picks it up and brings it back. She watches the floating twig intently until small minnows swim up to it, and she then seizes one by a rapid thrusting grab with her long, sharp bill. Another green-backed heron from the same colony carries bits of material to a branch extending out over the pond and tosses the bait into the water below. When minnows approach this bait, he flies down and seizes one on the wing.

Must we reject, or repress, any suggestion that the chimpanzees or the herons think consciously about the tasty food they manage to obtain by these coordinated actions? Many animals adapt their behavior to the challenges they face either under natural conditions or in laboratory experiments. This has persuaded many scientists that some sort of cognition must be required to orchestrate such versatile behavior. For example, in other parts of Africa chimpanzees select suitable branches from which they break off twigs to produce a slender probe, which they carry some distance to poke it into a termite nest and eat the termites clinging to it as it is withdrawn. Apes have also learned to use artificial communication systems to ask for objects and activities they want and to answer simple questions about pictures of familiar things. Vervet monkeys employ different alarm calls to inform their companions about particular types of predator.

Such ingenuity is not limited to primates. Lionesses sometimes cooperate in surrounding prey or drive prey toward a companion waiting in a concealed position. Captive beaver have modified their customary patterns of lodge- and dam-building behavior by piling material around a vertical pole at the top of which was located food that they could not otherwise reach. They are also very ingenious at plugging water leaks, sometimes cutting pieces of wood to fit a particular hole through which water is escaping. Under natural conditions, in late winter some beaver cut holes in the dams they have previously constructed, causing the water level to drop, which allows them to swim about under the ice without holding their breath.

Nor is appropriate adaptation of complex behavior to changing circumstances a mammalian monopoly. Bowerbirds construct and decorate bowers that help them attract females for mating. Plovers carry out injury-simulating distraction displays that lead predators away from their eggs or young, and they adjust these displays according to the intruder's behavior. A parrot uses imitations of spoken English words to ask for things he wants to play with and to answer simple questions such as whether two objects are the same or different, or whether they differ in shape or color. Even certain insects, specifically the honeybees, employ symbolic gestures to communicate the direction and distance their sisters must fly to reach food or other things that are important to the colony.

These are only a few of the more striking examples of versatile behavior on the part of animals that will be discussed in the following pages. Although these are not routine everyday occurrences, the fact that animals are capable of such versatility has led to a subtle shift on the part of some scientists concerned with animal behavior. Rather than insisting that animals do not think at all, many scientists now believe that they sometimes experience at least simple thoughts, although these thoughts are probably different from any of ours. For example, Terrace (1987, 135) closed a discussion of "thoughts without words" as follows: "Now that there are strong grounds to dispute Descartes' contention that animals lack the ability to think, we have to ask just how animals *do* think." Because so many cognitive processes are now believed to occur in animal brains, it is more and more difficult to cling to the conviction that this cognition is never accompanied by conscious thoughts.

Conscious thinking may well be a *core function* of central nervous systems. For conscious animals enjoy the advantage of being able to think about alternative actions and select behavior they believe will get them what they want or help them avoid what they dislike or fear. Of course, human consciousness is astronomically more complex and versatile than any conceivable animal thinking, but the basic question addressed in this book is whether the difference is qualitative and absolute or whether animals are conscious even though the content of their consciousness is undoubtedly limited and very likely quite different from ours. There is of course no reason to suppose that any animal is always conscious of everything it is doing, for we are entirely unaware of many complex activities of our bodies. Consciousness may occur only rarely in some species and not at all in others, and even animals that are sometimes aware of events that are important in their lives may be incapable of understanding many other facts and relationships. But the capability of conscious awareness under some conditions may well be so essential that it is the sine qua non of animal life, even for the smallest and simplest animals that have any central nervous system at all. When the whole system is small, this core function may therefore be a larger fraction of the whole.

The fact that we are consciously aware of only a small fraction of what goes on in our brains has led many scientists to conclude that consciousness is an epiphenomenon or trivial by-product of neural functioning, as discussed by Harnad (1982). But the component of central nervous system activity of which we *are* consciously aware is of special significance, because it is what makes life real and important to us, as discussed in detail by Siewert (1998). Insofar as other species are conscious, the same importance may well be manifest. Animals may carry out much of their behavior quite unconsciously. Many may never be conscious at all. But insofar as they *are* conscious, this is an important attribute.

Although nonconscious information processing *could in theory* produce the same end result as conscious thinking, as emphasized by Shettleworth (1998) and others, it seems likely that conscious thinking and emotional feeling about current, past, and anticipated events is the best way to cope with some of the more critical challenges faced by animals in their natural lives. As pointed out by the philosopher Karl Popper (1987), what he termed “mental powers” are very effective in coping with novel and unpredictable challenges. This is especially true of many animals under natural conditions, where mistakes are often fatal. The effectiveness of conscious thinking and guiding behavioral choices on the basis of emotional feelings about what is liked or disliked may well be so great that this core function is one of the most important activities of which central nervous systems are capable.

The nature of animal minds was a major subject of investigation until it was repressed by behaviorism, as discussed in chapter 2. Darwin, Romanes, Lloyd Morgan, von Uexkull, and many other scientists of the nineteenth and early twentieth centuries were deeply interested in animal mentality. This history has been thoroughly reviewed by Schultz (1975), Wasserman (1981), Boakes (1984), Dewsbury (1984), and R. J. Richards (1987), and especially cogently by Burghardt (1985a, 1985b). What is new is the accumulated results of a century of active and successful investigation of animal behavior. These discoveries have now provided a wealth of data about the complexities and versatility of animal behavior under natural conditions, and what they can learn to do in the laboratory. We can therefore return to the investigation of animal minds with far better and more extensive evidence than what was available to nineteenth-century biologists.

I will take it for granted that behavior and consciousness (human and nonhuman) result entirely from events that occur in their central nervous systems. In other words, I will proceed on the basis of emergent materialism as analyzed by Bunge (1980, 6), Bunge and Ardilla (1987, 27), and Mahner and Bunge (1997, 205–12), and assume that subjective consciousness is an activity of central nervous systems, which are of course part of the physical universe. Just what sort of neural activity leads to consciousness remains a challenging mystery, as I will discuss in detail in chapter 8. But there is no need to call upon immaterial, vitalistic, or supernatural processes to explain how some fraction of human or animal brain activity results in conscious, subjective thoughts and feelings.

Defining Consciousness

No one seriously denies that we experience conscious thoughts and subjective feelings, even though we cannot describe them with complete accuracy and therefore no one else can experience them *exactly* as we do. The question under consideration in this book is the extent to which nonhuman animals also experience something of the same general nature as the subjective feelings and conscious thoughts that we know at first hand. The content of an animal's conscious experience may be quite different from any human experience. It may ordinarily be limited to what the animal perceives at the moment about its immediate situation, but sometimes its awareness probably includes memories of past perceptions or anticipations of future events. An animal's understanding may be accurate or misleading, and the content of its thoughts may be simple or complex. A conscious animal must often experience some feeling about whatever engages its attention. Furthermore, any thinking animal is likely to guide its behavior at least partly on the basis of the content of its thoughts, however simple or limited these may be.

In their *Dictionary of Ethology* Immelmann and Beer (1989) define animal consciousness as "immediate awareness of things, events, and relations," but they hasten to add the conventional behavioristic claim that "statements about the nature of this awareness in comparison to that of humans are dismissed by tough-minded scientists as idle speculation." But speculation is where scientific investigation often begins, and I hope to stimulate new and enterprising inquiries that will significantly reduce our current ignorance. Consciousness is not a neat homogeneous entity; there are obviously many kinds and degrees of consciousness. Many scientists (for example, Hauser 2000, xiii) feel that terms such as *consciousness* are too vague and slippery to be useful in scientific investigation, and they are very reluctant even to begin talking about the possibility of nonhuman consciousness without first settling on clear-cut definitions. On the other hand, Francis Crick (1994, 20) is not so easily inhibited:

> Everyone has a rough idea of what is meant by consciousness. It is better to avoid a *precise* definition of consciousness because of the dangers of premature definitions. Until the problem is understood much better, any attempt at a formal definition is likely to be either misleading or overly restrictive or both. If this seems like cheating, try defining for me the word *gene*. So much is now known about genes that any simple definition is likely to be inadequate. How much more difficult, then, to define a biological term when rather little is known about it.

Semantic Piracy

An unnecessary confusion has arisen concerning the meanings attached to terms such as *awareness, emotion, mind,* and *conscious.* In ordinary usage they denote conscious mental states, but many scientists justify their aversion to these terms and the concepts they designate by arguing that they cannot be defined with the precision necessary for scientific analysis. This has led to a sort of semantic piracy committed by defining mental states in essentially behavioristic terms. Such terms as *positive* or *negative emotion* replace *like* or *fear* when a living or even a nonliving system is more likely to respond in one way than another, ruling out implicitly by choice of terms subjective experiences such as liking or fearing. Crist

(1996, 1998, 1999) has lucidly analyzed the pervasive effect of this "mechanomorphic" terminology. Thinking is often redefined explicitly or implicitly as information processing. Calling it cognition has become a popular way to study animal mentality without recognizing that at least some of the cognition is probably accompanied, and influenced, by conscious subjective experiences.

The Content of Animal Consciousness

Animal thoughts and emotions presumably concern matters of immediate importance to the animals themselves, rather than kinds of conscious thinking that are primarily relevant to human affairs. Consciousness is not an all-or-nothing attribute. It varies widely within our species, and it would be remarkable if the content of every animal's consciousness were identical. Conscious thinking and strong emotional feelings can ordinarily deal with only one or a very few things at a time. Large, complex animals must also be able to organize and retain information about innumerable perceptions and potential actions of which only one or a very few can be the focus of conscious awareness at any one moment.

Recognizing that an animal's consciousness may be quite different from any human thoughts and feelings makes the problem of identifying and analyzing it more difficult. We may, however, tend to exaggerate this difficulty, because many basic concerns are likely to be very similar for most animals that have any conscious experiences at all. Hunger and desire for food, fear of dangers, affection, and hatred may well have much in common for any animal that experiences them. Once we give up the implicit assumption that any conscious experiences of other animals must be a subset of human experiences, we are faced with the difficulty of determining, or indeed even clearly imagining, what such nonhuman experiences are actually like to the animals who experience them. For instance, the range of odors that various animals can distinguish, and to which they react in different ways, suggests variations in subjective emotional feelings that odors may elicit. If a particular species of animal is capable of some type of conscious experience that our brains cannot generate, how can we ever hope to learn what it is? Although total and perfect understanding seem at first to be beyond our reach, enterprising investigation can probably achieve significant if incomplete understanding.

Sensory and perceptual capabilities are an obvious category in which interspecies differences are probably large and important and in which we can anticipate a beginning of such investigation. As discussed in chapter 2, the philosopher Thomas Nagel (1974) aroused great interest by raising the question "What is it like to be a bat?" He chose this example because bats' reliance on echolocation, detecting obstacles and capturing rapidly moving prey by hearing echoes of one's own sounds, seems at first glance so remote from any human experience as to be truly beyond our ken. But this specific example may not be as telling as it seems on first hearing. For blind people also carry out a form of echolocation, detecting objects by hearing echoes of sounds they make. To be sure, human echolocation is severely limited compared to that of bats and dolphins with respect to the types of targets that can be detected and discriminated from each other, and especially with respect to the ability to distinguish relevant echoes from competing sounds. But the fact remains that despite enormous differences in resolution and practical usefulness many blind people, and also well-practiced blindfolded subjects, do make good use of echolocation, as reviewed by Griffin (1958) and Rice (1967a, 1967b).

The case of human echolocation has an intriguing and perhaps significant further ramification. Many blind people succeed quite well at echolocation but do not realize that they are detecting objects by their sense of hearing. Yet if they are prevented from making any sounds, if their hearing is blocked, or if they are subjected to masking noise, they lose almost completely their previous ability to detect obstacles and avoid collisions with them, as demonstrated by the meticulous experiments of Supa, Cotzin, and Dallenbach (1944). But this specialized type of auditory discrimination is not wholly divorced from consciousness. Many other skillful users of echolocation are well aware that they detect obstacles by hearing some difference in the sound of their voices, footsteps, cane taps, or other sounds they have found to be helpful in finding their way. Echolocation is of course only one case in which sensory or perceptual input channels to the central nervous system differ from those with which we are familiar. But it is an instructive example of how critical scientific investigation can lead to understanding of behavioral and subjective experiential phenomena that at one time appeared mysterious and inexplicable.

It is important to distinguish between perceptual and reflective consciousness. The former, called "primary consciousness" by Farthing (1992), Lloyd (1989), and others, includes all sorts of awareness, whereas the latter is a subset of conscious experiences in which the content is conscious experience itself. Reflective consciousness is thinking, or experiencing feelings, about thoughts or feelings themselves, and it is often held to include self-awareness and to be limited to our species. I will return later in this chapter and in chapter 14 to some of the important ramifications of this distinction. But most of the evidence reviewed in this book suggests perceptual rather than reflective consciousness, and to avoid tedious repetition of *perceptual* I will use the term *consciousness* to mean this rather than implying that in particular cases the animal is capable of both categories of conscious thinking.

Recognizing our ignorance is a necessary first step toward reducing it. The customary view of animals as always living in a state comparable to that of human sleepwalkers is a sort of negative dogmatism. Because we know far too little to judge with any confidence when animals are or are not conscious, the question of animal consciousness is an open one, awaiting adequate scientific illumination. There is of course no reason to suppose that other animals are capable of the enormous variety of thinking that our species has developed, largely through the use of our magnificent language—especially written language, which allows the dissemination and preservation of knowledge far beyond what can be achieved by direct communication and individual memories. The principal difference between human and animal consciousness is probably in their *content.*

Approaches to Animal Awareness

In spite of all the reasons advanced to justify a neglect of animal consciousness, a few scientists have recognized that we cannot adequately appreciate animals without understanding their subjective experiences. For example, the psychologist Walker (1983) edged closer to doing so than most psychologists when he reviewed extensive experimental evidence concerning animal learning and problem-solving in a book titled *Animal Thought.* He summarized his conclusions as follows: "Some kind of mental activity is being attributed to the animals: that is, there is considered to be some internal sifting and selection of information rather than simply the release of responses by a certain set of environmental conditions. Knowledge of goals, knowledge of space, and knowledge of actions that may lead to goals seem to be independent, but can be fitted together by animals when the need arises" (81). "Our organ of thought may be superior, and we may play it better, but it is surely vain to believe that other possessors of similar instruments leave them quite untouched" (388). Although Walker tip-toed cautiously around the question of nonhuman consciousness, he did push at the behavioristic fence.

The comparative psychologist Roitblat (in Roitblat and Meyer 1995, 3) advocates "a perspective on animal behavior . . . that views animals not as passive reflex devices, but as active information processors, seeking information in their environment, encoding it, and using it for their benefit in flexible and intelligent ways." But he holds back from explicit recognition that seeking information and using it might sometimes be conscious processes. On the other hand, Burghardt (1997) has recently recommended that scientific ethology be expanded to include the private experiences of animals. He proposes that understanding the subjective experiences of animals be recognized as a fifth basic objective in addition to the four that Tinbergen advocated (proximal mechanisms of causation and control, ontogeny, evolutionary history, and survival value).

Because private experiences include subjective feelings as well as cognition, Burghardt (1997) feels that "cognitive ethology" is not a sufficiently inclusive term. When I first suggested this term (Griffin 1976, 102) it seemed appropriate to emphasize that animal behavior could not be adequately understood without considering animal cognition. But over the past quarter-century we have come to appreciate more fully that in most cases subjective emotional feelings are at the heart of human mental experiences, and very likely those of animals as well, as discussed by M. Dawkins (1993) and Damasio (1999). To the extent that animals make conscious choices about their actions, this will usually entail deciding to do what they perceive as helpful in achieving a desired goal, and the desire often seems to be intense, whether the animal wants something or strives to avoid it. Subjective feelings and emotions are thus an important part of cognitive ethology. Limiting it to nonconscious information processing, or to thoughts devoid of emotional importance to the animal itself, is an obsolete relic of behaviorism.

An increasing part of the renaissance of scientific concern with consciousness is renewed attention to emotional feelings as well as "factual" thinking about objects and events. For example, Panksepp (1998), Damasio (1994, 1998, 1999), and LeDoux (1996) have reviewed recent analyses of the neural correlates of emotion. As usual, most of the scientific attention has been directed at human emotions. But Cabanac (1999, 176) has boldly proposed an objective and evolutionary approach based on "emotional fever." Because gentle handling produces an increase in body temperature and heart rate in rats, mice, and lizards but not in frogs or goldfish, he suggests that "the first elements of mental experience emerged between the amphibians and reptiles." This is a sweeping conclusion to base on a few simple physiological measurements, but the search for physiological correlates of mental experiences must begin somewhere.

It not at all clear, however, whether increase in heart rate or the set point for body temperature is a reliable correlate of emotional experience. And the conclusion that emotional feelings are totally absent in all amphibians and fishes seems rather premature. There is a tendency to take some well-studied reflex, such as insect-catching by frogs, as representative of all behavior of members of an entire group of animals. Sensory and neural mechanisms of insect-catching by frogs have been analyzed in detail, and because frogs will also catch small inedible moving objects this behavior has come to be seen as an inflexible reflex. But the courtship behavior of frogs or nest-building fishes involves more versatile behavior than catching flies.

The ethologist Marian Dawkins (1993, 1998) has contributed significantly to the recognition of the central importance of emotions in cognitive ethology. She has developed procedures by which an animal's preferences can be evaluated by allowing it to choose between different environments. More recently she has critically analyzed evidence for animal consciousness and the difficulties of determining whether animals have conscious experiences. Her basic conclusion is that at least mammals and birds are probably conscious at times but that there are many pitfalls that must be carefully avoided in scientific attempts to determine the existence and content of animal consciousness. She sums up the balance of evidence as follows: "Our near-certainty about (human) shared experiences is based, amongst other things, on a mixture of the complexity of their behavior, their ability to 'think' intelligently and on their being able to demonstrate to us that they have a point of view in which what happens to them *matters* to them. We now know that these three attributes—complexity, thinking and minding about the world—are also present in other species. The conclusion that they, too, are consciously aware is therefore compelling. The balance of evidence (using Occam's razor to cut us down to the simplest hypothesis) is that they are and it seems positively unscientific to deny it" (1993, 177).

Three contributors to a recent symposium volume (Dol et al. 1997) have argued that many animals have conscious experiences of some sort, although other contributors disagree, as will be discussed in chapter 2. Van der Steen (1997) considers consciousness a heterogeneous "umbrella concept" but concludes that although the subject is a difficult one it can be studied scientifically and that some animals are probably conscious at times. Meijsing (1997, 57) reviews the diverse views of scientists and philosophers and concludes: "From an evolutionary point of view, as soon as there is locomotion there is perceptual awareness and as soon as there is perceptual awareness there is self-awareness (meaning awareness of an animal's own body)." A somewhat similar view has been developed by Sheets-Johnstone (1998). And Wemelsfelder (1997, 79) argues that subjective experience is not "hidden" but is expressed in behavior: "Attention is not a by-product of the ability to process information, it forms the very condition for that ability, enabling the animal to evaluate and apply acquired information in flexible and adaptive manner." These views, however, fail to recognize that much complex human behavior takes place without any awareness of what our bodies are doing, so that additional evidence is needed to distinguish between complex and adaptive behavior and behavior that indicates consciousness. Recognizing that we cannot be certain which animals are conscious, R. Bradshaw (1998, 108) and others recommend that when issues of animal welfare are concerned it is best to "assume animals do have consciousness in case they do; if they do not it does not matter."

The taboo against considering subjective experiences of nonhuman animals has become such a serious impediment to scientific investigation that it is time to lay it aside and begin the difficult task of investigating the subjective experiences of nonhuman animals, as recently advocated by Burghardt (1997). To accomplish this we will have to overcome the effective indoctrination—often accomplished by nonverbal signals of disapproval—that has inhibited students and young scientists from venturing into this forbidden territory and subjected those who do so to criticism and ridicule, as exemplified by the strong opinion of Boakes (1992) quoted in chapter 2. One result is that students of animal behavior are inhibited from reporting versatile behavior that suggests conscious thinking, and scientific journals sometimes refuse to publish data or interpretations that support the inference of animal consciousness, as described by Searle (1990a, 1990b), Whiten and Byrne (1988), and Heinrich (1995, 1999).

Probabilities of Consciousness

One helpful approach to the very challenging problems of investigating nonhuman consciousness is to think in terms of likelihood or probability. We can define pA as our estimate of the probability of conscious awareness in a given case. This might be the probability that a given animal was aware of a particular object, event, or relationship, for instance, the likelihood that it sees another animal in the underbrush or that it is a predator stalking potential prey. Or we might think in terms of the probability that a given species is capable of experiencing consciously a certain type of thought or feeling. For example, many scientists believe that animals can be aware of the behavior of others but cannot think about their beliefs and intentions.

If we are absolutely certain that a particular animal is consciously aware of something, pA would be 1.0. If we are equally sure it is not, we would find pA to be 0. If we believe that it is absolutely impossible to make any such judgment, which seems to be the position of many behavioral scientists, we should set pA at precisely 0.5. A cognitive ethologist would ideally like to determine whether pA is 0 or 1 in particular cases. An adamant behaviorist would find it is impossible to assign any value other than 0.5. Realistically, however, we can escape from this dilemma by recognizing that although we cannot yet assign any meaningful quantitative values to pA, we can make reasonable inferences about the general range of values that are most likely to be correct. For instance, when the bonobo Kanzi uses his keyboard to answer spoken questions as to where he wishes to go, and then goes there enthusiastically but resists efforts to lead him in another direction, it is reasonable to estimate that pA is close to 1. And in many other cases observational or experimental evidence makes it very likely indeed that pA is 0.

At present we can only make rough estimates of pA for most intermediate cases. The simplest judgment is whether pA is below or above 0.5, and opinions will of course differ, depending on the relative weight we assign to various types of indicative evidence. But thinking in terms of pA and its plausible value can help clarify our analysis of these extremely difficult issues. It may also be helpful to extend this approach by recognizing that our judgments of pA are necessarily subject to much uncertainty, so that any numbers we choose should be taken as approximations, and such numbers might well always be expressed with some indication of this uncertainty. Attempts to assign quantitative values to pA are rather premature at this time, because we have little firm evidence on which to base them. Perhaps the best we can do in most cases is to think in terms of only three ranges of values for pA: below 0.5, about 0.5, and above 0.5.

Evidence Suggesting Animal Consciousness

There are several types of scientific evidence that provide promising insights into what life is like for various animals. One category of evidence is the versatility with which many animals adjust their behavior appropriately when confronted with novel challenges. Animals encounter so many unpredictable challenges under natural conditions that it would be very difficult if not impossible for any combination of genetic instructions and individual experience to specify in advance the entire set of actions that are appropriate. But thinking about alternative actions and selecting one believed to be best is an efficient way to cope with unexpected dangers and opportunities. In theory such versatility might result from nonconscious information processing in the brain. But conscious thinking may well be the most efficient way for a central nervous system to weigh different possibilities and evaluate their relative advantages.

A second major category of promising evidence about animal thoughts and feelings is their communicative behavior, which will be discussed in detail in chapters 9–12. And a third type of evidence is available from neuropsychology. For what little is known about the neural correlates of conscious, as opposed to nonconscious, thinking does not suggest that there is anything uniquely human about the basic neural structures and functions that give rise to human consciousness. Chapter 8 will analyze this evidence in some detail.

The "Hard Problem"

The lack of definitive evidence revealing just what neural processes produce consciousness has led Chalmers (1996) to designate the question of how brains produce subjective awareness as the "hard problem." He and others claim that it is such a difficult problem that normal scientific investigation is unable, in principle, to solve it, and that consciousness must be something basically distinct from the rest of the physical universe. But this view, fortunately, has not seriously interfered with a striking renaissance in the scientific investigation of consciousness. In the 1990s numerous neuroscientists, psychologists, philosophers, and others have taken up active investigation and discussion of consciousness. Even distinguished molecular biologists such as Edelman (1989), Edelman and Tononi (2000), and Crick (1994) have joined the quest. One of the most inclusive of several international conferences devoted to consciousness and related subjects led to two massive volumes edited by Hameroff et al. (1996, 1998). General reviews of this reawakened concern with consciousness have been published by Crook (1980, 1983, 1987, 1988), Baars (1988, 1997), Chalmers (1996), Flanagan (1992), and Searle (1992, 1998).

Although much of our behavior takes place without any awareness, and this includes most of our physiological functions and the details of such fairly complex actions as coordinated locomotion, the small fraction of which we *are* aware is certainly important. It is a completely reasonable and significant question whether members of other species experience anything, although the content of their conscious experiences is likely to be quite different from ours. Perhaps we can never discover *precisely* what the content of nonhuman experiences are, because scientific understanding is seldom complete and perfect. But it seems probable that we can gradually reduce our current ignorance about this significant aspect of life.

The Comparative Analysis of Consciousness

Despite the renaissance of scientific and philosophical interest in consciousness, one of the most significant and promising approaches to the general question has been largely neglected. This is what biologists call the comparative method: analyzing an important function in a variety of species in which it occurs, sometimes in simpler forms, in which it can be studied more effectively without the many interacting complications that obscure its basic properties in the more complicated animals. Crick and Koch (1998), leaders in the renewal of scientific studies of consciousness, take it for granted that monkeys are conscious. But they prefer to defer investigating nonhuman consciousness because they claim that "when one clearly understands, both in detail and in principle, what consciousness involves in humans, then will be the time to consider the problem of consciousness in much simpler animals" (97).

Restricting scientific investigation to the most complex of all known brains may be unwise, however, for insofar as consciousness can be identified and analyzed in a variety of animals, certain species might turn out to be especially suitable for investigating its basic attributes. Obvious analogies are the use of fruit flies for investigations of genetics, squid giant axons for analyzing the biophysics of nerve conduction, laboratory rats and pigeons for studies of learning, and *Aplesia* for detailed analysis of the cellular and molecular basis of learning. It would have been unwise for the early investigators of genetics or learning to limit their research to primates, and the same may be true for contemporary and future studies of consciousness.

Perceptual and Reflective Consciousness

The psychologist Natsoulas (1978) emphasized a major distinction that is often overlooked. One widespread and important meaning is what he designates as Consciousness 3, following the *Oxford English Dictionary:* "the state or faculty of being mentally conscious or aware of anything." This Natsoulas calls "our most basic concept of consciousness, for it is implicated in all the other senses. One's being conscious, whatever more it might mean, must include one's being aware of something" (910). Another important meaning is what Natsoulas, again following the *OED,* calls Consciousness 4, which he defines as "the recognition by the thinking subject of his own acts or affections. . . . One exemplifies Consciousness 4 by being aware of, or by being in a position to be aware of, one's own perception, thought, or other occurrent mental episode" (911). The other shades of meaning analyzed by Natsoulas (1983, 1985, 1986, 1988) are less important for our purposes, but these two impinge directly on the issues that will be discussed in this book.

Natsoulas' Consciousness 3 is similar to conscious perception, although its content may entail memories, anticipations, or imagining nonexistent objects or events, as well as thinking about immediate sensory input. An animal may think consciously about something, as opposed to being influenced by it or reacting to it without any conscious awareness of its existence or effects. It is convenient to call this *perceptual consciousness.* Consciousness 4, as defined by Natsoulas, entails a conscious awareness that one is thinking or feeling in a certain way. This is conveniently called *reflective consciousness,* meaning that one is aware of one's own thoughts as well as the objects or activities about which one is thinking. It is a form of introspection, thinking about one's thoughts, but with the addition of being able to think about the thoughts of others. The distinction between perceptual and reflective consciousness is important for the sometimes confused (and almost always confusing) debate among scientists about animal consciousness. Many scientists use the unqualified term *conscious* to mean reflective consciousness and imply that perceptual consciousness is not consciousness at all. On the other hand, the existence and distribution of relatively simple perceptual consciousness is important in its own right, and when we understand it better we will be in a much better position to investigate whether any animals are also capable of reflective consciousness.

Many behavioral scientists such as Shettleworth (1998) and philosophers such as Lloyd (1989, 186) believe that it is likely that animals may sometimes experience perceptual consciousness but that reflective consciousness is a unique human attribute. The latter would be much more difficult to detect in animals, if it does occur. People can tell when they are thinking about their own thoughts, but it has generally seemed impossible for animals to do so, although animal communication may sometimes serve the same basic function. The very difficulty of detecting whether animals experience reflective consciousness should make us cautious about concluding that it is impossible. Most of the suggestive evidence that will be discussed in this book points toward perceptual rather than reflective consciousness. Those swayed by a visceral feeling that some important level of consciousness *must* be restricted to our species may cling to reflective consciousness as a bastion still defended by many against the increasing evidence that other animals share to a limited extent many of our mental abilities. The related question of self-awareness will be discussed in chapter 14.

The relation between these two general categories of consciousness can be illustrated by considering a class of intermediate cases, namely, an animal's awareness of its own body—for example, the appearance of its feet or the feeling of cold as a winter wind ruffles its fur. This tends to become an intermediate category between perceptual and reflective consciousness, for an animal might be consciously aware not only of some part of its body but also of what that structure was doing. It might not only feel its teeth crunching on food but also realize that it tastes good. Or it might not only feel the ground under its feet but also recognize that it is running in order to escape from a threatening predator. Furthermore, an animal capable of perceptual consciousness must often be aware that a particular companion is eating or fleeing. This means that it is consciously aware of both the action and of who is performing it. These would all be special cases of perceptual consciousness.

This leads to inquiring how likely it is that such an animal would be incapable of thinking that it, itself, was eating or fleeing. If we grant an animal perceptual consciousness of its own actions, the prohibition against conscious awareness of who is eating or fleeing becomes a somewhat strained and artificial restriction. Furthermore, a perceptually conscious animal could scarcely be unaware of its own enjoyment of eating or its fear of the predator from which it is trying desperately to escape. One could argue that perceptually conscious animals are aware of their actions but not of the thoughts and feelings that motivate them. But emotional experiences are often so vivid and intense that it seems unlikely that when an animal is conscious of its actions it could somehow be unaware of its emotions. We might well pause at this point and ask ourselves whether it is really plausible to claim that an animal can consciously experience emotional feelings and simple thoughts but never be aware that it, itself, is having these experiences. Animals must often feel afraid, but are they incapable of thinking about their fear? Are they sometimes conscious of their actions but never of the thoughts and feelings that motivate them?

Consider the case of an animal that barely escapes from the attack of a predator. It was surely frightened at the time, and if it later sees the same predator it presumably remembers both the event and how frightening it was. If so, this memory of its experience of fear would be a case of thinking about one of its emotional experiences. If we grant such an animal the ability to experience perceptual consciousness of a remembered event, how reasonable is it to rule out the possibility of simple reflective awareness of its remembered fear? To continue with this example, suppose that in its previous narrow escape this animal succeeded only because at the last moment it remembered that it could squeeze into a particular cavity. In this emergency it had been perceptually conscious that a certain action, squeezing between the roots of a particular tree, would get it into a safe retreat. Suppose on the following day it sees the same predator in the same area. Our animal would probably remember not only its fear but the tactic that had saved its life the day before. This would be recalling not only the emotional state of fear but also the simple thought of how to escape. If we insist that all animals are incapable of even such simple sorts of reflective consciousness, we are in effect postulating that a perceptual "black hole" encompasses their most intimate and pressing experiences.

Summary

It is self-evident that we sometimes think about our situation and about the probable results of various actions that we might take; that is, we plan and choose what to do. This sort of conscious subjective mental experience is significant and useful because it often helps us select appropriate behavior; thus mental experiences are "local causes" of behavior, although of course they, in turn, are influenced by prior events.

Animals are best viewed as actors who choose what to do rather than as objects totally dependent on outside influences, although their choices are often constrained within quite narrow limits. Especially when they try to solve newly arisen challenges by adjusting their behavior in versatile ways, their choices are probably guided by simple conscious thoughts, such as fear of dangers, a desire to get something good to eat, or a belief that food can be obtained in a certain place or by a particular activity.

The difficult but important questions about animal mentality can best be approached from the viewpoint of a materialist who assumes that mental experiences result from physiological processes occurring in central nervous systems. These processes, and relationships among them, are neither tangible objects nor immaterial essences. They appear to be roughly analogous in this respect to homeostasis, which is an important physiological process but one that cannot be pinned down to a specific structure. In the scientific investigation of animal minds there is no need to call on immaterial factors, vitalism, or divine intervention. We know that our central nervous systems are capable of producing conscious experience on occasion, and nothing yet learned about neuroscience precludes the possibility that other nervous systems can achieve something of the same general kind. Of course, this does not mean that all consciousness is identical or that the experiences of other species come close to rivaling the versatility, breadth, and complexity of ours. The great reluctance to consider the possibility that a wide range of animals may experience at least simple conscious thoughts stems in part from a general opinion that consciousness is something immaterial. This helps explain why cognition has come to be accepted but consciousness has not.

Communicative behavior of animals can serve the same basic function as human verbal and nonverbal communication by expressing at least some of an animal's private experiences. Therefore cognitive ethologists can gather verifiable, objective data about some of the private experiences of communicating animals by interpreting the messages that they convey to others. It is often claimed that although language allows us to obtain significant (though imperfect and incomplete) evidence about the thoughts of our human companions, animals lack language and therefore this source of information is not available. But, as discussed in detail in chapters 9 and 10, animal communication is much richer than we used to believe, and this supposed barrier crumbles once we are prepared to listen.

In view of the likelihood that all or at least a wide range of animals experience some form of subjective conscious awareness, it is both more parsimonious and more plausible to assume that the difference between human and other brains and minds is the *content* of conscious experience. This content of consciousness, what one is aware of, surely differs both qualitatively and quantitatively by astronomical magnitudes. Rather than an absolute all-or-nothing dichotomy between human brains uniquely capable of producing conscious experience, on one hand, and all other brains that can never do so, on the other, this hypothesis is consistent with our general belief in evolutionary continuity.

Out of all these multiple cross-currents of ideas three categories of evidence stand out as the most promising sources of significant, though incomplete, evidence of conscious thinking by nonhuman animals, and they will be reviewed in the following chapters:

1. Versatile adaptation of behavior to novel challenges will be considered in chapters 3–7.
2. Physiological evidence of brain functions that are correlated with conscious thinking are discussed in chapter 8.

3. Most promising of all is the communicative behavior by which animals sometimes appear to convey to others at least some of their thoughts; this will be the subject of chapters 9–12.

Finally, after all this suggestive evidence has been reviewed, it will be appropriate to consider in chapters 13 and 14 several general questions that are relevant to the question of animal mentality.

CHAPTER TWO

Objections and Their Limitations

Many people find it difficult to understand why so many behavioral scientists are so adamantly reluctant to consider animal consciousness. The historical reasons, thoroughly reviewed by Boakes (1984), Burghardt (1973, 1985), Dewsbury (1984), and Rollin (1989, 1990), have involved a reaction against excessively generous nineteenth-century interpretations of animal behavior, such as those of George Romanes. Coupled with this rejection of scientific investigation of mental experiences was an equally fervent rejection of the idea that genetic influences, as opposed to learning, had a significant effect on human or animal behavior. Although this is logically a wholly different matter from the significance of subjective mental experiences, the two issues have tended to be linked in theoretical consideration of animal behavior and mentality, as discussed by Burghardt (1978).

In psychology these distinct but closely coupled trends were combined in behaviorism, as advocated most influentially by Watson (1929) and Skinner (1974). During the same period a strong reductionist tradition also developed in biology, typified by the views of Loeb (1912), who, although he believed in the existence of human consciousness, argued that all animal and even much human behavior could be explained in terms of relatively stereotyped "forced movements," or tropisms. In their zeal for objective proof of any claims about animal behavior or mentality, most of the psychologists who established the long-dominant behaviorist movement insisted on three major points:

1. Learning and other factors operating during an animal's individual lifetime account for almost all behavior not directly controlled by its structural capabilities. Questions about genetic influences on behavior often used to be deflected with the quip "Of course rats can't fly."
2. Only external influences and directly observable behavior should be considered in explaining what animals do; behavioral scientists

should limit their concern to observable inputs to and outputs from the black box called an organism. And

3. Subjective mental experiences, especially conscious thinking, should be ignored for two reasons:
   A. They are unmeasurable "private" phenomena, perceptible only by the one who experiences them, so that statements about them cannot be independently verified, and
   B. They have no influence on behavior, and are thus incidental by-products of brain function, or epiphenomena.

Claim 1 of behaviorism has been largely abandoned, although it was often vigorously defended by many behavioral scientists with much the same fervor as that currently directed against suggestions that mental experiences may occur in animals and exert some influence on their behavior. Extensive evidence shows that what animals learn is strongly constrained by species-specific capabilities; some behavior patterns are learned much more easily than others that the animal is quite capable of performing, as reviewed by Shettleworth (1972). Furthermore, the widespread interest in sociobiology and behavioral ecology has led to a strong emphasis on the adaptive value of behavior: how it increases the likelihood of an animal's survival and reproduction and hence its evolutionary fitness.

Claim 2 has also been greatly modified owing to the cognitive revolution in psychology. As summarized by Roitblat et al. (1984, 1): "Animal cognition is concerned with explaining animal behavior on the basis of cognitive states and processes, as well as on the basis of observable variables such as stimuli and responses. For a time it appeared, at least to some, that discussion of cognitive states was not necessary, either because they were exhaustively determined by environmental events, or because they were epiphenomenal and without any causal force. In any case, it was assumed that a sufficiently detailed description of overt events would suffice for explanation. A great deal of the research into animal behavior has made it clear, however, that such cognitive states are real and necessary components of any adequate theory that seeks to explain animal behavior." These cognitive states "include learning, remembering, problem solving, rule and concept formation, perception, [and] recognition" (Roitblat 1987, 2).

Animal cognition has thus come to be accepted as real and significant, and its investigation is recognized as important. But claim 3A still appeals to many behavioral scientists. Thus animal consciousness is still taboo, among both biologists who study animal behavior and psychologists such as Terrace (1984, 7), who asserted that "both in animal and human cognition it is assumed that the normal state of affairs is unconscious activity and thought." Although many biologists have always dissented from claim 1 of behaviorism, they have generally accepted claims 2 and 3. Therefore it will be convenient to refer to both psychologists and biologists who deny the significance of consciousness as "inclusive behaviorists." Despite renewed interest in animal cognition, the scientists who are willing to venture into this difficult area have tended to cling tightly to the security blanket of conventional reductionism.

Most philosophers have long since abandoned the argument of logical positivists that only objectively measurable attributes are significant, and cognitive psychologists now reject the negative dogmatism of the strict behaviorists. But students of animal behavior are still severely constrained by a guilty feeling that it is unscientific to inquire about subjective feelings and conscious thoughts (Colgan 1989, Heyes 1987, Latto 1986, Yoerg and Kamil 1991, Vauclair 1996, Roberts 1997, Kamil 1998, and Shettleworth 1998). Although ethologists have recognized more complexity and versatility in animal behavior, many have lagged behind the cognitive psychologists and continue to try doggedly to fit all the new knowledge about animal behavior into the same old pigeon holes that seemed sufficient years ago to Pavlov and Watson. Thus the ghost of Jacques Loeb (1918) still makes its cold and clammy influence felt when animal behavior is described solely in terms of stimuli, responses, and adaptive advantages.

Animal Cognition without Consciousness

Cognition in the sense of information processing in nervous systems obviously must occur in virtually every animal. External influences can affect behavior only by stimulating sense organs and neurons or other tissues, and these processes necessarily entail transfers of information. Furthermore, some of this information produces internal representations that interact with subsequent stimulation, with motor systems, and with each other. The exact nature of such representations is poorly understood, and we are often unaware of them even when they affect our behavior. Therefore it is almost trivially obvious that animal nervous systems process information. What is not obvious is the extent to which such activity of nervous systems leads to conscious mental experiences. Because so much of the coordinated and smoothly functioning information processing in human brains takes place without any conscious awareness, many scientists have tended to conclude, or at least strongly imply, that in nonhuman nervous systems *none* of the information processing is conscious. In addition to the nature of information processing in animal nervous systems, one of the central problems of cognitive ethology is the distribution and content of conscious awareness—subjective emotions, desires, beliefs, and behavioral choices intended to achieve certain results or avoid others.

The prevailing viewpoint of most behavioral ecologists was clearly expressed by Wittenberger (1981, 48):

> Cost-benefit analyses [are discussed] *as if* [emphasis in original] behavior results from a conscious decision-making process. . . . This procedure is just a shorthand logic used for convenience. We cannot assume that animals make conscious decisions because *we cannot monitor what goes on inside their heads.* Nevertheless, *it really does not matter* [emphasis supplied] what the proximate bases of those decisions are when evolutionary reasons underlying the behavior are our principal concern. . . . The question of whether those choices are conscious or unconscious need not concern us, as long as we remember that our tacit assumptions about purposiveness are just that. . . . Particular stimuli or contexts elicit particular behaviors. An animal need not know why those stimulus-response relationships exist. It need only know what the relationships are. This knowing need not involve conscious awareness, though in many cases animals are undoubtedly conscious of what they are doing; it need only involve the appropriate neurological connections. . . . Animals can be goal-directed without being purposeful, and they can behave appropriately without knowing why.

This perspective is widely shared. For example, Shettleworth (1998, 10) reviews and analyzes animal cognition from the stated viewpoint that "some animals may be conscious in some sense but we cannot know." This claim of inaccessibility is taken to justify a lack of interest in animal consciousness. On the other hand, cognitive ethology broadens one's perspectives to include not only information processing but also seeking to understand what life is like to animals.

Dukas (1998, 1) maintains the customary fire wall against consciousness by opening *Cognitive Ecology* as follows: "Animal cognition is a biological feature that has been molded by natural selection. . . . Cognition may be characterized as the neuronal processes concerned with the acquisition and manipulation of information by animals." This concern with animal cognition is a marked change from behaviorism, which limited psychology to the analysis of external influences that were held to control virtually all behavior. Yet a striking imbalance has resulted, with extensive investigations of nonhuman information processing, but an implicit denial that any of this cognition is accompanied or influenced by conscious thoughts or feelings.

One of the strongest statements of this denial was expressed by Kennedy (1992, 31), who concluded that "although we cannot be certain that no animals are conscious, we can say that it is most unlikely that any of them are." Livet (1992), like Kennedy, argues strongly against attributing mental states such as desires to animals, calling any such attribution "popular experimental psychology." For example, he argues that when birds return to their nests in a homing experiment there is no need to assume that they search for their nest.

Kamil (1998, 22) has objected strongly to any consideration of conscious experience of animals and advocates redefining cognitive ethology to exclude it from scientific consideration. After recognizing that the analysis of animal cognition in nature is appropriately termed cognitive ethology, he objects that I and others have preempted the term by

> defining cognitive ethology in terms of subjective experience: awareness, consciousness, etc. Many of us working on animal cognition find this term objectionable for many reasons; so objectionable that we completely avoid using the term. One of the reasons it is objectionable is that much of the argument has focused on attempting to prove the existence of these internal states and the nature of some of these arguments is unacceptable. In particular, many of Griffin's arguments are so weak and anecdotal that they remind one of nothing more than Paley's (1851) arguments for the existence of a creator, natural theology based on the argument from design.

Blumberg and Wasserman (1995) also likened the suggestion that an animal's conscious thoughts might influence its behavior to Paley's argument from design. This claim implies that consciousness is something immaterial. But as pointed out in chapter 1, this makes sense only if one is a dualist, in philosophical terms, and holds that consciousness belongs to a different realm from the physical universe. Perhaps these critics are not dualists and agree that consciousness results from activity of a central nervous system. But they obviously believe strongly that it is unsound to infer anything about nonhuman conscious thoughts. They seem to adhere to the behavioristic view that all behavior can be understood without consideration of conscious experiences such as beliefs, desires, hopes, fears, likes, or dislikes. Strict and consistent behaviorists must deny that their own thoughts have any effect on their behavior, but few, if any, really believe this, and they would presumably agree that we can make reasonable inferences about human thoughts from human behavior, especially communicative behavior. If so, their aversion to the possibility of animal consciousness is apparently based on a conviction that there is a difference in kind between humans and all other animals, that our thoughts sometimes influence *our* behavior but that no animal's thoughts can affect *its* behavior.

Paley's type of argument from design was rendered unnecessary by the recognition that natural selection can produce intricate living mechanisms in the course of biological evolution. Behaviorists from Watson and Skinner to Shettleworth (1998) attempt to explain behavior without admitting that conscious experiences can have some effect on it. Many behaviorists used to say that virtually all behavior was caused by learning or other forms of environmental influence during the lifetime of a given animal. This was later modified to include genetic influences. But it is difficult to see how prior learning or genetic programming can account for effective coping with novel and unpredictable challenges, which many animals accomplish by means of versatile and adaptable behavior. Simple thinking about the likely results of alternative actions and selecting the ones believed likely to produce desired results can account for such versatility without the slightest need to appeal to immaterial factors or divine intervention.

Vauclair (1992, 1996, 1997) has criticized cognitive ethology because it does not restrict itself to representations in the brains of animals. Even though he recognizes that animals may be conscious of such representations (1996, 30), he insists that "even if one accepts that representations presuppose conscious experience, it is not the subjective quality of the experience itself that is under investigation . . . without at least a system of mutual understanding, it would be difficult for us, to paraphrase Nagel (1974), to envision what it is like to be a bat. Fortunately the principal aim of a scientific study of the minds of other animals is not to find out what it is like to be a certain type of animal, but rather to clarify how mental states cause observable behaviours." This self-imposed restriction of concern overlooks the significant degree of mutual understanding that is often achieved by those who come to know animals intimately—horsemen and their mounts, sheepdogs and their masters, circus trainers, and even thoughtful pet owners, as discussed by Hediger (1947) and Roberts (1997). And in chapters 9 and 10 I will discuss in detail how animal communication provides objective evidence about nonhuman thoughts and feelings.

When struggling with difficult but important scientific problems it is a mistake to ignore any relevant data. As Seymour Kety (1960) put it, "Nature is an elusive quarry, and it is foolhardy to pursue her with one eye closed and one foot hobbled" (1862). Of course, it is easy to overinterpret animal behavior and to infer more humanlike content than is warranted. But the difficulty of being certain of the content of an animal's conscious experience does not justify denying that it exists and can be investigated.

Elsewhere Vauclair (1992, 139) has claimed that the approach I have advocated "rests on the eminently debatable hypothesis according to which the animals under study share the same conscious experiences as those of mankind." Yet I have repeatedly emphasized (for example 1976, 67–68) that whatever conscious experiences occur in animals may be very different from any human thoughts or feelings. Vauclair seems to assume that human experiences are the only possible kind. Interpreting some example of animal behavior as indicating conscious awareness is often implied to be a claim that the animal is experiencing exactly what a human in the same situation would think or feel. Shettleworth (1998, 235) appears to make the same assumption when she denies that the experiments of Herrnstein and others discussed in chapter 7 demonstrate that pigeons "have a concept in the same way humans do." But there is no reason to expect mental experiences to be unitary all-or-nothing entities, especially since we know they can vary widely within our own species according to gender, age, prior experience, and cultural background.

A somewhat similar criticism of cognitive ethology has recently been expressed by Dyer (1998, 202): "In the most restrictive view, cognitive processes are those involved in human mental qualities such as self-awareness. This seems to be the view of many so-called 'cognitive ethologists' (Ristau 1991), who are largely concerned with investigating the role of humanlike mental processes in animals. . . . I use the term 'cognition' in a broader sense that refers to information processing in general by nervous systems and that encompasses simple perceptual and learning processes as well as more complex mechanisms used to acquire internal representations of the world." Again we see the tacit assumption that consciousness is "humanlike," with the implication that it is absent or unimportant in other species. Perhaps Dyer is simply expressing the widespread aversion of ethologists to entanglement in the complexities of questions about mental experience, or perhaps he assumes that consciousness is limited to our species.

In short, most students of animal behavior agree that animals may be conscious, but scientists such as Vauclair, Shettleworth, Dyer, and others avoid considering this possibility and imply that to consider it is unscientific. Cognitive ethology is a large area about which we know very little. But it should certainly include all varieties of mental functioning, conscious or not, and we should remain open to the distinct possibility that the experiences of other species are quite different from any of ours.

An especially striking example of extreme reluctance to consider animal consciousness is the recent advocacy by Owings and Morton (1997, 1998) of the idea that animal communication should not be viewed as a transfer of information. Instead they insist that it is assessment and manipulation or management (A/M) and that information is an unfortunate concept that should be discarded in analyses of animal behavior. For example: "Assessment/Management is founded upon the fundamental biological process of regulation, rather than the anthropomorphic concept of information exchange. . . . To call 'information transfer' what animals, organs, or even cells do when they interact is implicitly to use human communication as a model for all forms of interaction among organismic entities. . . . In the A/M approach, the central question about communicative behavior from the perspective of management, is not 'what does it stand for?,' but 'what does it serve to accomplish?' " (Owings and Morton 1997, 359, 362, 367). Engineers and physicists would be surprised to find that the concept of information transfer is considered anthropomorphic.

Owings and Morton claim that "two meanings of cognition can be found in the literature. One treats cognition as a subset of input-processing activities, involving relatively complex mental phenomena such as intentionality, self-awareness and possession of a model of the mental states of others (e.g. Seyfarth & Cheney, 1994). This is not our meaning. We use cognition broadly here to refer to all input-processing activities beyond perception, including storing, retrieving, computing, and integrating as a basis for behavioral decisions . . . even simple processes like Pavlovian and operant conditioning are also cognitive processes; . . . We do not confine the term [cognition] to processes that approximate human cognitive sophistication; . . . The term intentions . . . refers to signals that are statistically predictive of future behavior, not to cognitive processes" (1998, 60, 71, 73). The claim that cognitive ethologists confine their interest to mental processes with "human cognitive sophistication" is an attempt to discredit by exaggeration.

It is difficult to see how excluding conscious experience *broadens* rather than narrows the scope of animal cognition. It is quite reasonable to emphasize the *effects* of animal communication. But these effects can only be achieved by transfer of information, so emphasis on assessment and management does not require claiming that no information is transferred in animal communication. Does an unstated motivation to avoid animal consciousness underlie this confusing claim that the assessment/management approach is an alternative to recognizing that information is transferred by communicating animals?

Anthropomorphic Objection, and Conceit

For many years any consideration of animal consciousness was strongly discouraged by the accusation that it was anthropomorphic. This widespread attitude resulted from the recognition that many earlier ascriptions of human thoughts to animals were wholly unjustified. But the charge of anthropomorphism has often been inflated to include even the most tentative inference of the simplest kind of conscious thoughts by animals. When one carefully examines such opinions, it turns out that they entail the implicit assumption that whatever it is suggested the animal might do, think, or feel really *is* a uniquely human attribute. This assumption begs the question being asked, because it presupposes a negative answer, and is thus literally a confession of prejudgment, or prejudice.

This simple point is often overlooked or ignored. To be more explicit, consider a specific example: a male bear is observed to approach a bush filled with ripe blueberries, and a cognitive ethologist infers that it consciously desires and intends to eat them. Such an inference might be incorrect in two general ways: (1) The bear might be conscious, but the content of his consciousness might be different from what was inferred. For example, as he approaches the bush he might have no interest in the blueberries but smells a sexual attractant left at the base of the bush by a female bear. This would be an error in inferring what the *content* of the bear's consciousness was. Or it might not be conscious at all even though we, mistakenly, inferred from his behavior that it is. (2) In most cases, when scientists criticize such an inference as mistakenly anthropomorphic they mean that animals are incapable of the type of conscious experience that has been inferred. It is this usage of the term *anthropomorphic* that tends to be circular and illogical, because, once the question whether an animal may be conscious is raised, to deny this possibility out of hand with no supporting evidence amounts to no more that a reiteration of a prior conviction that the animal in question is incapable of such a conscious experience. Of course, such prior convictions are likely to differ across species: what seems plausible for a chimpanzee may seem preposterous for an earthworm. This general point was clearly made by Bennett (1964, 11) and Griffin (1977). And Fisher (1987, 1990, 1991) has spelled out in considerable detail why there is no fundamental philosophical basis for the taboo against anthropomorphism as it has been perceived by most scientists, and he concluded: "The idea that anthropomorphism names a widespread fallacy in commonsense thinking about animals is largely a myth . . . and the use of the term as a critical cudgel ought to be given up" (1991, 84). Furthermore, when applied to the suggestion that animals might think about simple things that are clearly important to them, this charge of anthropomorphism is an unsupported and conceited claim that only our species is capable of even the simplest conscious thinking.

Mechanomorphic Indoctrination

Boakes (1992, 22) clearly expounded the behavioristic view that nonhuman consciousness is hopelessly inaccessible and that any consideration of it should be banned from science:

> Attributing conscious thought to animals should be strenuously avoided in any serious attempt to understand their behaviour, since it is untestable, empty, obstructionist and based on a false dichotomy. . . . By the age of 3 years most children can describe some arbitrary event or scene in the objective world that the listener cannot directly perceive. No such ability has been found in any other species. . . . This unique human ability can then be used to report on a subjective world. In its absence there is no way of knowing what this might be like for a nonhuman creature. . . . For many psychological students their first, and only, contact with animal psychology is a practical class in which they shape a rat to press a lever. Their almost universal way of explaining what happens is to talk in terms of the rat's beliefs, expectations, intention or boredom when it stops doing very much. Recently such talk has become more acceptable as long as imaginary quotation marks are placed around the mentalistic terms. Griffin would have us all remove the quotation marks. His book left me with the resolution to go back to abolishing such talk the next time I run such a practical. It clearly produces a false sense of complacency and gets in the way of looking carefully at what the animal is actually doing and thinking hard about what may be going on in its brain.

One could scarcely ask for a more explicit description of the indoctrination to which students have been subjected. In chapters 9–12 I will review numerous specific examples of animals that communicate about events or scenes that recipients of the communication cannot perceive directly. Furthermore, it should be obvious that thinking hard about brain function is seriously impeded rather than facilitated by adamantly ignoring some of the most important processes that occur in central nervous systems. Behaviorists have been insisting for decades that the only appropriate scientific view of animal behavior is one that treats the animals as nearly as possible like mindless robots, a view that Crist (1999) has termed "mechanomorphic."

The Inverse Clever Hans Error

Students interested in animal behavior have long been haunted by the specter of "Clever Hans errors." Suggestions that an animal might be consciously aware of the likely results of its behavior routinely elicit a sort of knee-jerk accusation that they result from Clever Hans errors (Sebeok and Rosenthal 1981). This hazard refers to the performance of a trained horse named Hans. His devoted trainer believed that Hans could add, subtract, multiply, and even divide written numbers. Hans gave his answers by tapping with his forefoot, 8 taps when 2 × 4 was displayed, 27 taps in response to 3 × 9, and so forth. Some skeptics were troubled by the fact that Hans solved difficult arithmetical problems about as rapidly and apparently as easily as simple ones (Washburn 1917). But he tapped correct answers even when others besides his familiar trainer presented the problems, and many scientists were convinced that he understood the problems and arrived at correct answers by a sort of mental arithmetic.

A psychologist, Oskar Pfungst, showed by careful experiments that Hans was not watching the written numbers but the person who presented the problem. If no one was visible, or if the person did not know the correct answer, Hans tapped his foot at random. What Hans had actually learned was to detect the small and inadvertent motions that people made while watching to see whether he stopped after the correct number of foot taps. Even when observers tried to avoid doing so they could not help revealing by subtle motions or facial expressions when the right number of taps had been produced. In his detailed account of these experiments Pfungst (1911) described many other complex patterns of behavior learned by Hans and showed that they all depended on inadvertent cues from human companions.

The saga of Clever Hans has been almost universally accepted by scientists as a definitive example of mistaken inference of complex mental abilities in animals. But this enthusiastic application of scientific caution has mushroomed into denial that animals experience even the simplest conscious thoughts. Inability to do arithmetic has been taken as evidence for the absence of any thinking whatever. What has been almost totally overlooked is the real possibility that Clever Hans was consciously thinking something simple but directly relevant to his situation—perhaps something like: "I must tap my foot when that man nods his head, and then I'll get some sugar." The horse's behavior was quite consistent with this interpretation, but scientists' efforts to debunk unjustified inferences have trapped them for generations into a dogmatic dismissal of the plausible alternative that animals, though quite incapable of complex mental arithmetic, may experience *simple* conscious thoughts.

The perceptual discrimination needed to detect inadvertent counting gestures has been recognized as remarkable and significant, and these abilities are probably important under natural conditions. The detailed observations by ethologists reviewed in chapter 4 have shown that many animals monitor the behavior of predators and react to the most minor changes in posture or behavior that signal a likelihood to attack. Predators are also very adept at noticing the slightest indications that potential prey are weak or sick, and these appear more likely to be attacked. The same basic neurophysiological mechanisms that underlie such finely tuned discriminations are probably employed by animals that learn to respond to inadvertent counting movements of their human trainers.

Discredit by Exaggeration

Scientists often dismiss suggestions of animal consciousness by overinterpreting them to include complex levels of thinking that are clearly beyond the capabilities of nonhuman animals. This attitude was parodied on the cover of a popular weekly magazine that included a balanced review of the current revival of interest in animal thinking and intelligence. The cover picture showed a dog with a balloon rising from his head and in the balloon was $e = mc^2$. This is a typical exaggeration of the suggestion that animals may think about simple matters that are important to them into an implication that they share the more complex levels of human thought. Another example of this tendency to exaggerate that which one wishes to deny is Ingold's implication (1988, 90) that those who suggest animals are sometimes conscious are claiming that *all* nonhuman animals *always* are.

Can *Any* Behavior Demonstrate Consciousness?

McFarland (1989, 126) disagreed strongly with my suggestion that versatile adjustment of behavior to cope with novel challenges is evidence of conscious thinking: "I fail to see the logic of this argument unless it is based on some notion of evolutionary continuity. Even here, it seems to me, we are treading on dangerous ground. . . . An evolutionary argument would lead us to suppose that each species has to solve the problems appropriate to its niche. . . . Thus, as a justification for supposing that animals have mental experiences similar to our own, the evolutionary argument cuts both ways." McFarland thus seems to be arguing, like Kennedy (1992) and Kamil (1998), that conscious mental experiences are uniquely human attributes and that they are adaptive only in the human niche. But recognizing that they probably differ widely according to the animal's way of life seriously weakens McFarland's objection. A residual aftermath of behaviorism is the widely held view that anything an animal does is probably done unconsciously. But we all know that we are sometimes conscious, so that the behaviorist qualms are clearly not applicable to one species. Thus only if one assumes that matters are different with other species does it make sense to claim that nothing any animal might do would qualify as convincing evidence that it is conscious.

Paralytic Perfectionism

In the provocative and influential paper "What is it like to be a bat?" the philosopher Thomas Nagel (1974) argued that bats are so different from us that we are inherently unable, in principle, to answer this question. But on close examination of his arguments it appears that what he really means is that we cannot hope to know *fully* and *exactly* what it is like to be a bat or any other animal. This conclusion should be tempered, however, by recognizing that we can never know the feelings and thoughts of our human companions with total completeness and perfection. Thus the unqualified version of Nagel's philosophical argument, and similar reservations of many scientists about nonhuman mentality, tend to inhibit any investigation at all unless totally perfect data are clearly in view. Such perfectionism severely impedes, or even prevents, investigation of animal mentality. Figuratively speaking, it paralyzes scientific exploration, and for convenience it can be termed "paralytic perfectionism."

The impossibility of ever knowing absolutely everything about something does not diminish the significance of partial but useful understanding. There is probably nothing that we can ever know with complete perfection and total certainty. But in other areas of inquiry this fortunately has not discouraged scientists from learning what they can, and we are quite content to investigate most other scientific questions with full and balanced recognition of the importance of partial answers, tentative conclusions, and competing hypotheses. Yet when questions about animal mentality arise paralytic perfectionism has provided a potent excuse for neglecting a difficult but important scientific problem. For example, Shettleworth (1998) repeatedly claims that no evidence of animal consciousness is completely convincing, and she goes on to dismiss attempts to study it as misguided for that reason. But other difficult scientific questions are discussed with normal scientific open-mindedness: "Rarely is the kind of evidence available that would allow more than a tentative hypothesis about the evolutionary history and adaptive value of any cognitive mechanism" (35). And: "It is not always apparent how cognitive abilities or the opportunity to use them can be manipulated to test whether they serve their hypothesized function in the field" (42).

A clear example of the demand for perfect evidence is the argument advanced by Heyes and Dickinson (1990, 87) that observations of animal behavior under natural conditions cannot provide convincing evidence of "a mental life built on the rational interactions of intentional states such as beliefs and desires." They use the term *intentional* to mean rational action based on a belief that the action will attain a desired result. They then cite laboratory experiments in which animals do not behave in a fully rational manner when confronted with quite unnatural situations. Their first example is an experiment by Hershberger (1986) in which young chicks that had previously obtained food from a bowl were placed in an apparatus where, when they moved toward the bowl, it moved away from them at double the chick's speed of approach. And conversely, if a chick moved away, the bowl approached at twice its speed of movement. Despite this bizarre situation the chicks did learn, albeit slowly, to get some food from this remarkably fickle source. After seven days with 10 trials per day they succeeded in getting food about half the time. In other words, they displayed incomplete but effective learning in coping with this extremely odd food source. But because they did not learn to drastically change their normal food-getting behavior Heyes and Dickinson imply that their behavior was not rational. Insofar as evolutionary selection has molded the operation of central nervous systems, it difficult to imagine how it could have prepared a bird to deal with a food source that behaved like the one contrived for this experiment.

Heyes and Dickinson's second example is based on experiments by Holland (1979) in which hungry rats were placed in an apparatus provided with a bowl that contained food only some of the time. A tone sounded from time to time, and when it was turned on the bowl was empty. Ideally, rational rats should have learned to stay away from the bowl when the tone was sounding, but in fact they did often come to it when the tone was turned on, and as a result their rate of food intake was lower than it would have been had they approached only during the periods of silence. Heyes and Dickinson interpret this lack of perfect rationality as evidence that when a hungry rat approaches a familiar food source this does not suffice to prove that it desires the food or believes that it is available in the bowl. They generalize this view as follows: "Many behaviours, that appear to be intentional on the basis of simple observation, fail to change in appropriate ways under the influence of new environmental contingencies, and therefore, if our analysis is correct, naturalistic observations of behaviour provide no reliable information about the intentionality of animal action" (94). It seems plausible, however, to interpret the result of Holland's experiment in terms of simple and rational, but imperfect, thinking on the rat's part. The bowl *was* a source of food; the rat merely failed to learn that there was a particular condition under which it was empty. But it is not a wholly foolish strategy for a hungry animal to keep looking for food where it has been found previously even though it was not always available.

Heyes (1987, 108) criticized my discussion of cognitive ethology because I appeared to be allying myself "with the layman, the affectionate pet owner and curious huntsman" and because I was willing to consider a wide variety of suggestive evidence that animals may be thinking consciously in particular situations. She seems to be satisfied only with data from precisely controlled experiments. The examples discussed above show some of the difficulties with this position. Based on a rigid definition of criteria adequate to demonstrate desires or beliefs, Heyes doubts that certain animals desire food or believe it can be obtained where they found it previously, because when the conditions are changed drastically they do not solve the problem in a fully rational fashion. But beliefs and desires do not necessarily entail perfect rationality, so the application of her rather rigid criteria tends to constitute a modified form of paralytic perfectionism.

Weiskrantz (1997, 78) discusses the idea that attributing consciousness to animals is useful because it helps us predict their behavior but rejects this as "not a convincing argument for the attribution being *correctly* made. We attribute human-like qualities to objects in computer games, to servo-controlled missiles, to mechanical toys and robots" (emphasis in original). Because we sometimes attribute consciousness to things that are not conscious, this argument goes on to imply, consciousness should *never* be inferred from any behavior whatever. It is helpful to consider a simple analogy: On seeing a bright light in the night sky we often think it is a star or planet, but then we notice that it moves and realize that it must be an airplane. Should we then conclude that we can never under any circumstances make the inference that a bright spot in the sky is a celestial body? Of course, we do not know of any comparably simple way to discriminate a conscious from an unconscious animal, but that is no reason to despair of ever learning how to do so.

Mentophobia

As Searle (1990a, 585) put it, "If one raised the subject of consciousness in cognitive science discussions, it was generally regarded as a form of bad taste, and graduate students, who are always attuned to the social mores of their disciplines, would roll their eyes at the ceiling and assume expressions of mild disgust." The intense fervor with which many psychologists and biologists insist that nonhuman consciousness is a totally inappropriate subject for scientific investigation appears to be so deep-seated and emotional that it verges on an irrational aversion, or mentophobia. Although it is most intense when the question of animal consciousness arises, many scientists are also uncomfortable when faced with questions of human consciousness.

Weiskrantz draws on his experience as a neurologist and points out that some human patients with various types of damage to their brains can react appropriately to stimuli that they deny perceiving consciously. The clearest case of this sort is "blindsight" (discussed in detail in chapter 8), in which patients with damage to parts of the visual cortex are blind to visual stimuli in the affected parts of their visual field. But when asked to guess about these stimuli they perform at much better than chance levels in identifying an object presented in their "blind" area. Other parts of the visual system are apparently functioning well enough to permit mostly correct guesses, but these patients are not consciously aware of the stimuli. Weiskrantz and others are troubled by the possibility that any example of animal behavior that suggests consciousness might be accomplished by brain mechanisms comparable to those that permit the blindsight patient to make correct guesses about things he says he cannot see.

Weiskrantz continues with a thoughtful discussion of the difficulties in determining with complete certainty whether under particular conditions an animal is aware of what it is doing. He stresses the need for some sort of commentary by which the animal could report its conscious experience, and considers that this can be accomplished by arranging laboratory situations in which the animal responds in a way that indicates what it is thinking. But he overlooks the abundant evidence of communicative behavior by which animals do at times appear to be reporting their feelings and simple thoughts. Chapter 9 will describe this category of evidence.

All of these criticisms and objections seem to boil down to a basic reluctance to believe that nonhuman animals can think rationally in even the very simplest ways. Cognition, internal representations, assessment, and manipulation are considered appropriate for scientific study, but whether the animals in question have any understanding of what they are representing or manipulating is not. This amounts to a basic assumption that only our species is capable of consciousness, thus ruling out a priori any inquiry into the fundamental question of what life is like for nonhuman animals. Such investigations are extremely difficult and challenging because we have only fragmentary evidence on which to base any conclusions about the occurrence, extent, and nature of animal consciousness. But this does not justify the blanket denial that animals are ever conscious. It is therefore appropriate to continue by considering some examples of animal behavior that suggest, although they do not rigorously prove, that the animals in question are consciously aware of objects and events that are important in their lives.

CHAPTER THREE

Finding Food

Locating suitable food is one of the most widespread and pressing problems faced by animals. Unlike pets and laboratory rats or pigs in factory farms, most animals must spend a large fraction of their waking hours locating food and extracting it from their environment. In the case of herbivores, this may seem simple at first thought, but it is seldom an easy matter of wandering about nibbling whatever vegetation is encountered. Not all plants are equally nutritious, and some contain distasteful or even toxic substances. Even grazing animals that appear to need nothing but abundant grass do pick and choose just which patches are most worth cropping and pay considerable attention to signs that food is available from particular plants. Active predators face more obvious challenges, because they must not only locate but pursue and capture prey animals that seldom wait passively to be eaten but devote considerable effort to avoiding that fate.

Foraging behavior varies widely, and its versatility is not closely correlated with the phylogenetic group to which the animals belong; so-called lower animals often display ingenuity comparable to that of mammals. This review of feeding tactics that suggest thinking on the animal's part will be divided rather arbitrarily into categories that can be roughly characterized as feeding on passive and active prey. This chapter will consider the former category, in which the food consists of plants or of animals that are relatively inactive, so that the principal problems are how to locate suitable food and handle it. Chapter 4 will concentrate on predation upon prey that exert effective efforts to escape and can be taken only by means of actively versatile tactics. In some significant situations these tactics include coordinated action by two or more individual predators.

Foraging Decisions by Bumblebees

Bumblebees would not seem likely to employ a high order of thoughtful decision-making, but when Pyke (1979) analyzed in detail the ways in which a particular species of bumblebee *(Bombus appositus)* gathered nectar from clusters of monkshood flowers in the Colorado mountains, he found that they followed fairly complex rules. These flowers vary considerably in nectar content, depending in part on whether an insect has already removed nectar from a particular flower. Pyke marked bumblebees so that individuals could be distinguished and recorded their behavior when visiting clusters of flowers that had not been visited by other insects. No one flower held enough to fill a bumblebee's stomach, so several had to be visited before she flew off to her nest. The lowest or next-to-lowest flower was almost always visited first, and the bumblebee then moved upward, usually selecting the closest flower she had not already visited. Out of 482 observations, the same flower was visited twice on only four occasions. Either the bumblebee remembered for a short time which flower she had already visited, or else she left a scent mark or some other indication that enabled her to avoid wasting her time on empty flowers.

Pyke described the simplest plausible rules to account for the foraging tactics of these bumblebees: "Start at the lowest flower on a given inflorescence, then move to the closest flower not already visited, unless the last movement had been downward and was not in fact the first switch from one flower to another on a particular inflorescence. In the latter case, move to the closest higher flower not just visited" (1170). Expressing these rules in English may make them seem more complicated than they actually are, but even if simplified into a set of actions within the capabilities of a foraging bumblebee they are not the simple, stereotyped sort of reactions we are accustomed to expect from insects.

Many other insects engage in equally complicated foraging behavior, but most are difficult to study because they rely so heavily on olfaction, and because it is extraordinarily difficult to monitor and experimentally manipulate the chemical signals that guide their behavior. Locating food may entail searching for the odors that signal its availability, following gradients in the concentration of such odors, or moving upwind when they are detected. But odors carried by winds do not spread in simple patterns, because turbulent flow near the ground produces an odor plume with a complicated and shifting shape, so that a simple stereotyped movement upwind would cause the insect to fly in a much more indirect path than is actually the case. For instance, male gypsy moths attracted to a female pheromone fly zigzag paths, often heading 90° from the wind direction, and apparently assessing the shape of the odor plume, as described by Zanen and Cardé (1999).

In cases where insects use vision to locate food sources, as when bees and other insects that feed on pollen and nectar search for flowers, they are so small and move so rapidly that it is very difficult to determine just what searching movements they employ. But I suspect that when cognitive ethologists become sufficiently disinhibited from the mindset that all insects are genetically programmed clockwork, they will devise effective methods to monitor how they go about searching for food. This in turn might disclose that many species are at least as versatile as the bumblebees studied by Pyke.

Many recent studies of foraging behavior of animals have attempted to define quantitative rules governing foraging decisions. A representative example of this approach is developed in a series of papers by L. A. Real and his colleagues (Real 1981, Real et al. 1982, Real 1987, Dukas and Real 1993, and Real 1994). In a typical experiment a tent of mosquito netting was set up around a colony of bumblebees *(Bombus sandersoni)*. The plastic floor of the tent was studded with wells 3 mm deep, spaced 2.5 cm apart, which served as "flowers," and each well was marked with a colored square of paper. Two hundred food wells were arranged with equal numbers of the two colors located randomly with 20 μl of sugar solution in each well. The bees didn't find them at first but were led to filled wells by a trail of honey, and they learned to forage on the artificial flowers in two days. They visited them singly, and to avoid complications caused by odors the plastic floor of the tent was removed after one bee had returned to the colony nest in order to remove any olfactory cues that might guide subsequent visits by foragers.

When both yellow and blue "flowers" held a constant 2 μl of sugar solution, 90–95 percent of the visits were to wells marked with yellow. Then the conditions of the experiment were changed; all of the blue wells now contained 2 μl and one-third of the yellow wells held 6 μl, while the other two-thirds were empty. Blue now marked a reliable source and yellow an unpredictably variable one, although purely random sampling of a large number of wells marked with either color would have yielded equal amounts of sugar. In this changed situation the bumblebees' preferences reversed, and 80–90 percent of the visits were to the "reliable" blue flowers. They preferred a reliable source to one that occasionally yielded more food but at other times provided none. In later experiments the difference in the sugar content in the yellow wells was varied to find how much more rewarding they had to be, on the average, before the bees were equally likely to visit either color. This type of experiment was then used to determine quantitative trade-off functions, and these were found to be affected in turn by ecological factors.

This general type of experiment with risk sensitivity has been elaborated in several more recent investigations described by Bateson and

Kacelnik (1998), Kacelnik and Bateson (1996), and other papers in a special issue of *American Zoologist* (36:389–531, 1996). The term *risk sensitivity* is used by foraging theorists to mean preference for either variable or constant food sources depending on the circumstances. Animals are considered risk prone if they prefer the variable type of food source and risk averse if they prefer the constant type. Animals tend to be risk averse when the average amount of food is greater than they need. But if the average is much lower than needed they tend to be risk prone. On the other hand, when the difference is in the time needed to obtain food rather than the quantity available, animals tend to be risk prone even if the food is ample.

For present purposes, however, the important result of these experiments was that bumblebees, wasps, and many other animals learn to make clear choices between food sources marked by different colors, and they adjust these preferences according to the circumstances, tending to choose variable sources when the overall yield is low and constant ones when food is abundant. Real (1987, 401, 409) summarized these findings as follows: “An organism’s perceptions of the environment may not correspond to the objective reality but to some subjective estimation of the probability of an event’s occurrence. . . . The subjective and not the objective probability distribution always governs the organism’s choice. . . . The organism’s responses to variability in resource distributions is a compound effect of both the perceived (subjective) probability of the occurrence of different reward levels or benefits and the utility that results from such benefits.” Real and others analyzing the results of these experiments refrain from mentioning the possibility that “subjective” preferences entail conscious choices for color and for reliable and predictable, rather than variable and uncertain, rewards. This reticence may be cautious deference to the prevailing taboo against considering the possibility that simple perceptual consciousness might be involved in making these important decisions.

Prey Selection by Starlings and Wagtails

Birds rely primarily on vision to locate and capture food, and they are easier to observe than many other animals. As a result we know more about their behavior in general and their foraging in particular. In some cases they are obliged to make choices in their search for food that would seem likely to be facilitated by a little simple thinking about the possibilities available to them and the probable results of various alternative courses of action. For example, a thorough study of a group of starlings in the Netherlands by J. M. Tinbergen (1981) revealed that when feeding nestlings they concentrated primarily on two species of caterpillars that were to be found in opposite directions from their nests. This made it possible to tell from their initial flight direction which of the two they intended to gather. One of the two species was preferred under most conditions, but the parents switched to the other when there was a pressing need for food, and especially when their broods were experimentally increased by placing additional nestlings in their nest. These choices were made at or near the nest, where neither type of caterpillar was visible; the starlings had to remember in which direction to fly for each type.

Krebs and Inman (1994) have reviewed several controlled laboratory experiments in which starlings were obliged to spend varying amounts of "travel time" to move between patches of food when the food available in the patches decreased with time. This meant that to obtain the most food the bird had to leave a partly depleted patch after a time interval that varied with the travel time to the next patch, that is, it would pay to remain longer in a diminishing patch if more time was needed to get to another patch. When travel time itself was experimentally varied in an irregular and unpredictable fashion, the starlings ordinarily acted as though expecting the most recent travel time they had experienced. But in other situations they also showed that they remembered the last several travel times.

In these experimental situations the birds were obliged to make trade-off decisions based on what they had reason to expect about food availability and travel time. They did quite well in general, adjusting their foraging behavior so as to approximate the most efficient use of their time and energy. Quantitative models have been constructed to work out what behavioral trade-offs are in fact most efficient, and for the most part the birds approximated the optimal behavior. Because these trade-offs can be specified by quantitative models, mechanical analogs or robots could theoretically be constructed that would follow the same rules of behavior, resulting in efficient foraging.

A clear and reasonably representative example of the choices and decisions involved in feeding behavior stems from the studies of two species of wagtails feeding on Port Meadows along the banks of the Isis River in Oxford (Davies 1977; Krebs and Davies 1978). The pied wagtail *(Motacilla alba yarrelli)* is a year-round resident of southern England, and the yellow wagtail *(M. flava flavissima)* is a migrant present only during the summer months. They were studied early in the season before they started to breed. At this time they were gathering food only for themselves, but they were probably also putting on weight that helped prepare them for the nesting season that would follow in a few weeks. They were easily observed because the grass was heavily grazed by cattle and horses, and the numerous dung pats provided food for the flies on which the wagtails fed. Only one bird at a time fed on insects from a single dung pat, but small groups often hunted at the pools where several kinds of aquatic insects were abundant. At a particular dung pat the wagtail would usually capture only one of the larger flies, and this disturbance would cause the others to scatter into the grass. The bird would then search for and catch many flies in the immediate vicinity.

At the start of a feeding session each bird had to decide where to search for food—whether to join a flock of wagtails or hunt by itself—and whether to concentrate on dung pats or on aquatic insects at the shallow pools of water on low-lying areas. The wagtails made their choices with considerable efficiency, so that they obtained approximately the maximum possible amount of food with minimum expenditure of time and effort. This entailed concentrating their efforts where food was most plentiful and moving on when it became depleted. These shifts were not rigidly programmed; the birds did not wait until every last fly had been captured but moved to richer sources when the effort required to catch another fly became greater than that needed to move on. But the shifts were not random; the wagtails moved to other areas where insects were plentiful. These decisions seemed to be based on seeing the larger flies on fresh dung pats. But the birds may well have also been influenced by memories of locations where they had found plentiful food in the recent past.

Behavioral ecologists who analyze feeding tactics such as those of these Oxford wagtails, and the many other kinds of foraging behavior that have been analyzed quantitatively as reviewed by Real (1994), ordinarily avoid speculating about any possible thinking on the bird's part as it makes decisions that are important for its survival and reproduction. For example, after discussing quantitative analyses of foraging behavior in some detail, Shettleworth (1998, 422) concludes that "optimal foraging theory depicts animals as selected to behave as if they are looking ahead. . . . But general principles of reinforcement theory seem to account for how animals actually do behave in the situations of interest to foraging theorists. One way to look at what's going on in the standard sort of risk-sensitivity experiments, and indeed any similar choice experiments, is that associative strength developed to each of the options in the forced trials and the free trials reveal the relative values of those associative strengths." This implies that there is no need for even the simplest sort of perceptual consciousness. But the multiple factors that must be evaluated, and the widely varying details of feeding situations, would seem to render a little simple thinking helpful and therefore adaptive. This is especially probable when the animal is faced with novel and unpredictable challenges for which previous experience and genetic programming have not prepared it.

Blackbirds' Decisions about Feeding Ecology

A detailed study of the behavioral ecology of marsh-nesting blackbirds by Orians (1980) has revealed how many subtle factors influence both the selection of insect prey and the choice of mates and nesting territories. The red-winged blackbird *(Agelaius phoeniceus)* and the somewhat larger yellow-headed blackbird *(Xanthocephalus xanthocephalus)* are abundant breeding birds in the marshes of the northwestern United States and western Canada, and they nest in areas that are sufficiently open that most of their feeding behavior can be observed relatively easily. Orians and his colleagues concentrated on the nesting season, when the parents are under great pressure to obtain enough food for their nestlings. This of course is a situation where natural selection operates powerfully on the birds' behavior, for the number of healthy young that can be raised depends directly on the amount and quality of food their parents capture and bring back to the nest.

Both species of blackbirds nest in vegetation growing in shallow water. The redwing is strongly territorial; in the spring the males arrive first and establish territories that include an area of marsh or adjacent upland. Females arrive later, and after visiting several male territories each female settles in one, mates with the territorial male, and builds her nest within his territory. The females do all the nest-building and incubation of the eggs and almost all the feeding of the nestlings, although after the young have left the nest the males also feed them. The yellowhead males provide more help in feeding their nestlings, but otherwise the habits of the two species are similar. Both feed heavily on adult forms of aquatic insects that have just emerged from the water, but the redwings also feed on insects they find on the dry upland areas. The larger yellowheads exclude redwings both from their nesting territories and from the richest sources of aquatic insects.

Marshes vary greatly in the abundance of the insects on which the blackbirds feed, and the density of nesting birds is roughly correlated with insect abundance, although other factors also play a role. For example, the yellow-headed blackbirds avoid areas with a continuous stand of trees extending more than about 30° above the horizon or, in one case, a marsh where tall cliffs rose abruptly from the water's edge. The tendency to avoid nesting in such areas, even when insect food is abundant, is probably related to the danger that hawks may select such places for their nests.

It is important to recognize that the male blackbirds make extremely important choices about nesting territories well before their young hatch and require an abundant source of insect food. At this time very few aquatic insects have emerged, so the choice of territory must be guided by something other than the contemporary abundance of insects. Somehow they do ordinarily make appropriate choices, selecting out of extensive areas of marsh the localities that later produce the richest harvest of aquatic and terrestrial insects. One might suppose that these choices are guided by memory and tradition, the blackbirds simply remembering where they nested last year or where their parents raised them. But these marshes change rapidly from year to year owing to ecological changes such as variation in water level and invasion of lakes by carp that drastically reduce the populations of aquatic insects. Newly arrived blackbirds often forage at the air-water interface when they are selecting territories. When the females arrive they seem to ignore the vigorous displays of the males and instead spend a great deal of time at the edge of the water. Perhaps they are looking for the aquatic larvae and nymphs of insects that will later emerge as adults.

It seems likely that the patterns of open water and emergent vegetation may be utilized, since marshes that will later provide abundant emerging insects have certain properties. For example the density of stalks of aquatic vegetation is important. If they are too closely spaced, relatively few insect larvae will be present, but many will emerge at the outer edge of this vegetation. This almost certainly improves foraging at the outer edge of vegetation but makes it less productive elsewhere. Of the two species the larger yellowheads tend to occupy the more open vegetation, while the redwings nest in denser vegetation, usually closer to the shore, where food is less abundant. This may result from actual territorial exclusion of the redwings by the larger yellowheads, but it may also reflect the fact that the redwings do more of their foraging on upland areas.

After the young blackbirds have hatched, the mother, and in the case of the yellowheads sometimes the father as well, have an extremely demanding task of finding, catching, and carrying back to the nest a sufficient number of insects to feed their hungry young. The actual selection and capture of insect prey is difficult to study in detail, because many of the insects are small and the birds cannot always be approached closely enough while they are feeding to see just what they are doing even with the aid of binoculars. Orians and his colleagues used several ingenious methods to determine what quantities of different insects were taken. One tactic was to place around the neck of a nestling a loose collar formed from a soft pipe cleaner, not tight enough to prevent breathing but sufficient to prevent swallowing of insects. The accumulation of insects in the nestling's mouth was then removed for analysis after it had been fed. This procedure showed that as many as ten insects might be delivered to a nestling on one return visit by its mother.

Orians and his colleagues also learned by tedious observation and long practice to identify through binoculars many species of insects as they were captured by the blackbirds. Sometimes a bird carrying a large load dropped what it had been pursuing and caught an additional insect, but in such cases it always picked up the previously gathered prey and then carried the whole lot back to its young. The foraging behavior of these blackbirds conformed at least approximately to expectations based on optimal foraging theories. For example, when gathering food close to the nest birds should return more often with smaller loads than when they are obliged to search for food at greater distances. In the former case the return trips require less time and energy.

The adult blackbirds also had to feed themselves, and they usually swallowed the first few insects captured on any one foraging trip before beginning to gather food for the young. The utilization of specific types of insect prey differed to some extent according to the circumstances. When they were not feeding young these blackbirds often ate dragonflies. The available dragonflies were quite large and provided excellent nutrition for the young, and even when they were the first insect caught on a particular sortie the parent bird did not swallow them but carried them back to the nest. What, if anything, do such busy parents think as they devote most of their waking hours to gathering food for their hungry young? Perhaps "Those youngsters need food" or "That dragonfly will stop their squawking for a while." We cannot say, as yet, but these are plausible inferences that should be kept in mind as hypotheses awaiting an adequate test.

Behavioral ecologists tend to assume that some genetically determined action pattern guides these choices. They seldom allow themselves to speculate about whether these decisions are too complex and too dependent on unpredictable circumstances to permit detailed preprogramming and whether any simple perceptual consciousness facilitates the examination of several marshy areas and the choice of one only after devoting a considerable amount of time to assessing the situation. Might the birds think that certain types of marshy vegetation are more likely than others to provide abundant insects a few weeks later?

How could cognitive ethologists hope to test this hypothesis? It might turn out after appropriate investigation that part of the male's display behavior communicates to females some such message as "Lots of insects here." If so, this would become a case where interpreting the communication provided evidence of the thought that was being communicated. Another speculative possibility would be to employ experimental procedures comparable to those that show which of several stimuli pigeons recognize as similar or different, as will be discussed in detail in chapter 7. Although such experiments have so far been limited to restricted laboratory situations, they might be modified and extended to inquire of blackbirds whether certain aspects of a male's displays were correlated with insect abundance.

Oystercatchers' Mussel-Opening Techniques

The oystercatcher *(Haematopus ostralegus)* is a large shorebird with a conspicuous red bill; it is related to the sandpipers and plovers. Mussels exposed at low tide are one of their principal foods, but they also feed on other shellfish and on earthworms exposed in plowed fields. The English behavioral ecologist Norton-Griffiths (1967, 1969) discovered that on the coast of Cumberland they use two principal techniques for opening mussel shells. When the mussels are fully exposed at low tide the birds seize them with their sturdy bills, pull them loose from the substrate, and carry them to a patch of sand where they turn the shell so that its flat ventral surface is uppermost. Even though this is not the most stable position, the oystercatcher maneuvers the mussel so that it remains in this position while hammering open the shell. To determine what forces were necessary for this operation, Norton-Griffiths built a mussel-cracking machine, using a close copy of an oystercatcher bill as the pick. The flat ventral surface of the shell proved to be the most easily broken part. Each oystercatcher learned where the sand was suitably hard for this operation and brought numerous mussels to the same spot.

When mussels are covered by shallow water the oystercatchers open them in an entirely different way. They search for slightly open shells and stab their bill into the opening. They do this in such a manner that they cut the large abductor muscle that closes the shell. After thus rendering the fleshy body of the mussel accessible, the oystercatcher tears the shell loose from the substrate and carries it to some convenient spot where it picks out the body of the mussel and eats it. At first this difference in feeding behavior appeared to be an adaptive adjustment to circumstances and opportunity; if the mussel shell is tightly closed, as it is when fully exposed, it must be hammered open, but when the shell is slightly opened underwater the stabbing technique is easier. But when Norton-Griffiths marked individual oystercatchers and observed their feeding behavior, it turned out that each one specialized in one or the other procedure. Further study strongly indicated that the young oystercatchers learned which technique to use when they began feeding with their parents.

Darwin's Finches

The finches that inhabit the Galápagos Islands provide a classic case of evolutionary diversification. Sometime within the past million years or so an ancestral population established itself on these dry volcanic islands, and its descendants evolved into thirteen species that range in size from less than 10 to more than 40 g. Collected by Charles Darwin, their nearly continuous variation left him somewhat puzzled, so it is not clear just how influential they were in his discovery of evolution by natural selection. The history, ecology, and evolutionary biology of these birds has been recently reviewed authoritatively by Grant (1986) and in more popular style by Weiner (1994) There are three groups of Galápagos finches, which can conveniently be categorized as ground finches of the genus *Geospiza,* tree finches of three genera, and the warbler finch *Certhidae olivacea.* The species differ most conspicuously in the size and shape of the beak, and these differences are clearly correlated with feeding habits. Finches with short, thick beaks feed on seeds, many of which are too hard and tough to be cracked by the more slender beaks of other species that specialize on insects.

Darwin's finches live in a harsh environment where food is often very difficult to obtain, except after the occasional rains, when vegetation and insects become much more abundant than during the usual dry periods. Their diet tends to be opportunistic; as Grant summarizes it:

> As a group Darwin's finches rip open rotting cactus pads, strip the bark off dead branches, kick over stones, probe flowers, rolled leaves, and cavities in trees, and search for arthropods on the exposed rocks of the shoreline at low tide. They consume nectar, pollen, leaves, buds, a host of arthropods, and seeds and fruits of various sizes. . . . By virtue of their deep beaks, and the masses and disposition of the muscles that operate them, ground finches crush seeds at the base of the bill. In contrast, tree finches apply force at the tips of their bills to the woody tissues of twigs, branches, and bark, and thereby excavate hidden arthropod prey. . . . The warbler finch, cactus finches, . . . woodpecker finch, and mangrove finch have relatively long bills which they use to probe flowers for nectar or holes in woody tissues for arthropods (393).

In addition to these general tendencies to specialize in foods for which their beaks are adapted, some of the Galápagos finches have highly specialized feeding habits. Woodpecker finches *(Cactospiza pallida)* and mangrove finches *(C. heliobates)* hold twigs, cactus spines, or the petioles of leaves in the beak and use them as tools to pry arthropods of various kinds out of crevices. On two small islands, Wolf, or Wenman, and Darwin, or Culpepper, about 100 km from other islands of the archipelago, the sharp-beaked ground finches *(Geospiza difficilis)* have developed the most unusual habit of feeding on the blood of boobies (genus *Sula*). This habit probably began as a mutually advantageous feeding on ectoparasites; a related species of ground finch commonly eats ticks from the skin of marine iguanas, some of which have ritualized displays that solicit tick removal by the birds. This general type of behavior is not unique to Darwin's finches; in other parts of the world several species of small bird have developed the habit of feeding on parasites that they pick off the skin of large mammals.

The "vampire finches" direct their vigorous pecking selectively at the base of boobies' feathers, most often near the elbow of the folded wing, drawing enough blood to drink. The boobies try to dislodge them, but the finches usually succeed in obtaining a blood meal by repeated attempts. This habit is well established on Wolf Island, but on other islands where the same species has been thoroughly studied numerous boobies also nest without any sign of its occurrence. On Wolf Island, *G. difficilis* also "push and kick seabird eggs against rocks, widen cracks that form in the shell, and then consume the contents" (Grant 1986, 393).

Galápagos finches are not the only birds with diversified and ingenious feeding habits, but they demonstrate how much versatility is required to make a living under difficult conditions. The use of twigs as probing tools and the selective pecking at the base of booby feathers to obtain a drink of blood are especially suggestive. It is important to recognize that the same individual finches employ a wide variety of food-gathering techniques according to the circumstances, although no one species exhibits the full range of feeding specializations displayed by the Darwin's finches as a group. We can only speculate about the origin of feeding on booby blood, but when a finch pursuing ectoparasites that crawled deeper into the thick feathers perhaps accidentally pecked hard enough to break the skin, was it pleasantly surprised to find a source of nutritious fluid? Did this perhaps remind it of the fluids to be found and eaten from inside cracked seabird eggs?

This calls to mind the extensive experiments of Tolman (1932, 1937) with laboratory rats that seemed to be surprised when food they had every reason to expect to find at the end of a well-learned maze was not forthcoming, as discussed in chapter 7. Here we would be dealing with pleasant surprises rather than disappointments, but if and when methods are developed to test the hypothetical inference of such simple but probably vivid mental experiences, we will have learned something important about the animals concerned.

Searching Images

When a hungry animal is searching for food under natural conditions it would waste a great deal of time and effort to scrutinize every detail of its surroundings. Both evolutionary selection and learning must exert a strong influence on searching behavior, as discussed in detail by Shettleworth (1998). Animals concentrate their attention not only on things that look, sound, smell, or feel like food but also on quite different things that are signs showing where food may be available. Specialized sensory systems are sometimes employed in searching for food. For example, some sharks detect the weak electric currents from the contractions of the heart or other muscles of prey animals that would otherwise be very difficult to locate (Kalmeijn 1974). And insectivorous bats distinguish the sonar echoes of edible insects from the many other echoes returning to their ears (Schnitzler et al. 1983; Ostwald et al. 1988). But most searching is based on vision, olfaction, or hearing.

Detailed studies of foraging birds have shown that they look for particular patterns that reveal where food is to be found, such as the barely perceptible outline of a cryptically colored moth resting on the bark of a tree trunk (Pietrewicz and Kamil 1981). In a wide variety of laboratory experiments, rats or pigeons learn that visual patterns which ordinarily have nothing to do with food are now a signal that food can be obtained. Somewhere in the animal's brain there must be a mechanism for recognizing what are termed *searching images.*

One of the most thorough and significant studies of searching images was carried out by Harvey Croze (1970). On a sandy beach where carrion crows were gathering mussels at low tide he laid out a row of empty mussel shell halves, convex side up, and beside each shell he placed a small piece of beef. After five hours the crows had taken them all. The next day Croze laid out 25 mussel shells, and under each one he hid a similar piece of beef. When the crows returned they turned over 23 of the shells and ate the meat. They had learned quickly that mussel shells lying on the sand, which would ordinarily be empty, had suddenly become signs of tasty food. On the third day the meat was buried in the sand underneath the shells. The crows turned these shells over, but finding nothing directly under them, dug with their bills in the sand until they found and ate the meat. Although olfaction is not well developed in birds, one cannot rule out the possibility that the crows could smell meat at close range, especially since pigeons can discriminate between different odors in laboratory experiments (Schmidt-Koenig 1979; Walraff 1996; Wiltschko 1996; and Able 1996). But regardless of the sensory channel employed, these carrion crows had obviously learned quite rapidly that food might be buried in the sand under empty mussel shells.

Croze continued to place similar mussel shells on the beach, but now he added no meat. For some time the crows continued to turn over these shells, but they gradually paid less and less attention to them. When they were only occasionally turning over mussel shells, Croze placed bait under some. When a crow found one that was thus baited, it began turning over many more. Under natural conditions it is common for something to be a sign of food only some of the time. Animals learn that it pays to inspect such objects even though they yield food only occasionally, and when they do yield food, to search for similar objects. Similar behavior has also been observed in other animals, as reviewed in the symposium edited by Kamil and Sargent (1981).

The signs of food availability may be difficult to recognize, for it is obviously advantageous for potential prey to have properties that impede easy detection. But locating food is so crucial that many animals have developed not only efficient sensory mechanisms for distinguishing signs of food from very similar objects, but an ability to learn what is a fruitful searching image. This is well illustrated by the experiments of Pietrewicz and Kamil (1981), who applied to blue jays instead of pigeons the type of operant conditioning procedures developed by psychologists. Naive blue jays were adept at learning how to pick out cryptically colored moths resting on backgrounds very similar to their own appearance.

Animals cannot ordinarily predict what objects are likely to indicate the presence of food, and an ability to learn about novel signs of food is useful to many species. An interesting example grew out of the marking of nest locations by ethologists. Ground-nesting birds often lay eggs that resemble the substrate on which they are laid, and some conceal their nests very effectively under vegetation. Having laboriously located such nests, ethologists often mark their location by placing stakes a short distance to one side, in order to facilitate finding them again, assuming that the stakes will not affect the behavior under study. But in one study of nesting phalaropes and semipalmated sandpipers near Churchill, Manitoba, many nests were marked by 50-cm stakes placed 2 to 3 m southeast of each nest (Reynolds 1985), and at least one sandhill crane learned that these novel objects were signs of tasty food. Reynolds observed this crane searching diligently near nest markers, and although nest predation was not observed directly, eggs were missing from nests near which there were perforations in the ground almost certainly made by the bill of a probing crane. This use of completely novel searching images indicates how versatile birds can be in learning what to look for when foraging.

Tit Tactics

The genus *Parus* includes the North American chickadees and titmice, those acrobatic and entertaining visitors to thousands of bird feeders, along with several European species known in England as tits. Because they do well in captivity and display ingenious foraging behavior, J. R. Krebs and others have studied the nature and efficiency of their feeding tactics when they search for insect prey. The stated purpose of these investigations was to test mathematical theories about optimal foraging behavior. Although the investigators did not admit to any interest in whatever thoughts and feelings the birds might have experienced, their findings provide suggestive hints of perceptual consciousness.

In one set of experiments Krebs, MacRoberts, and Cullen (1972), Krebs and Davies (1978), and Krebs (1979) studied how great tits *(Parus major)* coped with the challenging problems of foraging for concealed food. In order to standardize experimental conditions they did not study the capture of normal insect prey hidden on natural vegetation. Instead they used mealworms (larvae of the flour beetle), which are about 3 mm in diameter and about 25 mm long. Many insectivorous animals eat them avidly in captivity. To analyze both how the tits would learn to find mealworms hidden in different ways and how several hungry birds foraging in the same area would interact, four types of hiding places were employed: Plastic cups called "hoppies," ping-pong balls cut in half, called "pingies," 7.5 × 4.5 × 4.5 cm blocks of wood termed "milkies" with 1.5 cm holes drilled 2 cm deep into their tops, and "barkies" consisting of small strips of masking tape stuck on trunks of artificial trees constructed from wooden dowels. Under the tape there might be a mealworm forming a small bump. The milkies were designed to simulate milk bottles, which wild tits had learned years earlier to open by tearing off the thin metal foil used to cover their tops, as discussed below.

The first three types of container were filled with sawdust or bits of paper, and the openings were covered with masking tape. Only some contained mealworms, so that the birds had to open them to find whether they were or were not sources of food. To prepare the birds for this new type of foraging they were first given uncovered mealworms, then mealworms buried in sawdust in the same containers but not covered with masking tape, and finally covered containers. They learned surprisingly quickly that they could sometimes find food by pecking through and tearing off the tape. Some of the barkies consisted of short pieces of thick string covered with tape, but the birds never learned to discriminate between bumps formed by mealworms and bumps formed by bits of string. All these experimental arrangements were designed to simulate the task faced by birds when they search through large areas of vegetation for the few spots where something edible can be uncovered by probing or pulling off dead leaves or layers of bark.

Except for failing to distinguish barkies concealing mealworms from those concealing bits of string, the great tits learned to forage in these new types of insect hiding place. If mealworms were provided in one type of container but not the others, they concentrated their foraging on the type that had yielded food. Individual great tits specialized in particular methods of extracting the mealworms. Some concentrated on the hoppies, turning over the pieces of paper much as wild birds turn over leaves lying on the ground and peering into the container. Others wallowed in the container and threw out the pieces of paper. One discovered that by pecking through the masking tape he could peer into the container and see whether a mealworm was present. Some opened the milkies by hammering through the tape, while others pulled away one edge of the tape.

These individually varying patterns showed that foraging is not a fixed, stereotyped pattern and suggested that each bird was trying various actions in its efforts to find concealed food. When one bird was allowed to discover that there was a mealworm in a particular container, others housed in the same cage also began to look in similar places or similar containers. In short, these great tits learned a great deal about where to find concealed food in this novel situation both by remembering where and in what sort of hiding place they had found mealworms, and by observing other birds finding them. Once a food source had been identified, the dominant bird of a group would chase others away from it and take more of the food.

The versatility of insectivorous birds in finding hidden food items led to a spectacular development in the 1930s when two species of tits discovered that milk bottles delivered to British doorsteps could be opened by pulling the metal foil off the tops (Fisher and Hinde 1943; Hinde and Fisher 1951, 1972). At that time the milk was not homogenized, and the tits could drink the thick cream from the top of the bottle. Careful studies were made of the gradual spread of this behavior throughout much of England until a change in the technology of covering milk bottles eventually deprived the birds of this source of food. Although the matter was not studied directly, it seemed at the time very likely that increasing numbers of tits took up this habit by observational learning—seeing that another bird had found food in a novel sort of place, just as the experiments of Krebs and his colleagues showed they could do under controlled conditions. Many behaviorists are suspicious of such suggestive evidence for observational learning, however, as recently reviewed by Galef (1988), who considers it "an onerous concept" and prefers the term *social facilitation,* advocated especially by Clayton (1978), which seems to avoid the implication of conscious imitation. Shettleworth (1998) has reviewed in considerable detail the whole subject of what she calls "social learning" and recognizes that many animals do learn new kinds of useful behavior when able to observe others carrying out such behavior. But she treats social learning as a form of conditioning, implying that if it is a form of conditioning it therefore need not involve even perceptual consciousness of what is learned. I will return in chapter 14 to a more detailed discussion of the relation between learning and consciousness.

The basic motions used by tits to open milk bottles were much the same as those used to pull layers of bark from vegetation. This has led many to feel that the birds' discovery that milk bottles were a source of food was not so novel after all because it did not require the development of a whole new motor pattern. But it was versatile application of a previously well-developed type of action to a new and wholly different situation. This is typical of the interplay of ideas about animal behavior that has characterized advances in ethology. Some see novel behavior as evidence of radical inventiveness, a sudden insight that some completely new type of behavior will yield a desired result. Others note that the same limbs are moved in ways that are not very different from other long-established patterns of behavior and pooh-pooh the animal's performance as nothing new.

Sherry and Galef (1984, 1990) and Sherry (1990) have studied the acquisition of very similar milk bottle–opening behavior in captive chickadees *(Parus atricapillus),* close relatives of the British tits studied by Fisher and Hinde. They found that 20 of 64 chickadees presented for 75 minutes with a laboratory version of a foil-covered milk bottle spontaneously learned to remove the foil and consume the milk. The 44 birds that had not spontaneously opened their bottles in 75 minutes were given further opportunity to do so, either alone or with another chickadee easily visible in an adjacent cage. Some but not all then began to open the milk bottles, but the surprising result was that about as many did so whether or not the bird in the other cage opened its bottle while visible to the chickadee under study. In related experiments chickadees were more likely to open a milk bottle if an opened bottle was present in the cage. This seems to support Galef's general view that learning by imitation is rare or nonexistent in animals and that social facilitation is a more parsimonious explanation of cases like the milk bottle–opening by British tits in the 1930s.

On the other hand, Palameta and Lefebvre (1985) showed that pigeons were more likely to learn to open and feed from a particular type of food bin if another pigeon had done so in their presence. But a basic question in all these cases is whether, when an animal learns a new type of feeding behavior because of the presence or actions of another animal, it thinks consciously about the new behavior. It is less important whether the new behavior is drastically different from previous feeding behavior or entails applying motor patterns used in other situations to a new type of food source. As in many other examples of versatile behavior, it is theoretically possible that it might be carried out unconsciously, but the likelihood of this seems intuitively to be inversely proportional to the degree of novelty of the behavior that the animal performs.

The tits that originally learned to open milk bottles on English village doorsteps certainly did not comprehend the nature of milk bottles, the properties of glass and metal foil, or the nature and source of cream. But in hungrily exploring the environment of village streets, sidewalks, and doorsteps, they found that pulling off a sheet of shiny, pliable stuff allowed them to get at a new and tasty sort of liquid food, and they went on to exploit this new food source. The other tits that picked up the habit may have done so because each bird individually discovered the new food source on its own, found milk bottles already opened by another bird, or was stimulated by watching its companions apply their customary food-searching behavior to this novel situation. This is probably typical of innovation by animals; they learn that applying familiar motor patterns to new objects or in novel situations achieves a desired result or avoids an unpleasant one. Langen (1996) described a similar case of social learning by magpie-jays.

To what extent does the novelty in applying well-established actions to new situations indicate conscious thinking? The degree of novelty is surely an important factor, for the essence of the distinction between social facilitation and true observational learning is that in the former case the presence, or sometimes the food-getting, of one or more other animals elicits a form of behavior already familiar to the animal in question. Although inclusive behaviorists are so averse to considering even simple perceptual consciousness that they seldom phrase the distinction in such terms, what they seem really to mean by true imitation, which

Galef finds "an onerous concept," is the conscious copying of another animal's behavior.

Dropping Shellfish on Hard Surfaces

An enterprising sort of behavior carried out quite often by gulls is to carry clams, whelks, or other shellfish to some hard surface and drop them from a height of several meters. Similar behavior has been observed in other birds including crows, as thoroughly reviewed by Switzer and Cristol (1999) and Cristol and Switzer (1999). This process is quite different from finding food, seizing it, breaking it up, and swallowing. A potential food item must first be recognized as something that becomes edible only when its outer shell is broken. Then the bird must pick it up, fly with it to a suitable place, and drop it from a sufficient height. In some areas ethologists have reported that gulls are not selective and drop shellfish on soft surfaces where they do not break. But it is a common observation that herring gulls drop the great majority of the shells they carry on rocks, roads, parking lots, or other places where the shells do break. The accumulation of broken shell fragments can often be found concentrated along particular stretches of a paved road. In tidal estuaries consisting of sand flats and salt marsh with very few rocks, the rocks are sometimes surrounded by a halo of clamshell fragments.

Beck (1980, 1982) studied shell-dropping by herring gulls on the shores of Cape Cod, Massachusetts. These birds picked up clams, whelks, or empty mollusk shells inhabited by hermit crabs that they found at low tide and carried them 30 to 200 m to rocky areas, sea walls, and paved roads or parking areas. Beck observed that the gulls usually flew quite low over the beaches so that during all but the last part of each flight they could not see the hard surfaces toward which they were flying. They had evidently learned where to take shellfish for this purpose. In one area 90 percent of the drops were directed at a particular sea wall that occupied only about 1 percent of the area over which the gulls flew.

Beck also observed that when dropping shells on this relatively small sea wall the gulls did so from lower altitudes than when using a large parking lot. Perhaps they realized that there was no danger of missing the larger target and that the shells were more likely to break from a greater height. Shells did not always break, and the gulls often picked up an unbroken shell and tried again. Beck also observed that young herring gulls were less successful than adults; they dropped rather few shellfish but often carried up and dropped other objects, apparently in play. We have no detailed data, however, concerning the development of shell-dropping behavior.

Other studies (Zach 1978) showed that a particular population of crows nesting on an island near the coast of British Columbia had developed the habit of gathering whelks at low tide and dropping them on particular rocky areas, where they often broke so that the crows could eat the soft parts. Each pair of crows foraged at one section of beach and dropped whelks on a particular rocky area. Not all rocks were appropriate, because if they were too close to the water the whelks would bounce off and sink. If the rocks were not fairly level, edible fragments fell into deep crevices where they were also lost. From a range of sizes of whelks the crows selected the larger ones, and they did not pick up dead whelks or empty shells. Zach also studied the selection process by stuffing empty whelk shells either with very light material or with something similar in density to the living mollusk. He then glued back in place the horny operculum that closes the shells and found that the crow selected shells of normal weight or density. Presumably a lightweight shell would ordinarily mean a shriveled corpse rather than a fresh and edible morsel.

These crows dropped whelks from heights ranging between approximately three and eight meters. Only about 1 in 4 drops broke the shell, so that many had to be picked up and dropped over and over again. Unless they were disturbed, the crows persisted until the shell finally broke, which sometimes took as many as 20 drops. Why didn't they fly higher so that shells would be more likely to break? Zach's observation suggested that the crows had some difficulty in seeing where the whelk had fallen, and after dropping one they descended in such a way that it seemed that they were watching the trajectory of the falling shell. When an occasional whelk was dropped from a greater height it also seemed more likely to shatter into several fragments, making it more difficult for the gulls to locate and pick up all the soft parts. Perhaps it was less pleasant to eat whelks containing small bits of broken shell. This was indicated by the fact that crows sometimes dipped broken whelks into freshwater pools before eating them, apparently removing fragments of shell.

Only a small fraction of the crows collected whelks and dropped them on rocks, so this specialized feeding behavior had presumably developed quite recently. The initial stages of crows' discovery of this type of feeding and its spreading to other birds have not been studied. But Zach did observe some individual differences. For example, one crow picked up and dropped two whelks simultaneously, although all the others carried only one at a time. Such individual variability suggests that the birds had tried different tactics and were gradually learning which ones were most effective.

Shettleworth (1998) analyzed another case in which birds obtain food by dropping hard-shelled walnuts on paved roads, sidewalks, and parking lots in Davis, California. Some observers had inferred that the crows selectively dropped walnuts where approaching automobiles would run over them, but later studies by Cristol et al. (1997) showed that the crows were no more likely to drop nuts when cars were approaching than when they were not. Although recognizing that these crows were taking into account many relevant variables and dropping nuts from heights and on surfaces that were most likely to yield edible food, Shettleworth emphasizes the debunking of the interpretation that those "who saw nutcracking as an expression of clever crows' ability to reason and plan were engaging in . . . anthropomorphism . . . but research often reveals that simpler, more mechanical-seeming processes are doing surprisingly complex jobs" (4). But she does not consider the possibility that simple perceptual consciousness might facilitate the adjustments of height from which nuts were dropped and the selection of suitable places to drop them.

Caches

Many animals store excess food when it is abundant, as thoroughly reviewed by Vander Wall (1990, 219–24) and Balda and Kamil (1998). Out of Vander Wall's long list of animals known to store food, it is especially impressive that shrews and moles sometimes store large numbers of prey animals. When they find abundant prey, the best-studied species of shrew, the short-tailed shrew *Blarina brevicauda,* usually eat the first few insects captured but then store the rest in or near the nest. These shrews secrete a venom from the submaxillary glands which often paralyzes prey animals so that they remain alive but immobile. They do not decay as rapidly as they would if killed and remain available as food during what must often, under natural conditions, be long periods of scarcity.

Food hoarding increases in winter, and under some circumstances these shrews move live snails they have stored in their burrows back and forth between the burrow entrance and a storage chamber. This shifting of stored snails occurs primarily in cold weather, and it may serve to keep the snails in the coolest places available. Moles store earthworms and often bite anterior segments so that the worms are paralyzed or at least move very little. European moles *(Talpa europea)* sometimes store earthworms and insect larvae in specially excavated chambers near their nests.

Curio (1976, 22–25) describes how numerous predators store captured prey for later consumption. Often when a hungry animal finds an abundance of food it stores more than when the same food is located after it has eaten its fill. This might be due to thinking about future needs even when satisfying its immediate hunger. Under natural conditions much of this stored food is recovered and eaten later, but by no means all. Other animals take some, and the animal that stored it in the first place does not always retrieve it. But a substantial fraction is recovered, by some species at least, often after weeks or months.

Squirrels are the most familiar animals that store food, but there has been considerable uncertainty as to how they recover buried nuts. Cahalane (1942) concluded that western fox squirrels retrieve buried nuts not by memory but by smelling them. Lewis (1980) reached similar conclusions, and it has also been suggested that, rather than remembering where they stored nuts, squirrels notice signs of the ground having been disturbed. More recently, however, McQuade, Williams, and Eichenbaum (1986), Jacobs (1989), and Jacobs and Liman (1991) found that captive gray squirrels did remember the location of some of the food they had stored in a large cage. It remains to be determined what role memory plays under natural conditions.

Birds have provided stronger evidence of recovery of stored food by means of a detailed memory (Kallander and Smith 1990). Experimental studies of marsh tits by ethologists at Oxford University have been especially revealing. These birds are very similar to North American chickadees, but at bird feeders they must compete with more species of birds that eat similar foods. As reviewed by Shettleworth (1983, 1998) and Sherry (1984), two other tits of the genus *Parus,* the great tit and the blue tit, tend to eat rapidly when peanuts or other desirable seeds are available. Marsh tits, on the other hand, tend to grab single seeds, fly away, store them, and quickly return for repeated storing of individual seeds in different places. They do return to the hiding places and recover many, but not all, of these seeds. Cowie, Krebs, and Sherry (1981) studied the caching behavior of marsh tits under natural conditions by radioactively tagging sunflower seeds and detecting their hiding places with a scintillation counter. They found that stored seeds disappeared much more rapidly than control seeds stored in very similar sites by the experimenters.

Turning to laboratory experiments, where conditions could be better controlled, Sherry, Krebs, and Cowie (1981) allowed captive marsh tits to store several hundred seeds per day in a large aviary. The birds recovered stored sunflower seeds at far higher than chance levels after being kept out of the aviary for 24 hours. To control for other possibilities, such as having preferred types of places to store seeds and looking for such places when searching, these experimenters took advantage of the fact that visual information is processed and stored almost exclusively on the opposite side of a bird's brain from the eye where it is received. They covered one eye of marsh tits and found that this treatment did not affect their rate of recovering stored seeds. But if the blindfold was shifted from one eye to the other after seeds had been stored, the birds located seeds only at approximately a chance level.

This greatly strengthened the evidence that they were remembering specific locations where seeds were stored even when dozens of such places in a large aviary had to be remembered. In further experiments of the same general type Shettleworth and Krebs (1982) found that marsh tits remembered quite well where they had stored seeds. Furthermore, when storing additional seeds they usually avoided the hiding places where they had recently placed a seed. Evidently marsh tits specialize in storing and retrieving seeds, and they are probably much better at this task than many other species.

Cached food is often taken by other animals of either the same or a different species. Therefore it is important for an animal to place any stored food in places where others are as unlikely as possible to find it. This often means carrying it some distance from its source or from watchful companions, who may nonetheless seek to follow and rob the cache. Heinrich (1999) has described in detail the complex caching and cache-robbing behavior of ravens, which devote considerable effort and ingenuity to hiding food or other items in ways that their companions are less likely to observe. It is important for both the original cacher and later cache robbers to remember where the food was placed, and this is often made as difficult as possible by the cacher. Nevertheless, ravens do often manage to hide and later retrieve surplus food, although other ravens are often equally ingenious in discovering what has been concealed.

An even more impressive example of memory for food caches has been provided by studies of Clark's nutcracker, a relative of the crows and jays, which lives in alpine environments in western North America. Food is very scarce during long, cold winters, but during the autumn these birds gather enormous numbers of pine seeds and store them in crevices or bury them in the ground. This behavior has been studied in detail by Balda (1980), Vander Wall and Balda (1981), Vander Wall (1990), Balda and Turek (1984), and Balda and Kamil (1998). When seeds are plentiful a single bird may hide as many as 33,000 during the fall. Each cache ordinarily contains two to five seeds. To obtain enough food to survive through a typical snowy winter it is estimated that the Clark's nutcracker must recover approximately a thousand of its caches.

In order to study this remarkable type of memory under better-controlled conditions, captive nutcrackers were allowed to bury pine seeds in sand spread over the floors of a large aviary cage. After being kept out of the cage for a month, a nutcracker recovered and consumed a substantial fraction of these seeds. To be sure, some seeds could be found by random searching, and when the experimenters buried numerous seeds themselves, the bird found a few. But the nutcracker found many more of the seeds it had stored itself. Had it been using odor or some other cue than its memory, it would presumably have been equally able to find both kinds of seed. The cage was provided with conspicuous logs and stones on the floor. The nutcracker buried most of its seeds in the sand near such landmarks, and its success at finding them was reduced when no such landmarks were available. And when the stones or logs were shifted during the month that the bird was not allowed in the cage, it usually searched near a certain part of a log as if it remembered the location of each cache in relation to this large and conspicuous object. Its recovery rate was very low after the landmarks had been shifted.

Under natural conditions the environment where seeds are hidden changes drastically during the winter as leaves fall and the ground is often covered with snow. The nutcrackers must remember at least approximate locations with reference to relatively large and constant landmarks such as trees or sizable rocks. Field studies have shown that they do search even under the snow in more or less the right general locations where they have stored seeds. Presumably they remember a large number of such locations, and they may look at them from time to time as leaves fall and snow arrives to reinforce their maplike memories. When it is important for birds and other animals to remember a large number of details, they are often able to do so. To what extent they think consciously about this task is another and more difficult question. They can be regarded as thoughtless memory machines, but, as in other cases of this sort, simple perceptual consciousness may well be helpful.

Balda and Kamil (1998) have recently reviewed similar studies of pinyon jays *(Gymnorhinus cyanocephalus)* in the southwestern United States. Not only do they store hundreds or even thousands of pine seeds, but field and laboratory investigations have demonstrated that they are remarkably successful in remembering where they stored seeds. They succeed in reclaiming roughly 80 percent, which in many cases provide their principal food during the winter months. Balda and Kamil close their review as follows: "We know, from personal experience, that the knowledge we have gained through our laboratory research about these wondrous birds has forever changed our perceptions of them. The 'kaws' and 'kraws' will never sound the same; the commotion around pine trees will never seem the same; the sight of a nutcracker or pinyon jay arriving at its nest laden with pine seeds will never look the same. Awareness of the cognitive abilities of these animals forever changes our perception of them and their place in nature, and ours" (60). Does it not also seem likely that our admiration of their cognitive abilities might also include at least a suspicion that these birds experience simple conscious thoughts and feelings?

CHAPTER FOUR

Predation

The complex and often dramatic interactions of actively mobile predators and their elusive prey provide many suggestive insights into what life is like for the participants in this all-important phase of their lives. For the predator, the location and pursuit of appropriate prey calls for tactics that can be rapidly adjusted to changing and often unpredictable circumstances. Although success or failure in a particular hunt is not ordinarily crucial for the predator, its survival and reproduction depend on succeeding reasonably often. Prey animals must adjust their behavior in many ways, especially to balance the need for food against the dangers of predation. For them each encounter is literally a matter of life or death. The challenges of catching prey and escaping capture are just the sort of situations where conscious thinking may be most helpful. Predator-prey interactions are very seldom studied in laboratories of comparative psychology, but when they are studied carefully under natural conditions it often turns out that the animals adjust their behavior with an adaptive versatility that suggests simple thinking about the likely results of various behavior patterns among which they must make split-second choices.

The behavior of predators and prey has come under strong selective pressure in the course of their evolutionary history. Out of the thousands of species of predators that capture even larger numbers of other animals under natural conditions it has been necessary to select for this discussion a few specific cases where the behavior of both predators and prey has been described thoroughly enough to provide a reasonably clear picture of their respective tactics.

Jumping Spiders

Spiders are a highly diverse group including mobile predators as well as the more familiar web-building spiders. Although we have been used to thinking of them, along with the insects, as genetically programmed mechanisms, the versatility of their hunting behavior rivals that of mammals. The salticid spiders of the genus *Portia* are especially ingenious in their hunting behavior, as reviewed by Wilcox and Jackson (1998). They often feed on web-building spiders, which they kill by invading their webs and deceiving them by mimicking vibratory signals that are part of the web builder's courtship behavior. They adjust the vibrations they make in the victim's web in a variety of subtle ways, and in some cases they begin with a wide variety of signals but when one of these causes a response from the victim, the hunting *Portia* repeats that type of signal which lures the victim to the edge of its web. These are only a few of the versatile hunting tactics employed by various species of the genus *Portia*. If monkeys did what these spiders do, we would be strongly tempted to conclude that they were acting intentionally.

Pike and Minnows

The predator avoidance behavior of small fish, especially species that aggregate into compact schools, has often been interpreted in terms of stereotyped instinctive reactions. The interactions between predatory fish and their prey have been reviewed by Noakes (1983) and Helfman (1986). As emphasized by Magurran (1986), there is actually a great deal of individual variability in fish behavior, and this often involves adaptation to specific local situations. Variation is the raw material on which evolutionary selection operates, and recent investigations have revealed alternative behavioral strategies that are employed and are effective and adaptive for different members of a given species. For example, sticklebacks from lakes where pike are present are much more timid in the presence of pike than others from waters that have no large fish predators (Huntingford 1982). In waters where herons are important predators the sticklebacks spend more of their time near the bottom (Huntingford and Giles 1987). Coho salmon also adjust their feeding behavior in accordance with the availability of food and the risk of predation (Dill 1983). Groups of goldfish search for food preferentially near where other goldfish are feeding, but not near groups that are not gathering food (Pitcher and House 1987).

The interactions of pike and minnows involve especially significant adaptive versatility, as revealed by the detailed field and laboratory investigations of T. J. Pitcher, A. E. Magurran, and their colleagues. These minnows are about 8 cm in length, and they spend much of their time feeding singly or in small groups spaced some distance apart. Pitcher (1986) advocated using the term *shoal* for groups of fishes that remain together for social reasons and *school* for shoals in which swimming movements are synchronized and polarized, that is, all fish head in nearly the same direction.

When a 20–30 cm pike was introduced into a 1.5 × 2 m tank where 20 minnows had been living for some time, the first response of the minnows was to stop feeding and come together into a compact school, each fish about one-half to two body lengths from its neighbors (Pitcher, Green, and Magurran 1986). Earlier investigations had demonstrated that minnows in such schools are less vulnerable to predators. But then, at intervals of three or four minutes, some individual minnows or small groups began inspection behavior. They swam away from the school and approached to within four to six body lengths of the pike, paused for about a second, and then swam back to rejoin the school, which usually then moved farther from the pike than it was before the inspection. When a single minnow conducted an inspection it usually moved from the edge to the center of the school on rejoining it. Earlier experiments had shown that minnows have some difficulty in recognizing pike; they approach both realistic and crude model pike but resume feeding if the model does not resemble a real pike. During inspections, the minnows avoided the most dangerous zone immediately around the pike's mouth. This behavior strongly suggests that the inspecting minnows are aware the pike is a danger and are seeking to find out whether it is actually a predator and also how likely it is to attack.

In these experiments the pike often began to stalk the minnow schools by slowly approaching them. In addition to simply swimming away from the pike, the minnows reacted to a stalking or attacking pike by employing several quite different tactics. One response was for one or more members of the school to "skitter" by accelerating rapidly for one to five body lengths, often moving away from the school and then returning to take up a new position. Sometimes there was a more vigorous and synchronized "group jump" during which the school moved rapidly and vertically to a new position. It seemed unpredictable to the investigators (and presumably also to the pike) which of these tactics would be employed in a given instance, a kind of unpredictably variable "protean behavior" discussed by Driver and Humphries (1988). Another form of antipredator behavior performed by large schools was to open up a space roughly five minnow body lengths in size around the pike as it moved into the school, then close around it as it entered the open space and thus form a minnowless "vacuole" surrounding the predator.

Still another common maneuver is for the school of minnows to swim rapidly ahead of the approaching pike, then separate into two groups that turn toward the predator's tail as it passes. This "fountain" maneuver may also occur if the school meets a diver or fishing gear. As

Magurran and Pitcher (1987) describe the behavior, "escaping fish keep one eye on the object or predator they are avoiding while maintaining a swimming track of approximately 155 degrees to its current position. This causes the school to split into two. By then reducing the angle of their swimming track to maintain visual contact with the object, the fish automatically come together in a shoal behind it" (453). When closely approached or actually attacked by the pike, the schools sometimes exhibit a "flash expansion" in which the fish appear to explode by swimming rapidly in all directions away from the center of the school. Finally, after a school has split up, individual minnows often seek cover between pebbles at the bottom of the tank or under vegetation. Pike counter this tactic by "vigorous attempts to flush minnows out from their hiding places by directing jets of water towards them" (460).

Similar interactions between other predatory fish and the smaller fishes on which they feed have been observed under natural conditions. For example, Helfman (1986) has reported a series of experiments on the responses of a small Caribbean reef fish, the threespot damselfish, to presentations of models of a local predator, the Atlantic trumpetfish. These experiments were conducted at depths of 2–10 m near coral reefs by scuba or snorkel divers who moved a large or a small model trumpetfish toward the damselfish in either a vertical or a horizontal orientation. Trumpetfish were observed to strike at prey more often when in a head-down vertical position. Large and vertically oriented models were more likely than smaller or horizontally oriented models to elicit escape behavior, including entering refuges in the coral. Helfman described several other cases where prey animals exhibit what he calls a "threat-sensitive predator avoidance." As a conventionally cautious ethologist he refrains from any speculations as to conscious intent, but the fish he and others have studied certainly behaved as though aware of differences between threatening and nonthreatening predators and reacted appropriately.

Wilson et al. (1993) and Coleman and Wilson (1998) have found that the boldness with which small sunfish carry out this type of predator inspection varies from one individual to another. Recently Coleman, Arendt, and Wilson (1997) have found that in an experimental tank where a largemouth bass was separated from sunfish by a glass partition, some of the sunfish spent much more time than others near the glass. Then when these individuals were removed, others that previously had not come close to the bass began to do so. This suggests that in such situations, when sunfish are aware that some of their companions are inspecting a predator they remain at a distance, but when no other sunfish is inspecting they may begin to do so.

The overall impression that can be drawn from this series of appropriate maneuvers by predatory and prey species of fish is that the former are actively trying to catch the latter, and that the potential prey are well aware of the danger. On seeing a real or model pike, minnows stop feeding and aggregate into schools; individual minnows approach and inspect the pike from what seems to be a relatively safe distance; and when approached or attacked the minnows adopt a variety of evasive maneuvers that are often effective. Damselfish also approach model trumpetfish on some occasions, but both prey species avoid the head and mouth regions of the larger fish. One can postulate a complex network of instinctive reflexes to account for the observed behavior, complete with random noise generators at strategic points to explain unpredictable sequences. But the "ad hocery" of such schemes increases in proportion to the completeness of our understanding of the natural behavior. It thus becomes increasingly plausible, and more parsimonious, to infer that both pike and minnows think consciously in simple terms about their all-important efforts to catch elusive food or to escape from a threatening predator.

Drowning Prey

Predatory birds strike or seize their prey and in almost all cases kill it with the beak or talons. But in a few rare cases when a prey animal is too large to be subdued easily it is held under water until it drowns. Meinertzhagen (1959) described such behavior by white pelicans *(Pelicanus onochrotalus)*. In the zoological gardens near Cairo "where many wild duck take refuge by day, this pelican can be seen sidling up to a teal, suddenly seizing it, holding it under water until drowned and then swallowing it" (48). Two first-hand observations of prey-drowning by hawks have been supported by published photographs. In the first case a female harrier flying low over a marshy area struck a common gallinule and fastened her talons in its back. As described by Fitzpatrick (1979), "Instead of carrying the gallinule off, she pushed it down into the water which varies in depth from 2 to 10 inches in this particular area. After submerging the gallinule, the Marsh Hawk remained upright with the water midway up her breast . . . for approximately the next ten minutes. She occasionally flapped her wings and brought the gallinule up out of the water. Each time it was brought up while alive, the gallinule thrashed about, whereupon the hawk would submerge it" (837). Only after it ceased moving did the marsh hawk drag her prey about four feet to an exposed stump and eat portions of it.

In the second case a Cooper's hawk had seized a starling but had difficulty in killing it. Grieg (1979) described his observations as follows: "I had watched the hawk struggle with the starling for several minutes with no apparent success in some underbrush. The hawk was aware of my presence and was moving away from me and staying well hidden, but then to my surprise it carried the violently struggling starling out into the open directly in front of me, no more than 40 feet away, and into a depression where several inches of rainwater had collected. Once in the water with the starling, the hawk merely stood on top of it, and when the starling would struggle to raise its head and a wing out of the water, the hawk would shift its feet so that it would push the starling's head back under the surface" (836). After the starling had ceased struggling the hawk carried it away. In this case the observer's presence may have disturbed the hawk to some extent, but not enough to prevent it from dispatching and carrying off its prey.

It is most unusual for hawks to drown their prey or submerge part of their bodies in water, although Green, Ashford, and Hartridge (1988) observed a sparrow hawk that dropped to the surface of the water after seizing a lapwing in flight. The lapwing was held under water for about three minutes and then carried off, apparently dead. Knowing nothing of the previous history of these hawks, we can only speculate about how such behavior might have arisen, or how, or whether, they came to realize that holding a bird underwater would quiet or kill it. But it does provide a thought-provoking example of versatile inventiveness. An anonymous reviewer of the first edition of this book raised the interesting question whether a hawk that customarily captures both fishes and air-breathing birds or mammals might know enough to subdue the latter but not the former by holding it underwater. I know of no data bearing on this possibility, but enterprising ethologists might keep it in mind for future investigations.

Predation on the East African Plains

The most abundant antelopes of eastern Africa are the Thompson's gazelles, or tommies, which are preyed upon by many carnivores, including leopards, lions, cheetahs, hyenas, and wild dogs. Walther (1969) and others have been able to observe many details of the behavior of the 10–15 kg tommies, because they live in open country where they can be watched throughout the day without serious alteration of their normal activities. During periods when they are migrating they live in mixed herds with roughly equal numbers of males and females. But when they remain in one area most herds consist of females and their young together with subadult males or "bachelors." The adult bucks defend individual territories roughly 100–200 m in diameter; these are approximately contiguous, with little unoccupied space between them. Groups of females move through these territories and are tolerated even when no courtship or mating is taking place. But bachelor males are chased away and spend most of their time around the edges of the adult males' territories.

Despite the fact that tommies are an important portion of the diet of several predators, they do not appear to spend their lives in a constant state of terror. Their principal means of escape is simply running away, and this is almost always effective, provided the predator is seen at a sufficient distance. But even when lions are plainly visible the tommies do not flee unless the lion seems likely to attack. Walther states that tommies seem less disturbed by predators at a reasonable distance than by heavy rainstorms. As reviewed by Elgar (1989), potential prey animals such as tommies spend more time apparently watching for predators when they are in small groups than when many are within view of each other. Elgar stresses that many other factors influence time spent in "predator vigilance," but it seems clear that such watchful behavior helps avoid predation and that other group members benefit when one notices a sign of danger.

When a tommy sees something unusual or suspicious it becomes obviously alert; its head is held high, the ears pointed forward, muscles tensed, and it looks toward whatever has aroused its attention. Sometimes a tommy in this situation stamps on the ground with a foreleg, and often it emits a soft snort that Walther describes as a "quiff." Quiffs seem to occur only or at least primarily when something has alerted the tommy, and they cause others to assume the alert posture and look in the same direction. When a serious attack does occur the tommies flee at a steady gallop that can attain speeds of 45–60 km per hour. This is faster than any of their predators can run except for a very short distance in an initial charge. But tommies often run in other ways, including stotting, in which they jump a meter or so into the air, holding their legs relatively stiff. Stotting usually occurs at the start of a chase, but only when the pursuer is not too close, and at the end when the predator has given up. When a herd of tommies is closely pursued most or all of them bounce about in an irregular fashion, and this behavior seems to confuse many predators and hinder their concentration on a single gazelle so that often they all escape. But when closely pressed the tommies always gallop. The type of gait also varies according to the type of predator; stotting is common when tommies are chased by hyenas or wild dogs but is employed rarely with lions, cheetahs, and leopards, which achieve higher speeds in their initial charge.

Caro (1986a, 1986b) has analyzed stotting in detail and concludes that its most probable function is to warn predators that they have been detected, or in the case of fawns, to alert the mother that the fawn needs protection. Fitzgibbon and Fanshawe (1988) have provided additional evidence that stotting shows the tommy to be in good physical condition and likely to outrun a pursuing predator. This sort of display directed at a predator occurs in many animals, as recently discussed by Zahavi and Zahavi (1997).

Just before tommies are actually captured they often change direction rapidly, doubling back like a hare, but in all such cases closely observed by Walther the predator caught the gazelle in the end. When a mother and fawn were approached by a jackal, which is about the size of a fox, they would at first run off together. But sometimes the mother would attempt to defend her fawn by charging at the jackal and striking it with her horns. Occasionally a second female, perhaps an older sister of the threatened fawn, would join in this kind of attack. Mothers with young fawns sometimes attack quite harmless animals and birds. But when the fawn is chased by a larger and more dangerous predator, such as a lion or a cheetah, the mother moves about excitedly at a considerable distance without intervening directly.

Hans Kruuk (1972) described a sort of distraction behavior by mother tommies directed at hyenas that were chasing their fawns. They would run between the two, crossing in front of the hyena and staying very close beside it but just out of reach. Although usually only one female, presumably the fawn's mother, would engage in this sort of behavior, sometimes as many as four would run close to the hyena at the same time. Yet all these efforts seemed to be ineffective. On 12 occasions fawns were aided by distracting efforts of adult females, but 6 of them were caught. In 19 cases observed by Kruuk the females made no attempt to distract the hyena, but only 5 of these fawns were taken. Possibly the females could discriminate between more and less serious dangers to a fawn and engaged in the distraction efforts only when the danger to the fawn was acute. If so, the two sets of data may not have been strictly comparable.

Such seemingly hopeless behavior as that displayed by an adult female tommy entails a real risk that the hyena will capture her as well as the fawn. In strictly evolutionary terms, these attempts to save the fawn appear to be maladaptive inasmuch as they do not seem to lower the likelihood that the fawn will be killed. Likewise, the final frantic efforts of a tommy to escape have no statistical survival value since they are ineffective. This apparent lack of any survival value leads inclusive behaviorists to disparage the significance of such behavior, because their concern is limited to ways in which behavior contributes to the inclusive fitness of the animals in question. But when we broaden our horizons by considering what life may be like to the animals themselves, it is not surprising to find that they make strenuous efforts to avoid being killed, or having their offspring killed, even when such efforts have little or no chance of success. This is an example of how our theoretical concepts can be broadened by considering how life may seem to the animals themselves.

Cow-sized wildebeest attacked by a pack of hyenas make surprisingly few and ineffective attempts to defend themselves. Kruuk (1972, 158) states that a wildebeest beset by a group of hyenas makes some attempts to butt its attackers but that "generally speaking, the quarry just stands uttering loud moaning calls and is torn apart by the hyenas. It appears to be in a state of shock." Although wildebeest sometimes try to escape by running into a lake or stream, the hyenas almost invariably manage to kill them. Do they hope, though vainly, that they can escape by running into the water?

Mother wildebeest often attack hyenas that are threatening their calves, but those without calves hardly ever do so. Yet Kruuk observed one striking exception on the part of a cow that was in the process of giving birth. Her calf's front feet had already emerged, but she nevertheless went to some lengths to attack a hyena that was walking past and that seemed not to be paying any attention. Shortly afterward this cow appeared to be avoiding the hyenas in the vicinity as she selected a quiet spot, surrounded by other wildebeest, where she lay down to complete the birth. Might such a cow realize that her soon-to-be born calf was in danger of attack by a hyena? We cannot tell, but perhaps it is best to keep an open mind and not dismiss such possibilities out of hand.

Mutual Monitoring by Predators and Prey

Potential prey animals are almost constantly on the watch for dangerous predators, and once one appears they keep it under close observation. Tommies are far from complaisant. They watch for dangers at intervals, although any one may concentrate on grazing for many minutes at a time. But the members of a herd look around at different times, so that there is almost always at least one watching for approaching danger. Only on rather rare occasions does the predator become serious about attacking, and only then, ordinarily, do the prey animals become seriously alarmed. When predators and prey can see each other, they spend most of their time monitoring each other rather than attacking or fleeing. This involves making subtle distinctions; predators notice slightly abnormal behavior that signals weakness or vulnerability. And prey animals notice the changes in the behavior of predators that signal likelihood of attack. Since it would be advantageous for predators to conceal such signals, they are probably unavoidable and inadvertent.

When Thompson's gazelles detect a predator, they often do not flee but move closer. They appear to be much interested and to be inspecting the dangerous creature. Walther sometimes saw a herd of tommies recognize a predator at 500–800 m and then approach within 100–200 m. Under these circumstances the herd contracted into a smaller area than when feeding, the individuals remaining closer to one another. When the predator moved, the herd followed it, evidently aware of the danger and ready to dash off at the first sign of an actual attack. The predators also seemed to understand the situation and rarely attacked a group of alert tommies. Predator monitoring by territorial males was especially evident. At the approach of a predator in daytime the females generally moved away, while the buck stayed in his territory and kept the predator under close watch. As it moved he usually followed at a safe distance until it reached the territorial boundary. Then one of the neighboring territorial males would take over the monitoring of the dangerous intruder. This sort of predator monitoring was so effective that predators captured only 1 of 50 territorial males that Walther studied intensively during a two-year period.

George Schaller (1972) described other examples of prey animals monitoring the behavior of predators and not appearing to be frightened unless the predator rushes at them directly. When a lion is walking along steadily, tommies, zebras, wildebeest, and other potential prey usually face the danger in an erect posture but do not run away. Wildebeest usually keep up an incessant grunting, but when a lion approaches they stop, so that the predator is surrounded by a zone of silence, which presumably warns others of the danger. A group of wildebeest may even approach a predator and line up to watch it pass. But if a lion stops and turns in their direction, the grazing animals usually flee for a short distance, then turn and stand watching again. Roughly 30 m from a lion seems to be considered a safe distance in open country, but when potential prey animals move into thick vegetation they behave much more cautiously. These sensible adjustments of predator monitoring behavior suggest understanding the relative dangers of various predators and how these dangers vary with circumstances, especially with the behavior of the predator itself.

Predators often seem to select for attack members of a herd that are weak, sick, very young, or very old, although an extensive review of the evidence bearing on this question showed that the results of many investigations failed to support this widely held belief (Curio 1976, 113–17). Even when all members of a herd appear to be healthy and vigorous, hungry predators still manage to capture some. But much other evidence reviewed by Curio (117–28) does indicate that unusual or conspicuous members of a group of prey are more likely to be taken by predators. Kruuk observed that hyenas appeared to be quite adept at noticing slight differences in an individual prey animal's posture, locomotion, or other behavior that indicate weakness or vulnerability. When a gazelle had been anesthetized for study or marking, he had to stand by in his automobile to protect it from attack by hyenas even after it seemed to have recovered completely and to be behaving normally. Thus, although the tendency to select more vulnerable prey may have only a small overall statistical effect, it seems reasonably clear that predators often try to identify and attack prey animals that are most easily captured.

This ability to perceive very minor changes in the behavior of another animal appears to be widespread among both predators and prey. The same basic neurophysiological mechanisms that permit such subtle discriminations are probably called into play when animals such as Clever Hans learn to respond to minor and inadvertent counting movements of their human trainers, as discussed in chapter 2. Recognizing how basically important such discriminative perception is to wild animals under natural conditions helps us understand how a horse or a dog can notice when a man stops counting as he watches the performance of an animal that is supposed to be solving arithmetical problems.

Kruuk observed that predators whose home area was close to that of territorial tommies never hunted these familiar neighbors. Some hyenas usually rested from midday until late afternoon in dens located within territories of male tommies. When they left their dens in the early evening they were presumably hungry, but, although they were often surrounded by tommies, they passed between them and hunted in other areas. Perhaps the hyenas knew that the local tommies were alert and difficult to catch. Regardless of that possibility, these hyenas were clearly distinguishing some tommies that were neighbors and were not attacked from others that were treated as prey. Thus their behavior was far from being a stereotyped set of responses to a particular species.

A special case of predator monitoring that involves an unusual sensory mechanism has been described by Westby (1988) in the course of extensive field studies of electric fishes. The electric eel *(Electrophorus electricus)* is well known for its powerful electric organs, which can stun or kill other animals. It has been customary to consider these as defensive measures, but it has also been known for some time that electric eel are predators and that their strong discharges help them catch prey. Westby had placed recording electrodes in a stream and was monitoring the social behavior of a small species of weak electric knifefish, *Gymnotus carapo,* which use their weak pulses both for orientation and for communication in courtship. While he was thus monitoring electric activity in the stream, the pulses of a 1.8-meter-long electric eel could be detected approaching the location of the knifefish. The latter stopped emitting its pulses most of the time but was detected and devoured by the electric eel, which probably located it by recognizing its occasional weak electric signals.

In a laboratory tank a captive electric eel emitted its strong "predatory" shocks when stimulated either by a live knifefish or by electrodes generating signals recorded from the same knifefish and played back into the water at normal intensity levels. Although such electric signals are unfamiliar to us, they are easily monitored and reproduced, so the investigation of social communication and electric shock predation can be conducted with more precise control of the signals than is possible with most other animals. This offers a potentially exciting opportunity for cognitive ethologists to analyze not only predatory shocks and efforts to avoid them but also a whole range of social communication of weakly electric fish by electrical recording, analysis, and playback, as is already being done by several investigators including Westby (1988), Bratton and Kramer (1989), and Kramer (1990, 1997).

Playing with Prey

Foxes, like many other predators, sometimes play with captured prey; while our human sympathies go out to the tormented victim, the predator seems to enjoy this sort of behavior. Henry (1986, 136–40) describes a highly suggestive incident in which a six-month-old red fox that he had observed extensively appeared to release a captured shrew intentionally and return it to the vicinity of its burrow. This fox had caught and immediately eaten one mouse, then caught another with which he "played vigorously for several minutes." After it had been killed the fox carried it some distance and cached it. Although this showed that the fox was no longer very hungry, he soon captured a shrew, which he carried some distance to an open roadway where he began to play with it.

Henry described the fox's behavior in his field notes: "The fox is leaping around, dancing about the shrew who runs over to one side of the road before the fox herds it back to the center. After 45 seconds of playing with this animal, the fox then does an extraordinary thing. He picks the shrew up in his mouth, walks back down the slope to where he captured the prey, and then with a toss of the head spits the shrew out directly at a small burrow. In less than a second, the shrew disappears into the hole and is out of view." Shrews have an unpleasant odor, and although hungry foxes eat them, they are more often cached. As Henry suggests, this fox may have returned the shrew to the burrow a few inches from where it was captured with the anticipation that it might provide food or fun on some future occasion. Of course we cannot be at all sure of such a suggestion, but as Henry concludes, "If we are going to understand red foxes completely, in all their depth and breadth, we cannot study and analyze just their common behaviors."

Cooperative Hunting

Many carnivorous mammals sometimes hunt in groups, although it is not clear how much this increases their total food intake over what each could capture individually. Sometimes hunting groups succeed in surrounding or concentrating prey that would otherwise be more likely to escape. This type of simple cooperation has been observed when pelicans, cormorants, South American otters, dolphins, or killer whales are pursuing fish (Curio 1976, 199–201). The behavior of large cooperating groups is quite different from that of individuals pursuing fish on their own or in smaller groups. For example, Bartholomew (1945) observed that double-crested cormorants *(Phalacrocorax auritus)* fished in a different type of formation when in very large groups of 500 or more than when only 50 or so were pursuing fish together. In such cases the individuals modify their behavior only enough to maintain an appropriate position in the formation. Cormorants in the Netherlands vary their fishing tactics according to the distribution of fish and the turbidity of the water, as analyzed by Van Eerden and Voslamber (1995) and Voslamber, Platteeuw, and Van Eerden (1995).

Meinertzhagen (1959) vividly described this type of cooperative hunting by Dalmatian pelicans *(Pelacanus crispus)* as follows: "There were over a hundred birds forming a line, diving towards shore in shallow water, two ends of the line advancing in perfect order and not very fast. A crescent was eventually formed, every bird keeping his correct station, and then, as water barely 18 in. deep was reached, the two horns of the crescent increased speed, closed in, and formed a complete circle, the birds almost touching. Within the circle the water boiled with small fish, heads were rapidly plunged in simultaneously and pouches filled with fish, the circle closing in all the time and feet paddling hard to prevent fish escaping below body-line" (47). More recently McMahon and Evans (1991) described how pelicans surround groups of fish in a semicircle formation.

Würsig (1983) has described how dusky dolphins herd anchovies in the open ocean, diving and swimming at them from below and from the sides while vocalizing loudly. This results in a tight ball of anchovies, and the dolphins take turns swimming into the aggregation and seizing fish while others continue the herding from outside the ball. Chimpanzees ordinarily feed on fruit, leaves, flower blossoms, seeds, and insects. But on occasion they actively hunt, kill, and eat monkeys, bushpigs, small antelope, rodents, and even human babies. In some cases they prepare to hunt monkeys by watching them for some time and apparently communicating with each other by low-intensity sounds and touching each other. Then they approach and pursue particularly vulnerable individual monkeys, such as females or young, with many signs of intentional cooperation, as described in detail by Goodall (1986), Boesch and Boesch-Achermann (1991), Boesch (1994), and Stanford (1998).

In many cases the advantage of group hunting seems to be that it permits the killing of large prey that would be able to defend itself against a single attacker. Hawks and eagles that prey on birds have been suspected of cooperative hunting, but when such hunting has been studied carefully it has often been unclear whether it was advantageous or whether the same birds obtained just as much food when hunting singly. In the case of Aplomado falcons *(Falco femoralis)* in Mexico, however, Hector (1986) found that there was some division of labor and simple coordinative signaling. Furthermore, Bednarz (1988) has both reviewed earlier studies suggesting cooperation and reported his observations of Harris's hawks in New Mexico that engaged in effective teamwork when hunting cottontail rabbits and jackrabbits.

These hawks, which average about 850 g in weight, hunted during the nonbreeding season in groups of two to six, probably members of a genetically related family. The full sequence of hunting behavior was observed in 13 cases. They captured both cottontail rabbits, somewhat smaller than themselves, and also jackrabbits that weighed about 2,100 g. To aid in individual identification Bednarz had previously attached radio transmitters to at least one hawk in each hunting group, and he could observe the others visually during most of 30 successful hunts in which a group of hawks captured 17 cottontails and 9 jackrabbits. Cooperative hunting seems to be advantageous for Harris's hawks, since single hawks were not seen to attack large prey during these observations. Especially with jackrabbits that are more than twice the size of the hawk, it seems unlikely that a single bird would have much success.

At the beginning of a hunt a group of five or six hawks would split up into smaller groups of one to three that appeared to search for prey during flights of 100–300 m from one conspicuous perch to another. In 7 of the 13 well-observed hunts, several hawks converged from different directions on a rabbit that was away from cover. In four captures of a cottontail rabbit "a flush-and-ambush strategy was employed. Here, the hawks surrounded and alertly watched the location where the quarry disappeared while one or possibly two hawks attempted to penetrate the cover. When the rabbit flushed, one or more of the perched birds pounced and made the kill" (1527). The two remaining hunts that could be closely observed were relay attacks, which Bednarz describes as follows: "This technique involved a nearly constant chase of a rabbit for several minutes while the 'lead' was alternated among party members. A switch occurred when the lead hawk stooped at the prey and missed, at which point the chase was continued by another member of the party. I recorded one relay chase that continued for at least 800 m and involved more than 20 stoops and hence switches in the lead" (1527).

Another intriguing example of cooperative hunting behavior was reported by Munn (1988), who studied the South American giant otter *(Pteronura brasiliensis)* in the Manu National Park of Peru. These otters are six feet long and weigh about 30 kg. Their principal diet consists of fish caught singly by individual otters, but they often hunt in groups when pursuing large prey. A common cooperative fishing tactic is to swim together in a loose phalanx and dive together for 10 to 20 seconds, with much thrashing and churning of the water. Sometimes groups of giant otters kill and eat black caimans from 0.6 to 1.5 m in length, and even anacondas up to 2.7 m long. They attack caimans almost simultaneously from several directions, and when attacking a large anaconda two or more otters would bite the snake at different points along its body, holding fast and bashing it against fallen tree trunks in the water. These group attacks are coordinated to at least a limited degree, and some quick and sensible thinking about where and how to bite in relation to where one's companions are attacking a large antagonist would seem helpful.

Kruuk (1972) summarized the results of his extensive observations of hyenas hunting wildebeest calves both individually and in groups. Only about one-third of 108 attempts by one or more hyenas to capture calves were successful. Single hyenas were almost always driven off by the mother, but when two hyenas were involved one often seized the calf while the mother was attacking the other. It is also clear from many observations that hyenas and wild dogs are more successful in hunting tommies in groups than singly. George Schaller's extensive studies of lions showed several cases in which a group of four or five lionesses spread out as they approached one or more gazelles, those in the center of the group approaching more slowly than those at the edges, thus creating a U-shaped formation that tended to surround the prey on three sides. Some prey animals escaping from lions in the center of the formation seemed to be more easily caught by those advancing on the edges of the formation. Yet both Schaller and Kruuk and more recently Scheel and Packer (1991) expressed caution about inferring conscious intent on the part of the cooperating predators. For example, Schaller's index does not include the word *cooperation.*

Schaller concluded that each lioness behaves more or less independently although of course quite ready to take advantage of another's hunting efforts, as when a gazelle fleeing from a companion runs close to the lioness in question. But Schaller did note that when hunting in groups the lionesses look at each other and seem to be trying to maintain an effective formation, such as the U-shaped pattern. Groups of lionesses that hunt together routinely for months at a time are often sisters or mother and daughters, and they certainly know each other intimately. Yet like many other ethologists, Schaller and Kruuk seem to be bending over backwards to avoid any suggestion that lionesses or other predators intentionally cooperate by coordinating their hunting tactics.

Recent studies of cooperative hunting have emphasized its adaptive advantages, and disadvantages, in terms of theories to explain how it is likely to have evolved and be maintained (Packer and Pusey 1982; Packer and Ruttan 1988; Caro 1995). But relatively little attention has been paid to the cognitive requirements of cooperation or the likelihood that the cooperating animals think about each other's behavior. Dugatkin (1997), in *Cooperation Among Animals,* recognized that cooperative hunting is common in mammals. But he avoids discussing the degree to which cooperators are likely to think consciously about coordinating their hunting efforts. Effective cooperation would benefit from at least a minimum of communication, and if this could be detected and analyzed adequately it might provide objective evidence of simple perceptual consciousness.

Stander (1992) has observed dozens of group lion hunts in very open country in Namibia. He studied these lions so intensively that he could recognize individuals, some by natural markings and some by tags and radio transmitters that were attached to them when they had been captured and anesthetized. In numerous group hunts of large prey particular lionesses took the same position within a roughly linear hunting formation night after night. Individual lionesses occupied the same position (center or one of the wings) on successive nights. These formations approached prey in a coordinated fashion before the group attacked at about the same time from different directions. But hunting tactics were varied according to the situation; for example, smaller prey were hunted by individuals in many cases.

I once had the good fortune to observe a group lion hunt involving more definite evidence of coordination. While I was briefly visiting Robert Seyfarth and Dorothy Cheney Seyfarth during their studies of vervet monkeys in the Amboseli National Park in Kenya, they drove me along a dirt road at the edge of a forested area bordering a large open plain. At a place where the woodland receded for a few hundred meters, the road ran between a semicircular area of grassland on one side and the open plain on the other. A large herd of wildebeest had split into two groups; 50 to 60 were grazing on the woodland side of the road, while the remaining 100 or so were feeding on the open plain about 150–200 m from the road.

As we paused to watch the wildebeest, four or five lionesses approached with a businesslike gait along the edge of the plain, roughly parallel to the road and within a few meters of it. Both groups of wildebeest obviously saw them, for they stopped feeding and watched the lionesses intently. Because the ground was irregular we could not see them all the time, but when about 200 m from the two groups of wildebeest two lionesses climbed slowly to the tops of two adjacent mounds, where they sat upright and remained stationary but conspicuous. After a few minutes had passed we could make out a third lioness slinking, her belly pressed close to the ground, along a ditch that paralleled the road. Although she was visible to us only occasionally, it was clear that she was moving toward a position roughly midway between the two groups of wildebeest. She soon crawled out of our view and for several minutes nothing seemed to be happening at all.

Suddenly a fourth lioness rushed out of the forest behind the wildebeest on the woodland side of the road. Although we had not seen her move to the woods, it seemed almost certain that she was a member of the same group and that she had moved into the wooded area out of our view, and presumably also that of the wildebeest. In any event this lioness charged directly toward the wildebeest that were located between the woods and the road, and they thundered off directly away from her toward their companions on the open plain. This route took them across the ditch close to where we had last seen the lioness slinking furtively a few minutes earlier. As the herd bounded over the ditch she leaped up and seized one of the 50 or so that were galloping all around her. As the dust settled she was busy killing this wildebeest by covering its mouth with hers as it lay on its back, legs kicking feebly in the still dusty air.

The two lionesses that had been sitting quietly on their mounds and the one that had chased the wildebeest toward the ditch walked slowly toward the downed prey but arrived only after its kicking had ceased. All four then began to eat their prey in a leisurely fashion, and after a few moments a jackal joined the scene. Meanwhile the whole group of wildebeest returned from the open plain where they had fled and stood in a line watching the lionesses and her victim from a distance of about a hundred meters.

A single observation such as this one cannot be taken as conclusive proof of intentional cooperation. It is what behavioral scientists sometimes disparage as an anecdote, but it is a very suggestive one. If this was not intentional cooperation, why did the first two lionesses climb to conspicuous positions and wait where the wildebeest could easily see that they presented no serious danger? Why did a third lioness sneak so furtively along the ditch to a position about midway between the two groups of wildebeest? And was it a pure coincidence that a fourth lioness disappeared from the group and that a lioness just happened at the appropriate time to rush out from a suitable point at the edge of the forest so as to chase the wildebeest over the ditch where one of her companions was waiting?

The individual elements of group hunting behavior I have described have been observed by others, but we were remarkably fortunate to see so much of this sequence. Yet the ethologists who report detailed observations of lion behavior have been so reluctant to infer conscious intent to cooperate that they seem to have refrained from reporting as much of this type of behavior as they have actually observed. As in so many other cases of versatile animal behavior, one can always argue that it is not absolutely necessary to infer even the simplest level of perceptual consciousness. But the weight of accumulating evidence tips the scales so strongly that the rigid exclusion of any possible consciousness has become stretched to the breaking point.

CHAPTER FIVE

Construction of Artifacts

A wide variety of animals display considerable ingenuity and versatility in the construction of shelters and structures that serve other purposes, such as capturing prey or attracting mates. These activities require adjustment of behavior to the local situation, to the materials available, and to the changing circumstances at various stages of construction. This necessary versatility often suggests that the animal is thinking about the results of its efforts and anticipating what it can accomplish; acting to attain an intended objective is a more efficient process than blindly following a rigid program. This chapter will describe selected cases where animals build more or less fixed structures that benefit them or their offspring, and chapter 6 will discuss the construction and use of tools. Although there is no sharp distinction between artifacts and tools, the details of the behavior involved in their construction and use make it convenient to discuss them separately.

The many types of structures built by animals are often characteristic of the species to which they belong, and in some cases such as termite mounds and the larvae of caddis flies the structures may be more useful for identifying species than the morphology of the animals that build them. This species specificity suggests genetic programming of a relatively stereotyped nature, but when analyzed carefully it often turns out that the animals display considerable versatility in adapting the details of the artifact construction to the particular situation. Karl von Frisch reviewed many significant examples in a delightful book, *Animal Architecture* (1974). It is no accident that the same brilliant scientist who discovered that honeybees communicate symbolically, as discussed in chapter 10, also turned his attention to the ingenious variety of shelters constructed by various animals, for it is difficult to repress one's admiration for the effectiveness and beauty of many of the structures built even by rather simple animals. One component of our admiration may stem from the suspicion that the animals may do their building with at least some rudiments of intention. More recently a very thorough survey of animal architecture has been provided by Hansell (1984); this adds much detail and many new discoveries to the more general and semipopular book by von Frisch.

Even some of the protozoa build simple cases, usually by secreting material from the cell surface. But a few, such as the amoebas of the genus *Difflugia,* add grains of sand to their outer surfaces, so that they come to resemble tiny croquettes. These particles are transported through the cytoplasm and distributed fairly evenly at the cell surface. But the processes involved appear to have more in common with cellular functions such as pinocytosis, in which the cell membrane bulges out and engulfs an external particle, than with active manipulation by the organism of objects outside of its body (Netzel 1977). Since protozoa lack anything at all comparable to a central nervous system capable of storing and manipulating information, it seems highly unlikely that they could be capable of anything remotely comparable to conscious thinking.

Hansell (1984) reviews several scattered examples of annelids and mollusks that build simple structures, such as the burrows of marine polychaete worms. Impressive structures are built primarily by arthropods and vertebrates. Several species of crabs dig burrows, and some fiddler crabs of the genus *Uca* add hoods near the burrow entrance as a part of their attraction and courtship of females (Crane 1975; Christy 1982). But some species of octopus also build shelters (Mather 1995; Hanlon and Messenger 1996). Mather (1994) has also described how some octopuses improve crevices in rocks where they take shelter, enlarging the cavity by removing stones and sand and bringing stones, shells, claws from molted crabs, and even bottles to partly block the entrance. Several species of fish build nests to shelter their eggs, the most thoroughly studied being the European stickleback, whose courtship behavior has been analyzed by Tinbergen and many others. A less well-known case is the weakly electric fish *(Gymnarchus niloticus),* the species used by Lissmann in his classic experiments demonstrating the use of weak electric fields for orientation. These relatively large freshwater fish build a nest from floating vegetation that has a corridor leading to a chamber at the end where the eggs are laid (Bullock and Heiligenberg 1986).

Ant Lions

Simple pitfall traps are constructed by two distantly related groups of insect larvae, the ant lions and the worm lions. Both dig shallow, roughly funnel-shaped holes in loose soil and wait with their bodies almost entirely buried at the bottom of the cavity for other small animals to fall in. The most common prey are ants, which generally attempt to escape from the pit by climbing up the steeply sloping walls (Wheeler 1930; Topoff 1977; Heinrich 1984, 141–51). When an ant or other small prey animal falls into the pit, the ant lion or ant worm seizes it with its mandibles and injects a poison that kills the victim. Often prey such as ants crawl about actively enough that they are not seized on the ant lion's first attempt. The ant lion then throws grains of sand in the general direction of the prey, and this seems to increase its chances of capturing the ant. Lucas (1982, 1989) has studied in detail one species of ant lion, *Myrmeleon crudelis,* in Florida. The pits are not perfectly conical but have walls that slope more gradually in the parts of the pit where the ant lion waits for prey. The walls are steeper in front of it, and these front walls also tend to be lined with finer grains of sand.

These predatory insect larvae thus construct a very simple artifact, the pit with somewhat different walls on different sides, and they also throw sand grains as a very crude form of tool. The details of their prey-catching behavior are difficult to study because everything happens too fast for the human eye to follow easily. It is not clear how adaptable the ant lions are, but they do alter their behavior to some degree according to the type of prey or the kind of soil in which their pit is dug, and occasionally they show interesting interactions. Wheeler reviewed earlier observations in which two ant lions dug pits so close together that an insect escaped from one pit only to fall into the other. In this case the first ant lion emerged from its pit and pursued the prey into its neighbor's pit.

Caddis Fly Cases

One of the more impressive examples of complex structures built by animals we usually consider to be very simple are the cases and nets built by the larvae of caddis flies. These are abundant animals in freshwater streams and ponds, and many of them construct shelters of various types attached to vegetation or to the bottom of the body of water they occupy. They develop from eggs laid in the water by winged female caddis flies, which emerge and mate after a long period as aquatic larvae. The larvae of North American caddis flies and the cases they construct have been described in detail by Wiggins (1977), who illustrates numerous variations in case and net construction characteristic of the numerous genera and species. The larvae themselves are rather nondescript and difficult to identify, but the shelters are so characteristic that they are often used by systematic entomologists for species identification. The cases built by caddis fly larvae may employ bits of leaves, particles of sand, or other available materials including the empty shells of very small snails. They are cemented together by silk secreted from glands on the larva's head. Sometimes caddis fly larvae crawl out of their cases, and they may fight over cases, or one may evict the occupant of a case and take it over (Otto 1987a, 1987b; Englund and Otto 1991).

Like other insects, these larvae grow through successive stages in which the exoskeleton is shed. In early stages the small larva uses minute and ordinarily homogeneous particles to form a roughly cylindrical case around its body. These cases may increase the flow of water over the gills (Williams, Tavares, and Bryant 1987) and also protect the otherwise vulnerable soft-bodied larva from small fishes or other predators such as immature dragonflies. The case is not totally impervious; it has an opening at the posterior end to allow feces to pass out, and water circulates freely from an anterior opening to the posterior one so that the gills can extract oxygen from the water. In many species the case is portable and is carried about as the animal moves by pushing its head and thoracic segments carrying the six legs out through the front opening. The case is held close to the body by hooklike projections from the abdominal segments. In constructing their cases, caddis fly larvae are somewhat selective about the materials used. Some species cut pieces from the leaves of aquatic plants, and others construct fine meshed nets that serve to strain minute animals and plants from the flowing water.

The case-building behavior of a few species of caddis fly larvae has been studied carefully in the laboratory by Hansell (1984, 1968a, 1968b, 1968c). He conducted detailed observations and experiments with *Silo pallipes,* which begins its larva life by constructing a simple cylindrical tube of sand grains cemented together. This species passes through five instars, between which the animal sheds its external skeleton and grows a larger one. In its first instar the *Silo* larva occupies a roughly cylindrical case composed of particles about one-half millimeter in diameter. But toward the end of this instar it adds two larger sand grains at the sides of the front end. During the second instar it adds two additional, still larger particles, and through the three succeeding instars it selects at each molt larger grains of sand for the anterior opening. At the end of the fifth instar the case is about ten millimeters long, and the larger anterior grains are two to five millimeters in diameter. Throughout this growth the larva enlarges its cylindrical case by adding more small particles.

Hansell's observation showed that the larvae reach out of their cases and feel particles in the immediate vicinity with their anterior appendages and reject many that are either too large or too small. Having felt one of appropriate size they move it into position and secrete silk to fasten it there. On a very small size scale this behavior is flexible and adapted to the available particles and the stage of case construction.

Hansell (1972, 1974) also studied another species of caddis fly larva, *Lepidostoma hirtum,* which cuts panels from bits of leaves to form a floor, roof, and two sides. All of the panels are approximately rectangular pieces, one or two millimeters in size, held together at the edges by secreted silk. The resulting case is strengthened by a staggered arrangement of the pieces. Each joint between two side plates intersects with the middle and not the edge of a roof plate. When Hansell cut away the front end of a case to form a continuous, smooth front edge, the larva cut leaves of different shapes from those normally used and glued them into place so as to restore the staggered arrangement. These simple insect larvae thus exhibit a considerable degree of versatility not only in the initial construction of their cases but in repairing them. Despite our customary assumption that insect larvae exhibit only stereotyped behavior, they have highly organized central nervous systems with hundreds of neurons and synapses that are quite capable of organizing relatively complex and flexible behavior.

In yet another species of caddis fly, *Macronema transversum,* the larvae construct a more elaborate case, which provides not only a shelter but a food-gathering mechanism, as described in detail by Wallace and Sherberger (1975). This case is a chamber considerably larger than the animal's body and roughly oval in shape. From the upstream end rises a tubular extension with a roughly ninety-degree bend and an opening facing the direction from which water is flowing. From the other end of the oval chamber rises a shorter outlet tube. The entire chamber is only two or three centimeters in length, but the flow of water by the surrounding stream serves to ventilate it fairly well owing to its construction. Opening into the side of the main chamber is a separate tubular structure corresponding to the normal caddis fly case. But both ends of this open into the larger chamber. The end where the larva places its head opens into the upstream side of the main chamber, and water flows through this side chamber past the larva's gills and out a posterior opening into the downstream part of the main chamber. Finally, the larva constructs a fine mesh net across the middle of the large chamber. Small animals and plants are caught on this net and serve as food. The entire structure is roughly comparable in complexity to the nests of many birds.

It has been customary to view such artifacts as caddis fly cases in much the same light in which biologists view the elaborate structures of animal bodies—patterns regulated primarily by genetic instructions. But on close examination it turns out that even simple creatures such as caddis fly larvae adjust their building behavior in ways that would seem to be aided by simple thinking about what they are doing. Such thinking might be limited to seeking to match some sensory or perceptual pattern that was itself produced by genetic instructions. But, as I have pointed out elsewhere (Griffin 1984), the fact that a central nervous system operates in a certain way because of genetic instructions does not necessarily mean that its operations may not also lead to cognition and perceptual consciousness.

Insect Nests

Many species of insects construct a wide variety of nests or shelters, as reviewed by von Frisch (1974). Often, as in the conspicuous nests of wasps and hornets, numerous females contribute by gathering bits of vegetation, chewing them, and mixing them with saliva and with silky secretions from specialized glands or with their own feces. The resulting daubs are then applied to some solid substrate, and other daubs are added to build up a multichambered nest much larger than the individual insects. In some ways the nests built by solitary bees and wasps are even more impressive, because they result from the work of a single female. When ready to lay eggs she searches widely for suitable materials and brings them from a considerable distance to prepare a place where she builds a nest. A female mason bee of the genus *Chalicodoma* moistens with her saliva particles of sand and dirt she has gathered and forms these into small oblong pellets. She then carries each such pellet to a spot on a large rock, where she cements them together into a roughly cylindrical cell open at the top. In each cell she lays one egg and regurgitates liquid honey around it. Then she closes the top with additional sand grains, cementing it securely with saliva. Next she applies dust particles to the outside of a group of such cells with the result that they come to look almost exactly like the surface of the rock to which they are attached. The mixture of fine dust particles and saliva dries into a structure almost as hard as the rock.

A different species of mason bee, *Osmia bicolor*, seeks out an empty snail shell and deposits in the narrow inner parts of the spiral chamber both her eggs and food for the larvae that will hatch from them. Some of the food thus deposited is a semisolid mixture of pollen and regurgitated stomach content known as "bee bread." After depositing numerous eggs and a quantity of bee bread the female *Osmia* fills the middle portion of the tapering spiral cavity of the snail shell with chewed-up pieces of leaf. Nearer the outside she deposits enough small pebbles to form a fairly rigid wall. Neither the leaf fragments nor the pebbles provide a totally airtight seal, and air can still circulate even after a second wall of leaf pulp is added outside of the pebbles. As though these preparations were not enough, the mason bee adds dry stalks of grass, tiny twigs, or pine needles, piling them over the snail shell in a large irregular dome.

These elaborate sequential actions may well be genetically programmed to a considerable extent, but experimental evidence is not available to indicate how much individual experience may affect this behavior. Might such an egg-laying female think consciously about what she is doing? She will die long before her offspring emerge, so she has no direct information about the eventual result of her complex efforts, although she did of course begin life as a larva in a similar nest. Although a female *Osmia* may have no possibility of understanding the long-term advantages of her elaborate efforts, she might nevertheless think consciously about the immediate present or very short-term future. She might strive to close the snail shell after laying eggs and feel good about hiding it, even though she has no understanding that offspring will eventually emerge.

Some wasps dig tunnels in the ground or into solid wood and lay eggs in them. The tunnel opening is then plugged, and some species select material for the plug that matches the surroundings so well that the burrow opening is very difficult to detect. In the sand wasps *(Ammophila campestris)* studied by Baerends (1941) the burrows are dug in sandy ground with an enlarged chamber at the lower end. These wasps close the opening with small stones, which they bring from some distance if none are available in the immediate vicinity. Then the wasp searches for and captures a caterpillar, which ordinarily is as large as herself. She paralyzes it by stinging it in several places and carries it on the wing to the burrow. There she opens the entrance, drags the caterpillar down to the nest chamber, and only then lays an egg. Paralyzed caterpillars survive well enough that the growing larvae can eat them.

As reviewed by Thorpe (1963) and described in detail by van Iersel and van dem Assem (1964), the sealing of a wasp burrow is accomplished very skillfully, so that after it is completed it is virtually impossible to see any signs of disturbance. When digging a burrow, the wasp throws sand grains some distance away so that they do not create a revealing pile of material. The female, on returning with a caterpillar, relocates her hidden burrow by remembering the appearance of surrounding landmarks. If these are experimentally altered she may be unable to find the concealed burrow entrance.

This type of complex burrow construction and provisioning has been studied experimentally by intervening in the process in ways that would not normally occur. For example, if a paralyzed caterpillar is moved after the wasp has dropped it near the burrow entrance while reopening the hole, she hunts about for the missing prey and drags it back to the entrance. But in many experiments she then repeated the entire behavior of depositing the caterpillar near the entrance and digging even though the burrow has already been opened. This has often been taken as evidence that the entire pattern is rigidly programmed genetically, and the further assumption is customarily added that for this reason the wasp cannot possibly think consciously about what she is doing.

Similar examples have received a great deal of emphasis. After some insects have constructed a nest they may fail to alter their behavior if an experimenter opens a hole in the wall or bottom of the nest so that the eggs fall out as soon as they are laid. *We* can immediately see what the insect ought to do to accomplish her general objective, and when she fails to do this, and instead repeats what would ordinarily be appropriate behavior in this abnormal situation, we tend to conclude that she is a mindless robot. But in other cases insects do behave sensibly and repair damaged nests. Undue emphasis has been placed by inclusive behaviorists on the examples of stupid behavior, and because many insects often fail to adjust their behavior after abnormal experimental disturbances, we have overgeneralized such inefficient behavior into a dogmatic conclusion that no insect ever thinks consciously about its activities.

The leaf-cutter ants of the genus *Atta* are in many ways one of the most successful species in the world if judged by the number of well-nourished individuals and their impact on the immediate environment. The general behavior of these ants has been well reviewed by Weber (1972) and by Hölldobler and Wilson (1990). They live in enormous underground colonies consisting of many chambers interconnected by tunnels. Each colony is founded by a single queen, but she lays eggs that develop into thousands of nonreproductive workers. Males and virgin queens are produced only after such a colony has grown to a substantial size. The workers are adapted in both anatomy and behavior for different functions such as caring for eggs, larvae, and pupae, gathering food, and defending the colony against intruding insects of other species as well as against vertebrate predators.

The food-gathering workers move out from the entrance over the forest floor in such enormous numbers that they quickly wear down the vegetation and form beaten paths. They climb plants of all sizes, cut pieces of leaf roughly the size of their own bodies, and carry these back to the colony. Often they can be seen by the hundreds walking methodically along their trails, each ant carrying a tiny green fragment. When they attack flower beds, the dismayed gardener may be startled by a busy trail of five-millimeter fragments of colorful flower petals being carried to the underground tunnel system.

Inside the colony the workers carry leaves or flower petals into special chambers containing masses of fungus. These chambers may be as large as a meter in diameter, and the fungus grows rapidly, nourished by the leaf fragments as well as by feces of the ants. The leaf cutters, like certain other species of ants, are in a very real sense engaging in a form of agriculture. They collect particular food plants, bring them into the colony, and manipulate the fungus in appropriate ways to facilitate its growth and the processing of the food, which the ants themselves could not digest unaided. In addition to gathering food and tending the fungus "gardens," the ants devote much effort to caring for eggs and larvae. Although much of the behavior of leaf-cutter ants is relatively predictable, the workers vary their activities according to the needs of the colony to some degree. It has not been practicable to rear such ants in isolation in a way that would serve to tease apart the relative contributions of genetic constructions and individual experience, but, as in so many other cases of complex animal behavior, it would seem that a little elementary conscious thinking would be helpful. For example, when suitable food has not been found in one area, the leaf-cutter ants shift their foraging activity to other places. They certainly exploit newly available food sources by massive invasions that can be totally destructive to human gardening efforts.

Another group of specialized ants, the African weaver ants *Oecophyla longinoda,* exemplify the capabilities of millimeter-sized central nervous systems (Hölldobler and Wilson 1990). Weaver ants live primarily in trees, but instead of digging a burrow or nesting cavity they construct nests by joining together leaves. The edges of the leaves are attached with sticky silk to form closed chambers considerably larger than the bodies of the ants that build them. Nest construction, like most of the activities of social insects, is carried out by nonreproductive female workers. The first problem in constructing a nest is to bend leaves out of their normal flat shapes and join the edges to form the walls of an enclosed cavity. Sometimes one ant can grasp one leaf with her rear legs and the other with her jaws and by bending her whole body and flexing her legs can pull the edges of the leaves together. But since the individual ants are much smaller than the leaves, numerous workers must line up along the edges of two leaves and all pull in a roughly coordinated fashion. Even this degree of cooperation is not sufficient when the leaves are separated by more than an ant's body length. This problem is solved by the ants forming chains, one seizing the edge of a leaf in her jaws, another holding her by the abdomen, while a third holds that ant's abdomen, and so on until at the end of a chain of several ants one grasps the second leaf with her hind legs. Numerous chains of ants pulling and bending leaves in rough coordination form them into an enclosure that will later be occupied as a nest.

This shaping of leaves and bringing their edges together by the cooperation of numerous individual weaver ants is only part of the process. Once brought together, the edges of leaves must be joined with silk. But the adult workers cannot secrete this silk themselves. They obtain it by carrying larvae of appropriate age and holding them first against one leaf edge and then the other. Thus the younger sisters of the workers are used as a sort of living tool in nest construction. The queen weaver ant lays her eggs, and larvae grow and are fed by workers that bring food into these nests. The food consists mostly of fragments of other animals the foraging workers have killed or scavenged. But one leaf nest is seldom large enough to hold a growing colony, so while the queen remains in the first nest the workers construct others nearby and then carry food, eggs, and larvae back and forth between them. Sometimes these colonies grow to consist of dozens of nests extending over several trees.

All this social behavior, which is necessary for the survival and reproduction of the colony, would seem to be facilitated if the workers were capable of simple perceptual consciousness. When starting nest construction they might intentionally work to pull the edges of leaves closer together. When leaves have been bent into approximately an appropriate shape, some workers might consciously realize that it is now necessary to glue them together and fetch larvae of a suitable age to secrete the necessary silk. Coordination of this building activity must require some sharing of information about what needs to be done. But this is probably carried out in large measure by chemical signals, which are very difficult to analyze. In the case of food-gathering, however, communicative behavior of weaver ants and other species has been studied in sufficient detail that we can understand at least the basic processes by which cooperative endeavors are coordinated, as will be discussed in chapter 10.

Bird Nests

The nests of birds are larger and more familiar to us than those of insects, but when one takes into consideration the difference in size of the builders many bird nests appear rather small and simple compared to the elaborate structures built by some of the social insects. Yet other bird nests, such as those of orioles, are covered structures with long entrance tubes. The elaborate nests of African weaver birds have been studied extensively by Collias and Collias (1962, 1964, 1984) and by Crook (1964); these and other complex bird nests are also described by von Frisch (1974) and Hansell (1984). The nest of the village weaver *(Textor* or *Ploceus cucullatus)* of Africa is an ingeniously complex woven structure, and the detailed observations and experiments of Collias and Collias have provided a more complete understanding of how it is built than is available for other species.

The finished nest of the village weaver is a roughly spherical structure with an opening leading downward from one side of the bottom. The nest is constructed from strips of grass or similar material that are woven by a series of fairly complex motions. When courting a female, the male weaver begins nest-building by forming a roughly vertical ring attached at two or three places to a small branch. He then enlarges this at the top, forming first a partial roof and later the walls of a roughly spherical chamber. The entrance is completed last. The actual building operation consists of grasping a strip of grass near one end and poking it with a vibratory motion, at first alongside some object such as a twig, and later into the mass of previously placed strips. When the end of the strip sticks in place, the bird releases its grip, moves to the other side of the twig or nest mass, and seizes the strip again with its bill. It then pulls it through or around whatever material is already present, bends or winds it about the original twig or another piece of nest material, and finally pokes the end of the strip back into the accumulating mass.

A constant feature of the stitching process is that, with successive pokings of the strip, the bird reverses the direction in which it is wound around some preexisting object. Different parts of the nest receive different amounts and kinds of material, so that the end result is a structure with a virtually waterproof roof, thicker than the walls, and a short, roughly cylindrical entrance tube. When a female accepts the male's courtship displays, and the nest he has constructed, she adds a soft, thick lining at the bottom of the egg chamber, first covering it with a thin layer of strips that she tears from leaves of tall grasses or palm fronds. After accomplishing this task, she inserts many soft grass tops with the stem end down, so that their feathery tops provide a soft cup. When feathers are available the female gradually begins to use them until there is a thick layer of feathers on top of the grass heads.

The tailor bird *(Orthotomus autorius)* of India constructs its nest by stitching together leaves to form a protective cup. Plant fiber or strips of spider silk are used as threads, and these are sometimes threaded through holes made by the bird at the edge of the leaves. Although many species of birds construct their nests by a variety of motor actions that serve to produce a structure that is sturdy, and often is camouflaged to make it less conspicuous, the weaving and stitching procedures are especially suggestive of conscious thinking about the process and its results.

It is customary to view nest-building by birds as genetically programmed behavior. In some species such as canaries, which build relatively simple cup-shaped nests, naive females that have never seen a nest do construct a reasonably normal one when ready to lay eggs. But in the case of the village weavers a considerable amount of learning seems to be involved. Young males build partial and irregular nests, but these are not accepted by females. Their nest-building skill is increased if they can watch adult males building nests. Thus the male that builds the relatively complex structures has ordinarily had a long period of practice and has also had abundant opportunity to watch older males build more complete nests. Although the general pattern of a village weaver nest is fairly constant, each male adapts his building efforts to the immediate situation, especially to the shape and position of the twigs where he starts building.

Birds generally repair damaged nests in a reasonably sensible fashion and do not go through the entire sequence of building unnecessarily. When Collias and Collias removed parts of a nearly completed weaver bird nest, the male rebuilt those parts and did not go through wasteful repetitions of other portions of the building sequence. The only partial exception to this rule involved the entrance tube. If this was damaged, the male generally built an abnormally long tube. Nesting birds behave in many appropriate and rational ways when raising young. Even simple nests are often concealed in vegetation, that is, are constructed in places where they are difficult to see. Furthermore, the parents arrive and depart from their nests quietly and inconspicuously. This is obviously advantageous, because many predators eat eggs and young, and the nest and the arriving and departing parent are obvious signs that food is available.

This is not to say that birds never do foolish things in the course of nest-building. Von Frisch (1974) described how blackbirds occasionally start building many nests in some artificial structure that has many similar-looking cavities. The birds apparently become confused as to just where the nest is to be and never succeed in completing any one nest. As in so many cases of this kind, we tend to infer a total lack of thinking when animals do something foolish and wasteful of effort. But we do not apply the same standard to members of our own species, and we never infer a total absence of thinking when people behave with comparable foolishness. It is important to realize that to postulate that an animal may engage in conscious thinking is by no means the same as to say that it is infinitely wise and clever. Human thinking is often misguided, and there is no reason to suppose that animal thinking always corresponds perfectly to external reality. Our thoughts may be inaccurate or quite different from what others see as correct and sensible. But error is not the same as absence of thought.

Events in our own lives show that stupidity does not preclude consciousness. If our usually dependable automobile fails to start, in our impatience many of us do something totally irrational, such as kicking the tires or swearing at the machine. We know perfectly well that such displaced aggression will not start the car, but we are probably thinking, "Why won't the damned car start this morning?" or "What will happen when I am late at the office?" or "I should have bought a new battery last fall."

What can we say about the possible conscious thoughts and feelings of nest-building birds? The conventional assumption of inclusive behaviorists has been that nest-building is a complex but predetermined form of behavior. The predetermination may stem from genetic information or learning; ordinarily it is some mixture of the two. One clear example of genetic influence on nest-building behavior stems from the experiments of Dilger (1960, 1962) with two species of lovebirds of the genus *Agapornis* that hybridize in captivity. One, *A. personata fischeri,* builds its nest from pieces of bark that it carries in its beak. Another, *A. roseicollis,* cuts strips of nest material from vegetation, which it carries by tucking them into feathers on its back before flying to the nest. In Dilger's experiments, hybrid birds cut paper strips but were inefficient in transporting them, apparently attempting both to tuck strips and to carry them in the bill. They improved with practice, however, and most ended by carrying strips in the bill after initial and incomplete tucking movements. This is a classic example of behavioral genetics where there is clearly a strong inherited tendency to behave in a certain way, and yet the animal does not follow the genetic instructions slavishly but gradually modifies its actions on the basis of experience and succeeds in obtaining an apparently desired and intended result.

This leads us to another challenging question about nest-building and other complex reproductive behavior of birds. When they are at an early stage of nest-building, do birds have any concept of the finished product they are working to achieve? We might go one step further and ask whether a bird beginning to build a nest has any idea of the eggs and young that will soon occupy the nest. Such speculation may be more plausible in species where the females do the nest-building, for they will lay the eggs and feed the young. Many biologists and psychologists, including Shettleworth (1998), seem to feel that it is outrageously far-fetched to suggest that a female bird might think about eggs and young as she begins to build a nest. Yet there might be some adaptive advantages to such thoughts, for even a simple concept of the function the nest will serve could help birds construct it appropriately. But again we are stymied by our inability to gather convincing evidence.

Despite the lack of completely conclusive evidence, it seems plausible that nest-building birds have some simple conscious desire and intention to produce an appropriate structure. Genetic influences may determine the general pattern of the desired result, but birds certainly vary their behavior as they work to achieve this goal. Although their efforts are not always totally efficient, and human observers can often imagine how they might do somewhat better, the birds give every indication of trying to achieve an anticipated goal.

Bowerbird Bowers

In many ways the most impressive structures built by any animals are the bowers constructed by male bowerbirds of Australia and New Guinea. Displaying males attract females to these structures, and mating takes place in them. Darwin placed considerable emphasis on the bowerbirds in his discussion of sexual selection, and more recent investigators have discovered many important facts about their behavior. The Australian biologist Marshall (1954) described and analyzed the behavior of bowerbirds in great detail, emphasizing physiological and endocrine influences on their courtship displays and bower-building. Gilliard (1969) pointed out that among bowerbirds and their relatives there is an inverse correlation between striking plumage and elaborate bowers. Species with conspicuously colored feathers used in courtship displays build simple bowers, if any, whereas the most elaborate bowers are constructed by species with relatively dull plumage. More recently, intensive studies of individually marked birds have added new dimensions to our understanding of their social behavior, beginning with the observations of Vellenga (1970, 1980) and Warham (1962) and continuing with increasingly ingenious methods in the studies of Borgia and his colleagues (1985a, 1985b, 1986, 1987, 1995), Pruett-Jones and Pruett-Jones (1994), and especially Diamond (1982, 1986a, 1986b, 1987, 1988).

Fourteen of the 18 species of bowerbirds build bowers or at least make small clearings on the ground that they decorate to some extent. Bower style varies with the species. It is convenient to begin with the relatively simple avenue bower built by the satin bowerbird *(Ptilonorhynchus violaceus)*. Adult males are glossy bluish-purple, and the females and immature males are green. This species has been more thoroughly studied than any other because it is common in eastern Australia and sometimes builds its bowers in suburban areas where it can be observed more easily than relatives that are found only on uninhabited mountains of New Guinea. Vellenga (1970, 1980) took advantage of this opportunity by color banding 940 satin bowerbirds captured in her garden over a period of five years. One of these individually identified birds could be traced to a park half a mile away where he maintained and defended his bower for fifteen years. Her observations of known individuals confirmed several aspects of bowerbird behavior that had previously been inferred by Marshall and others from less direct evidence.

As summarized by Diamond (1982, 99–100), largely on the basis of Vellenga's observations,

> The bower is a woven platform about 10 feet square, supporting an avenue of woven sticks, with walls a foot high nearly joining in an arch over a floor, and decorated with blue or green natural or man-made objects. Perishable decorations such as flowers are replaced daily. Some bowers are left unpainted, but others are painted daily either blue, black, or green by the male, using a wad of bark as a paint brush and using crushed fruit, charcoal, or (nowadays near civilization) stolen blue laundry powder as paint. The long axis of the bower is generally within 30 degrees of north-south, possibly so that the male and female can face each other during early morning displays without either having to stare into the rising sun. When Marshall picked up a bower and reoriented it, its architect promptly demolished it and rebuilt it in the correct orientation. . . . The bower-owning male Satin bowerbird continually tries to entice females into his bower by picking up in his bill an object such as a flower or snail shell, and posturing, dancing and displaying to the female.

The males also call loudly. As described by Borgia (1986), the male "faces the female while he stands on the platform. He gives a whirring call while prancing, fluffing up his feathers and flapping his wings to the beat of the call. Calls are punctuated with periods of silence, quiet chortling, buzzing, or mimicry of other birds. The female's initial response is to enter the bower and 'taste,' or nip, at a few sticks. Then she intently watches the courtship. If she is ready to copulate, she crouches and tilts forward" (98). Diamond (1982) emphasizes the fact that

> many courtships are interrupted at a crucial stage by the intrusion of other individuals. However, another reason for the low success rate may be the invariant sequel to successful mating: the male bowerbird savagely attacks the female, pecks and claws her, and chases her from the bower. Mating itself is so violent that often the bower is partly wrecked, and the exhausted female can scarcely crawl away. The courtship display can appear little different from the male's aggressive display. When a courted female is won over and starts to solicit

copulation, the male often changes his mind and chases her away. Thus, a female may have to make many visits to a bower before she overcomes her fear of the aggressive male. After mating, the female constructs a nest at least 200 yards from the bower and bears sole responsibility for feeding the young.

Bowers, like flags of possession, may serve as symbols of males' property rights in their wars with other males. An adult male spends much time repairing his own bower, protecting it from raids by rivals, and attempting to steal ornaments or destroy rivals' bowers. The battles and territorial shifts that Vellenga recorded among her 426 banded adult males make the European Thirty Years' War seem straightforward by comparison. . . . Dominant males directly prevent other males from wooing females by the destruction of their bowers. Young males continually try to erect rudimentary bowers in the territory of an adult male but the latter patrols his property several times a day and wrecks them. When Marshall placed 100 pieces of numbered blue glass into these rudimentary bowers one night he found 76 of the pieces transferred to the bowers of dominant males by noon the next day. Similarly, Vellenga observed a blue celluloid band to be transferred between bowers several times a day, until one male finally wove it into his bower. (100)

Borgia (1985a) and several collaborators observed the mating success of many individually marked male and female satin bowerbirds, using an automatic recording camera to monitor 207 copulations at 28 bowers during one season. The number of copulations at each bower varied from 0 to 33, with 5 of the 28 males achieving 56 percent of the matings. There was also a statistically significant tendency for males with well-constructed and highly decorated bowers to mate with more females. The most attractive bowers were more symmetrical, included larger sticks more densely packed, and were decorated with more blue feathers, yellow leaves, and snail shells. In another investigation of flower preferences Borgia et al. (1987) found that male satin bowerbirds selected primarily blue and purple flowers and never used those that were orange, pink, or red. Furthermore, they showed a preference for blue and purple flowers that were uncommon in the local environment.

Borgia (1985b) also studied the effects on mating success of bower destruction by other males. He found that the number of destructive attacks on bowers, and the amount of destruction, were inversely correlated with the quality of the bowers, that is, their symmetry, size of sticks used, density of stick-packing, and general quality as judged by the investigators. Bower destruction occurred only when the bower owner was absent, ordinarily to feed, and on his return the owner almost always drove off the raider. Since bower quality had been shown to be correlated with mating success, it is clear that the competitive bower raiding has a direct effect on the evolutionary fitness of the males. In addition to partial or complete wrecking of bowers, neighboring males often steal feathers and other decorations to improve their own bowers (Borgia and Gore 1986).

All this complex interaction of competing males using elaborate bowers to attract females suggests that some simple conscious thinking may be involved when bowers are being constructed or decorations gathered and set in place. Advocates of what I have called the "sleepwalker" view of animal behavior (Griffin 1985) may be tempted to argue that because all this behavior has such an obvious effect on the birds' evolutionary fitness it must be genetically fixed and therefore mindless. But as has been found to be the case with many behavior patterns, the generally similar actions performed by members of the same species do not preclude a major role for individual experience and learning. Early in the mating season, as summarized by Borgia (1986), "young males visit the bowers and the bower owners often display to them . . . (but later) . . . the bower owners become less tolerant of male visitors" (98).

Vellenga's earlier studies had shown, as summarized by Diamond (1982), that "many skills related to bower building and use have to be learned. Young males in green plumage spend about two years building rudimentary but increasingly complex bowers before acquiring blue adult plumage and building complete bowers. These 'practice bowers' are the joint efforts of several young males, which take turns placing and rearranging sticks, often clumsily and without success, occasionally with the cooperation of more skilled older males. The young males do not paint these bowers and are less discriminating than adult males in choice of color for bower decorations. Immature males spend much time watching the displays, mating and other bower activities of adults, and are displayed to by adult males" (101). Diamond (1987) also described how groups of two to four female bowerbirds visit bowers, presumably providing opportunities for younger females to learn from older females which features of bowers are most attractive.

Other species of bowerbirds build quite different types of bowers, but all seem to serve the same basic function of attracting females. For example, the stagemaker, or tooth-billed bowerbird *(Scemopoectes dentirostris),* of North Queensland clears a roughly circular area of the forest floor. As described by Gilliard (1969, 276), largely from the observations of Warham (1962) and Marshall (1954), "the male clears all the fallen debris from this space of earth as if with a broom. He then decorates it with fresh green tree leaves of one or more favoured species . . . the leaves may be up to twice as long as the male which carries them to his bower. These leaves are almost always placed upside down on the 'meticulously clean' court and are replaced with fresh leaves when they wither. The tooth-edged bill (for which the bird is named) serves as a specially modified tool for severing fresh leaves. . . . At one bower Warham found 56 leaves. He removed all of them [and after that] almost every day the male carried in from 2 to 10 leaves and placed them carefully on the ground court." Within a week 25 leaves had been gathered to replace those that had been removed.

Archbold's bowerbird of the New Guinea highlands *(Archboldia papuensis)* makes a display ground of flattened ferns and other vegetation but decorates it with piles of snail shells and sometimes black beetle wing covers. MacGregor's bowerbird *(Amblyornis macgregoriae),* also found in the New Guinea highlands, clears a roughly circular area and forms at its center a "maypole" consisting of numerous sticks piled around a thin sapling. Moss is gathered and placed on the cleared area around this vertical pile, which may be as much as meter in height. The golden bowerbird *(Prionodura newtoniana)* of Queensland constructs a double column of sticks, each somewhat like the "maypole" of MacGregor's bowerbird, but with a roughly horizontal stick connecting them. The male perches on this stick during parts of his displays. All of these styles of bower construction show considerable variation from place to place and from one individual to another, indicating that they are not fixed, stereotyped building behavior patterns.

Perhaps the most impressive of all bowers are some of those constructed by certain populations of the Vogelkop gardener bowerbird *(Amblyornis inornatus).* These have been studied by Diamond (1982, 1986b, 1987, 1988), who has not only described how they vary between populations of the same species, but has demonstrated by ingenious experiments how the birds select decorations of preferred colors. These birds are confined to rugged, uninhabited mountains of western New Guinea that are very difficult to reach because of both physical and political obstacles. In the Wandamen Mountains the bowers are truly impressive huts 40–80 cm high and 90–220 cm in diameter formed by weaving sticks together. On the downhill side of the hut is an opening 18–58 cm wide and 20–28 cm high. As described by Diamond (1987, 189), "the hut was built around a sapling at whose base was a green moss cone 20–23 cm diameter and 15 cm high, with a low stick tower joining the cone to the ceiling of the hut. The hut rested on a green moss mat."

In the Kumawa Mountains, about 200 km from the Wandamen range, the same species builds much simpler bowers, which lack any roofed hut but consist of maypoles, moss mats, moss cones, and stick towers similar to those located inside the huts of the Wandamen birds. In both cases small, conspicuous decorations are added by the birds. The moss mats are "woven tightly from fine, clean, dry, dead fibers of a moss that grows abundantly on trees" (Diamond 1987, 182). At some bowers these mats were almost perfect circles; when the diameter was measured at various orientations it differed by only a few percentage points. The "maypoles" were saplings 1–3 cm in diameter, without leaves or branches, rising up to 1.5–4 m above the ground. The moss cones were formed from the same moss used to make the mat. Stick towers consisted of hundreds of sticks 20–90 cm long piled against each maypole. Around the base of the tower the sticks lay horizontally, radiating neatly in all directions. In both areas there was also some variation in size and details of construction between individual bowers.

Most of the Kumawa bowers were decorated with dead pandanus leaves 20–150 cm long, often leaning against the maypole sapling. Some of these leaves weighed half as much as the bird, who must have dragged them for many meters from the nearest pandanus tree. Other common decorations were the brown shells of land snails, dark brown acorns, brown stones, beetle elytrae, and piles of 22–90 brownish sticks 8–100 cm in length. The Wandamen bowers were decorated quite differently. Most common were piles of black bracket fungi 20–50 cm high or groups of 4–32 dark brown or blackish beetle elytrae. Red and orange fruits were also present at most bowers, along with red leaves and black fungi other than bracket fungi, plus black fruits. Particular decorations were preferentially placed in certain areas relative to the bower; for instance, black and orange bracket fungi were placed downhill from the door. Colored flowers, red leaves, and red, orange, or green fruits were usually outside the bower on the mat, and objects chosen infrequently as decorations were usually inside the hut. These rare items included butterflies, beetle heads, amber beetles, acorns, and orange pieces of bark or jelly-like fungus. In some cases decorations of the same color were grouped more or less together. When Diamond altered these arrangements by shifting decorations to atypical locations the birds usually put them back where they apparently belonged.

All of these observations suggest that individual Wandamen bowerbirds were expressing preferences for certain patterns of decoration. Given the fact that young male bowerbirds practice bower-building and watch experienced adult males building bowers and courting females at them, it seems likely that these local and individual preferences in decorating bowers are learned manifestations of what Diamond (1986b, 3042) calls "culturally transmitted traits, like human art styles." To test this interpretation Diamond (1988) placed poker chips of seven bright colors in or near the bowers of both Kumawa and Wandamen bowerbirds. In the Kumawa area, where the birds do not naturally use colored objects to decorate their bowers, most males removed the poker chips placed in their display areas and did not bring any in from nearby.

At Wandamen bowers, however, the birds gathered many poker chips and arranged them in their bowers, showing a hierarchy of preferences, with blue collected in largest numbers, followed by purple, orange, red, lavender, yellow, and white in that order. But these relations differed among individual birds, and whichever color was preferred was gathered first, stolen from other bowers first, and removed from the bower least often. Poker chips of the same color tended to be grouped close to each other and also close to natural objects of similar color. Two or three poker chips of the same color were sometimes stacked on top of each other; occasionally a chip of a preferred color was piled on top of a less favored color. In general poker chips were treated in much the same ways as naturally occurring decorations.

This panoply of construction and decoration strongly suggests that the bird is consciously thinking about what it is building. Of course, determined behaviorists can always dream up complex sets of genetic instructions that might operate to generate whatever behavior may be discovered. In the case of bowerbirds, this exercise will require more ingenuity than in many other cases. If we allow ourselves to speculate as to what these birds might be thinking as they work at their bowers, the goal of luring a female ready for mating may well be prominent in the content of their thoughts. Marshall (1954) and others have argued that because bower-building and decorating are so clearly part of the birds' reproductive behavior, they were not accompanied or influenced by aesthetic feelings. Yet toward the end of this monograph, which emphasized the hormonal control of bowerbird reproduction and the behavior accompanying it, Marshall recognized that the birds may well enjoy the bowers they build. As von Frisch (1974) pointed out in his book *Animal Architecture,* it would be difficult to deny that impressing females motivates much human artistic creation. Unless committed a priori to an absolute human/animal dichotomy, we have no basis for rejecting out of hand the hypothesis that a male bowerbird thinks in simple terms about the bower he is building and decorating, the other males competing with him, and the females he hopes to entice for mating.

Beaver Engineering

When considering the kinds of animal behavior that suggest conscious thinking the beaver comes naturally to mind. These large aquatic rodents manipulate their environment in rather spectacular ways to obtain food and shelter. They fell trees and construct conspicuous lodges as well as digging less obvious bank burrows. They also deepen shallow stretches of water to form channels where they can swim and tow branches that would otherwise drag on the bottom. Beaver carry mud dug up from the bottoms of these channels to piles that may form small islands. This behavior may extend to digging canals through dry land that are also used to float branches to food storage piles. These canals and channels are not dug helter-skelter but along routes the beaver use to travel between a lodge or burrow and food supplies. Finally, as everyone knows, some beaver create ponds by building dams across small streams. Although they are far from being perfectly efficient engineers, their activities have a more obvious and substantial impact on their surroundings than those of most other mammals.

This review of beaver behavior is based primarily on the detailed studies of Wilsson (1971), Richard (1960a, 1960b, 1967, 1980, 1983), Hodgdon (1978), Patenaude (1983), Patenaude and Bovet (1983, 1984), and Ryden (1989). Wilsson and Richard studied captive beaver for the most part, whereas Hodgdon and Ryden concentrated on extensive observations of wild beaver under natural conditions. Patenaude succeeded in arranging an observation window that permitted videotaping of beaver behavior inside the lodge.

Beaver typically live in family units consisting of a monogamous pair of adults and their young. One litter of three or four kits is born each spring, and kits usually stay with their parents for two years. Both parents bring food to the young, and yearlings help by feeding and grooming their younger siblings, some staying with their parents for a third year (Hodgdon 1978). Although the adults do more dam- and lodge-building than their offspring, the yearlings participate to some extent. Many aspects of beaver behavior seem to vary from one situation or population to another, so that there are no absolutely fixed patterns of behavior that occur under all natural conditions. Scientists concerned with animal behavior have paid relatively little attention to beaver despite the impressive scale and scope of beaver works. One reason is the difficulty of observing a largely nocturnal animal that spends much of its time out of sight in a lodge or burrow. Furthermore, most of its body is submerged a great deal of the time, and some of its most interesting behavior occurs underwater, in burrows or lodges with underwater entrances, or even under the ice in midwinter.

As emphasized by Richard (1960b, 1967, 1980, 1983), it is difficult to repress the inference that beaver could scarcely accomplish what they do without some awareness of the likely results of their activities. He studied captive beaver in the fenced Parc à Castors, about 180 × 60 m in size, through which flowed a small stream. They built dams and lodges, and he was able to observe many details of their behavior and to perform experiments that tested their ability to solve problems quite different from the normal experience of their species. Some individual beaver learned to open puzzle boxes to obtain food, even when this required manipulating different types of latch. When confronted with food out of reach on a small platform at the top of a one-meter pole his beaver piled branches around the pole until they could climb up this pile to reach the food (Richard 1980, 97, fig. 55).

Richard's Parc à Castors contained numerous willows on which the beaver were free to feed, except for a few shade trees that were declared off limits "as in the Garden of Eden" to preserve the appearance of the park. These trees were protected by a cylindrical fence of heavy netting firmly anchored in the ground and wired securely to branches above the beavers' reach. But the beaver solved this problem by piling a pyramid of branches and mud around the tree and climbing up this pile to reach the unprotected trunk, which they then cut in their usual fashion. They did this to similarly protected trees on several different occasions.

Pilleri (1983) reports an ingenious response to a novel situation by two beaver confined in an enclosure where the water level of a 3.5 × 2.5 m pool 1.6 m deep was regulated by an outlet pipe equipped with a cap perforated by three holes 8 mm in diameter through which the water escaped. After about two weeks the beaver began to plug these three holes with "peeled twigs which had been gnawed off obliquely at both ends by the beavers and whittled down in such a way that they exactly fitted the holes. . . . The performance was repeated several times, always at night, after we had removed the sticks in the morning and restored the water in the pool to its normal level. Every night the beavers made new calibrated sticks and blocked up the holes. [Finally they] changed their technique. In addition to the sticks they used grass and whole piles of leaves mixed with mud" (99).

Many scientists disagree with Richard's conclusion that beaver know what they are doing, claiming that all their construction of burrows, lodges, canals, and dams results from genetically programmed action patterns that involve no conscious thinking or anticipation of the results of these activities. For example, after one of the most thorough studies of beaver behavior, involving both free and captive animals, Wilsson (1971, 240–54) concluded that even those actions that seem most intelligent are simply genetically programmed motor patterns. He specifically disputes the conclusion reached by Richard (1967) that beaver exhibit some degree of forecasting and "unformulated thought":

> The examples [Richard] gives as evidence for "forecasting," "intelligence" and "some kind of unformulated thought" can . . . just as well be interpreted as

> stereotyped phylogenetically adapted reactions. For example, the fact that a beaver often thoroughly investigates a leak in the dam, then leaves it for some hours and later brings different kinds of material with it when it returns to repair it, does not necessarily mean that it is able "to forecast in the choice he makes of building materials, which depends on the use made of them as well as the shape his construction will take." . . . Dam building behaviour is activated when the animal has received stimuli from the dam for a certain time and a delayed response is not unusual in phylogenetically adapted behaviour. (Wilsson 1971, 247)

Yet, despite this insistence that their behavior is stereotyped, Wilsson (1971, 187–89) describes how some of his captive beaver first piled material at a water outlet, but then, when this did not raise the water level in their tank, changed their tactics to a more appropriate placing of sticks and mud at the inlet. Thus, as pointed out by Richard (1983), dam-building behavior is sometimes modified as the beaver learns that one placement of material is ineffective in raising the water level.

A vivid example of the strong tendency to deny conscious intent was expressed by the anthropologist Ingold (1988). He points out that one of the earliest scientific studies of beaver engineering was conducted by Lewis Henry Morgan (1868), who was also, in Ingold's words, "one of the founders of the discipline of anthropology as we know it today" (86). After quoting Morgan's opinion that beaver fell trees and build dams with at least some rudimentary anticipation of the results of their activities, Ingold vigorously denies the correctness of any such inference. He cites approvingly the assertion by the American anthropologist Kroeber (1952) that (in Ingold's words) "the beaver *does not* and *cannot* construct an imaginary blueprint of his future accommodation; whereas this is something of which even the most 'primitive' human is capable. The human engineer constructs a plan in advance of its execution; the beaver lives merely to execute plans designed—in the absence of a designer—through the play of variation under natural selection" (90, emphasis in original). In all fairness to Kroeber, it should be mentioned that he made this statement originally in 1917, long before ethologists discovered how versatile animal behavior can be.

Ingold goes on to support his assertions by appealing to the authority of Karl Marx: "What from the very first distinguishes the most incompetent architect from the best of bees, is that the architect has built a cell in his head before he constructs it in wax" (90). Proving a global negative statement—that something never happens under any circumstances—is notoriously difficult, but Ingold and others deny that beaver might be aware of the results of their actions, without presenting any convincing evidence to support such a sweeping and dogmatic assertion. And Ingold reiterates his conviction that *"animals have no thoughts"* and that "rather than thinking without communicating, the animal *communicates without thinking;* so that the signals it transmits correspond to bodily states and not to concepts" (94, 95, emphasis in original).

It is difficult to reconcile this vehemently negative assertion with the evidence reviewed in later chapters, especially the experiments showing that pigeons can be taught something closely resembling concepts. Yet in a section titled "Thinking, feeling and intending" Ingold recognizes that animals are probably conscious of "doing and feeling" and writes: "Morgan in his time, and Griffin in ours, are suggesting that . . . beavers . . . plan things out, or envisage ends in advance of their realization. I do not think they do; but more than that, I do not think human beings do either, except intermittently, on those occasions when a novel situation demands a response that cannot be met from the existing stock-in-trade of habitual behavior patterns" (95–97). No one has ever suggested that animals *always* think consciously about what they are doing or about the likely results of their activities (although on p. 96 Ingold accuses me of that absurd claim, a typical example of discredit by exaggeration). What is so puzzling is how Ingold and others can recognize the likelihood that animals are sometimes, even if only rarely, conscious of what they are doing, yet feel so strongly committed to denying any sort of foresight, even for a short time into the future—denying that animals have any thoughts about the likely results of their own activities.

Given this basic uncertainty and disagreement, can we throw any light on these questions by considering the actual behavior of beaver? A good place to begin is their digging of burrows and construction of lodges. Under natural conditions most beaver dig burrows into the banks of whatever streams or lakes they occupy. Burrows start underwater and then turn upward until it becomes possible to construct a reasonably dry chamber. Beaver seldom if ever start digging burrows where the ground slopes too gradually to provide space for a beaver-sized tunnel above the water level. This implies that they recognize that some shores are too low to make burrowing worthwhile. As far as I have been able to learn from published accounts of beaver behavior or from my own observations, they do not start useless burrows. But aborted burrows would be difficult to locate, so lack of reports of their occurrence is not conclusive evidence of their absence. Perhaps they make this selection on the basis of steep underwater banks, but it seems more probable, pending appropriate investigation of the question, that beaver explore shorelines and select for burrowing only places where the ground rises far enough to provide space for a dry chamber within a reasonable distance of the water.

Burrow construction by beaver and other animals is an interesting sort of goal-directed behavior. It may even be directed toward the consciously perceived goal of providing a dry shelter. But the fact that burrowing occurs underground and sometimes underwater greatly restricts the sort of observational or experimental evidence that can be obtained without difficult and costly special procedures. Animals can be observed starting to dig a burrow, and completed burrows can be excavated and mapped, but the actual behavior of digging them, deciding in what direction to progress, and coping with obstacles such as rocks or tree roots remains almost totally unstudied.

When beaver burrows begun underwater turn upwards as they progress under and beyond the shoreline, they often reach the surface and break through to the air. Beaver sometimes pile sticks and mud over this opening and continue burrowing upward through the material they have added. Since their teeth can easily cut branches and shred woody material into soft bedding, they can burrow from below into a pile of their own making at least as easily as through a natural bank where the earth is often studded with roots and stones. This addition of new material may occur either after an actual opening to the air or shortly beforehand. As with many other aspects of beaver behavior, different populations seem to differ in this respect. But in either case the beaver brings mud and branches from a distance to a spot on the shore where its burrow is close to opening or has already opened to the air.

Bank lodges formed by adding material above a burrow often become island lodges when the construction of a dam raises the level of the water. But some island lodges are also built a short distance from the shore of lakes where the beaver build no dams and where the water level does not fluctuate appreciably. Hodgdon (1978) observed cases in which the beaver started an underwater burrow at a place where the water's depth varied steeply although all the immediate area was underwater. When the burrow broke through the earth, still below the water surface, they piled branches and mud over the opening, and thus created an island lodge.

The beaver that Hodgdon observed extensively at close range always began lodge-building by at least starting to dig a burrow underwater. When the bottom was too rocky for burrowing, these beaver went through the motions of digging before piling up material that eventually became a lodge. But Richard (1980, 95, fig. 53) describes a large lodge that had been constructed when a flood inundated the beavers' burrow and that was later exposed after the falling water level left it high and dry. The lower portion was entirely composed of sticks piled up by the beaver, through which an access tunnel had been excavated. This led into a chamber above the highest water level large enough for two persons to squeeze into. Since the actual construction of this lodge was not observed, there is no way of knowing whether the beaver began with an attempt at burrowing.

Aeschbacher and Pilleri (1983) found that when captive beaver built a lodge they "cut sticks of two different lengths from the branches provided, peeled them and inserted the longer ones into the roof of the main chamber and the shorter ones into the chamber entrance, in a more or less radial pattern which was afterwards plastered with mud, wood shavings and small twigs." Richard also describes cases where the beaver gathered appropriate material before starting to build, suggesting anticipation of the construction. Under natural conditions beaver add to and alter lodges and burrows, so these shelters are in a constant state of enlargement and modification, somewhat like medieval cathedrals. Lodges and burrows are often abandoned, even when the beaver remain in the vicinity and build new shelters. Even after they are abandoned beaver lodges are substantial piles of material, sometimes 2–3 m high and 5–10 m long, and their remains can be recognized years later even though grass, bushes, and even trees may have grown up on them. Abandoned lodges are sometimes reoccupied and reconstructed.

In many cases, except when there are very young kits in a family lodge, the beaver move about from one burrow or lodge to another in the same pond or stream, so that a shelter occupied one day may be empty the next. One pleasant day in mid-May I came upon a pair of adult beaver and at least one youngster in a sort of "bower" under a fallen tree partially screened by surrounding vines and small branches but clearly visible from my kayak as I paddled down a small stream in the New Jersey pine barrens. The nearest lodge was three or four hundred meters away, and the very gently sloping shoreline provided virtually no opportunity for burrowing. These beaver had presumably come out or stayed out after daybreak, although they must have had some other shelter. This "bower" was later converted into a typical small lodge by adding sticks, leaves, grass, and mud to the walls.

Beaver vary their lodge- and dam-building in different situations, and they use many sorts of material according to what is available. Some beaver, such as ones that built lodges inside caves (Grater 1936; McAlpine 1977; Gore and Baker 1989), adapt their building behavior to atypical situations. They may modify human artifacts, adding, for example, sticks, mud, and even stones to a concrete dam and thus raising the water level above what the human dam builders intended. Or they may build nesting chambers inside abandoned or even occupied buildings close to streams, such as unused mills (Richard 1980). These modifications of behavior suggest sensible use of available resources rather than the unfolding of rigid, stereotyped genetic programs.

One of the principal foods of beaver is the bark of trees, but they also eat other vegetation such as tubers of water lilies dug up from the bottom of their ponds. They may totally consume small branches up to about a centimeter in diameter, but from larger branches or the trunks of saplings they strip off only the bark and leave a shiny bare surface mottled by characteristic tooth marks. Their digestion is aided by symbiotic microorganisms, and like most herbivorous mammals the intestine has large branches, or caecae, where slow digestive processes take place. Beaver produce two kinds of feces, one of which they eat and thus recirculate some foodstuffs, presumably accomplishing in this way a more complete digestion of otherwise intractable materials.

Cutting trees is perhaps the best known of beaver activities. Most trees cut by beaver are no more that 15–20 cm in diameter, but occasionally a trunk as thick as 1 m is severed. When the tree falls, many but often not all of the branches are cut into smaller sections that the beaver can tow through the water. The bark may be eaten immediately, or sections of tree trunk and branches may be transported to underwater food piles near the lodge or burrow. Beaver tend not to use tree trunks or branches with edible bark for building lodges and dams. For these they usually employ sticks or logs from which the bark has been stripped, dead trees or fallen branches, and portions of trees whose bark is inedible or is eaten only rarely. When accumulating underwater food piles or building dams, beaver sometimes actively thrust one end of a stick into the mud or the previously accumulated tangle of branches; this prevents the stick from floating to the surface or being carried downstream by the current.

Although these activities seem directed toward an end result that is useful to the beaver, they are by no means perfectly efficient. Some trees are cut partway through and then abandoned; others lean against neighboring trees as they begin to fall, so that the beaver obtains no food at all. Beaver do not seem to realize in such cases that by felling the tree against which the first tree has lodged they could obtain a double supply of bark rather than nothing at all. Occasionally one finds tree trunks that have been cut partway through at two or more levels. Beaver of different sizes can chew effectively at different distances above the ground, or such multiple cuts may result from activity with and without an appreciable depth of snow. But they sometimes occur in areas such as the New Jersey pine barrens, where there is not enough snow to account for them on this basis.

Inspection of areas where beaver have been working shows that they expend more effort than we can see would be necessary. Even after branches or tree trunks are brought to ground level they may be cut into sections as short as 15–20 cm although sticks as long as 2–3 m can be transported; these short sections may be taken into the lodge, where the bark is eaten, or the whole piece may be shredded to form bedding. It is easy for us to judge from inspection of beaver works that they could have accomplished their apparent objective with much less work than was actually expended. But lack of perfect efficiency in accomplishing an operation does not prove the absence of any conscious plan.

Beaver dams are built by gradually adding sticks, mud, and occasionally stones to some relatively narrow portion of a flowing stream. I have weighed stones up to 3.3 kg from a small dam, and Richard (1983) reports use of rocks as heavy as 10 kg. Ordinarily the dam is begun where there is already some small obstruction to the flow of water such as a small stick or a rocky ledge on the bottom of the stream. As the water level rises, more material is added, usually with most of it placed where the water is flowing most vigorously. Richard points out that laying sticks roughly parallel to the direction in which the water is flowing gradually produces, as the dam grows higher, an array of sticks oriented at increasing angles to horizontal. But many sticks are oriented in other directions, so that the eventual dam is a tangle of branches and mud. Manmade objects such as pieces of sawed lumber or plastic are sometimes also incorporated into beaver dams. Although beaver apply mud and soggy vegetation to the upstream face of their dams, this does not produce a watertight structure, and usually more water trickles through or under a beaver dam than flows over the top. But beaver dams do commonly maintain a pond that may be a meter or more deep on the upstream side of the structure.

Wilsson (1971) and Richard (1967, 1980) have tested the hypothesis that the sound of running water is a stimulus to placing material on what will eventually become a dam. Playbacks of recorded sounds of running water did elicit piling of material close to the loudspeaker, but often this response occurred only after a considerable period of time. Richard (1967) found that playbacks of the sound of running water attracted the beaver's attention but that a lowered water level, and inspection of the leaking dam, led to the placement of most material where it was needed to stop the leak.

When people insert pipes through a beaver dam, as is often done to avoid having the beaver flood roads or basements, the beaver initially try ineffective measures such as placing mud on the dam itself near the noisy outflow from the pipe. But in many cases studied by Richard they eventually discovered the place where water was entering the pipe and plugged this opening, even when it was several meters upstream from the dam and near the bottom of the pond. In one experiment a long pipe with a strainer at its entrance was arranged so that water entered the pipe well above the bottom of the pond and far upstream from the dam. The beaver eventually piled material under the strainer to create a submerged island to which they then added enough mud and vegetation to plug the strainer and stop the escape of water. In recent years a considerable effort has been devoted to contriving "beaver bafflers" of ingenious design that prevent effective dam-building at culverts or other places where the resulting rise in water level would interfere with human activities.

Although sounds are clearly *one* stimulus to dam-building, these experiments show that adding material to a dam is by no means a rigid response to this particular stimulus. Most material is added on the upstream side of the dam even though the loudest sounds of running water are usually on the downstream face. Under natural conditions beaver pile material at only a few of many places where water is tumbling noisily. When beaver are adding small amounts of material to some of the places where water is flowing out of the pond they occupy, they often bring nothing at all to larger leaks where a much noisier flow is occurring. If the sound of running water were the only stimulus for dam-building, one would expect to see piles of sticks and mud at many places where the sound of running water was present, but this is far from being the case. I have tried playing back sounds of water flowing through leaks in beaver dams, placing the loudspeaker at the edge of the dam where a leak might occur, but so far the beaver have not piled any mud on my loudspeakers. The view that dam-building is a simple fixed action pattern released uniquely by the sound of running water seems to be a reductionistic oversimplification.

Beaver dams are often damaged by strong currents after heavy rains and snowmelts, and beaver usually repair breaks in actively maintained dams, but this may not happen for hours or days. Dam-building and dam-repairing are more prevalent in late summer and fall than earlier in the year. Occasionally, however, an especially disastrous break elicits strikingly energetic and appropriate behavior. One of the best examples of such emergency responses is described by Ryden (1989) in the course of extensive observations of a beaver colony whose members were cutting only a few trees because they were feeding primarily on the tubers of water lilies. These beaver were relatively habituated to human observers, whose presence did not seem to alter their behavior. For several weeks the adult male had been doing most of the dam maintenance. At nearly the same time every evening he inspected it and added small amounts of mud here and there, but his mate and the yearlings rarely visited the dam.

Then one day in late June human vandals tore open a large hole in the dam, causing a torrent of water to rush out of the pond. The water level dropped at a rate that clearly threatened to drain the pond within a matter of hours. Ryden and a companion were naturally outraged at this wanton destruction and sought to reduce the damage by piling large stones in an arc upstream from the opening in the hope that this would slow the flow of water and help the beaver repair their dam, even though almost all of the stones were underwater. When the adult male emerged from the lodge at the normal time in the late afternoon and made his customary visit to the dam he immediately responded to this emergency with drastically altered behavior. He first cut a few small branches and towed them to the gap in his dam, where he succeeded in pinning some into the newly placed rock pile, although others were washed downstream.

At this pond there were very few dead trees available for the beaver to gather as dam-building material, primarily because human picnickers had used them for firewood. In this emergency the beaver cut and brought to the dam green vegetation that he would otherwise have used for food if he cut it at all. Three other beaver from the colony joined in relatively fruitless efforts to fix branches to the top of the rock pile over which the water was cascading, although during many nights of observation they had seldom been observed at the dam.

When adding branches to the top of the largely submerged rock pile failed to slow the torrent that was draining their pond, the beaver changed their tactics within a few minutes. Instead of towing more branches to the hole in the dam they dove to the bottom of the pond, gathered mud and vegetation such as water lily stems, leaves, and roots, and used them to plug the underwater gaps between the rocks. This slowed the escape of water from the pond, and in time the combined efforts of the beavers and their human helpers stabilized the water level, but at a much lower level than that of the original pond. Beaver ordinarily gather mud and underwater debris and apply such material to the upstream face of the dam *after* they have piled sticks at one of many places where water is flowing. But in this case they seemed to recognize that stick piling was ineffective and turned instead to plugging underwater gaps between the stones, even though the noise of flowing water came from the top of the rock pile.

Ryden watched the beaver work for most of the night at this dam-repairing endeavor, leaving only in the early hours of the following morning for a brief rest. When she returned the beaver had retired to their lodge. Late in the afternoon, at his customary time of emergence from the lodge, the adult male's first act was to remove a large stick from the lodge itself and tow it 100 m to the dam for further repair efforts. Ryden had been watching the beaver pond continuously for two hours beforehand, and in all probability this beaver had not visited his dam since retiring roughly twelve hours earlier. Yet he performed this most unusual action of removing a branch from the lodge and towing it to the dam, presumably because he remembered the need for material to repair the damaged dam. Other beaver also removed sticks from the lodge when they emerged and brought them to the dam.

While the beaver were inside the lodge, Ryden and her companion had brought many dead branches to the pond to provide the beaver with material for dam repair. They did this because there was so little available near the shores of this pond. On finding this floating tangle of branches, the beaver began at once to use them for dam repair, piling many of these branches on top of the material they had added to the arc of stones the night before. In time they succeeded in restoring a functional dam. Meanwhile, the water had fallen so low that the entrances to the lodge sheltering a litter of kits had become exposed. Several days later the beaver reconstructed the entranceways so that these were again underwater.

Further dam-building with sticks and mud gradually raised the water level, but only slowly over many days. During this period, despite the lack of any very noisy flow of water, the beaver added material to the top of the dam and above the water level. This is something they do not ordinarily do when building a dam, although Hodgdon (1978, 209) describes other instances of "over compensation of repair efforts . . . where repaired dam segments became higher than the normal dam crest," and similar behavior is reported by Wilsson (1971) and Richard (1967). Ryden, however, has observed many repair efforts in which the top of the dam was kept almost perfectly level and no higher at the point where a break had been repaired. In these situations, when beaver adapt their behavior to changing circumstances, are they perhaps hopefully anticipating a higher water level that would restore their pond to its former depth? Of course, one cannot be at all sure of such a speculative inference, but it is a possibility worthy of further investigation. One way to learn more about this behavior would be to inquire whether such "overcompensation" occurs only, or primarily, when the break being repaired had lowered the water level below its former level. Or is it just as likely to occur after a relatively small break that has not resulted in a significant drop in water level?

Ingold (1988) claimed that only members of our species ever plan things out "on those occasions when a novel situation demands a response that cannot be met from the existing stock-in-trade of habitual behaviour patterns" (97). Such drastic breaks in a beaver dam as the one described by Ryden are fortunately rare events, and it is highly doubtful that these beaver had ever before encountered a problem of this magnitude. They certainly had never experienced human helpers placing large stones in an arc upstream from a major break in their dam. Their first response was to bring branches to the damaged dam and try to pin sticks into the stone pile. But they soon changed their behavior to digging up mud and vegetation from the bottom of the pond and using it to plug underwater spaces between the stones. Later they transferred material from lodge to dam. These were novel and appropriate responses to a wholly new situation without precedent in their experience.

Beaver often rework material they have built into their lodges, not only adding new branches and mud but shifting material to new positions. The taking of branches from the lodge sheltering young kits for use in dam repair observed by Ryden is an extreme example. Even the dams are sometimes modified by removal of material as well as by addition of sticks, mud, or stones. In winter beaver ponds in northern latitudes are often frozen over for many weeks. During this time the beaver must either remain in the lodge or swim underwater beneath the ice, holding their breath, to reach their food storage piles or to travel to other parts of the pond. Several observers, including Wilsson (1971), Hodgdon (1978), and Ryden (1989), have noted that beaver sometimes open holes in their dams, thus causing the water level to drop and creating an airspace under the ice. Such holes in the dam have not been reported under other circumstances, to the best of my knowledge. Otters also make openings in beaver dams, as described by Reid, Herrero, and Code (1988), but their openings are usually trenches at the top of the dam rather than tunnels.

Opinions differ as to why beaver cut holes in the dams they have so laboriously constructed and maintained in previous weeks or months. One possibility is that the resulting air space under the ice makes it unnecessary to swim underwater to reach their food stores or to travel farther from the lodge and perhaps reach openings in the ice through which they can come out on land to seek fresh food. Hodgdon (1978, 124–25, 209–10) observed 22 beaver families during long periods in midwinter when they were confined under thick ice for weeks at a time. Eighteen of these family groups cut 31 holes in their dams, six being "beaver-sized tunnels completely through dams." Most of these breaks were made in the late winter when rising water levels probably began to flood the lodge. They lowered the water level by as much as half a meter, although usually they produced a 10–15 cm air space under the ice. The important point is that the beaver reversed their normal behavior with respect to their dams, cutting holes rather than adding material. It seems reasonable to suppose that they had some objective in mind, unless, like Ingold, we refuse to consider the possibility of any sort of conscious intention on the part of beaver.

To account for such reversal of their customary behavior of adding material to dam or lodge as the thoughtless unfolding of a genetically determined program requires that we postulate special subprograms to cover numerous special situations such as rising water in midwinter when the pond is covered by ice. Such postulation of a genetic subprogram can always be advanced as an explanation of any behavior that is observed, but the plausibility of such "ad hocery" fades as their number and intricacy increases. A simpler and more parsimonious explanation may well be that the beaver thinks consciously in simple terms about its situation, and how its behavior may produce desired changes in its environment, such as deeper water when there is no pond, or an air space under the ice in midwinter under the conditions where Hodgdon observed that 18 out of 22 beaver families cut holes in the dams they had constructed.

CHAPTER SIX

Tools and Special Devices

The use of tools by nonhuman animals has often been considered a sign of intelligence, and at one time it was believed to be limited to our closest relatives, the apes and monkeys. But prolonged and intensive observation of animal behavior has revealed many instances of tool use, and even of preparation or construction of simple tools, in a wide variety of animals ranging from insects to birds and mammals. Some scientists, such as Hall (1963), have reacted against the trend to interpret tool use as especially telling evidence of rational behavior by arguing that it is not so special after all. Beck (1980, 1982) and Hansell (1984, 1987) have both reviewed the many known cases of tool use by animals, but they have also expressed the opinion that the use or even the preparation of tools does not necessarily indicate greater mental versatility than do many other kinds of behavior.

Other types of animal behavior, such as the shell dropping by gulls described in chapter 3, are just as strongly indicative of thinking as is tool using. But the latter is certainly *one* important category of behavior in which it would appear especially valuable for an animal to think consciously about what it is doing. Even in relatively simple cases the tool is a separate object from the food it helps the animal obtain or the process for which it is used. Therefore, selecting or preparing a tool indicates awareness of whatever it serves to accomplish. In an especially impressive type of tool preparation and use, a chimpanzee breaks off a suitable branch, strips it of twigs and leaves, carries this probe some distance to a termite nest, pokes it into termite burrows, and then pulls it out and eats with apparent relish the termites that cling to it. The details of this use of tools to obtain termites vary geographically and constitute an important part of chimpanzee culture, as discussed by McGrew (1994). Such tool use differs significantly from digging up burrowing prey or otherwise capturing it by direct action, because all but the final steps require seeking out, modifying, and manipulating something very different from the food that is thus obtained.

Although other types of behavior also entail acting on objects quite different from those involved in the final consummatory behavior, tool use, and especially the preparation of tools, constitutes an especially distinct separation of specialized behavior from the goal attained. To be sure, when a rat runs through a maze or a pigeon pecks a lighted key these actions also differ from obtaining food from the end of the maze or from the magazine of a Skinner box. But selecting a suitable object to use as a tool entails a more independent action on the animal's part—the adaptation of an otherwise unimportant object such as a branch or twig to a specific purpose. Therefore it is appropriate to retain much of the commonsense view that tool use, especially tool manufacture or preparation, is rather special, although of course not uniquely indicative of conscious thinking on the animal's part.

Such behavior is relatively rare, but it does occur consistently in many groups of relatively complex animals. The suggestive examples mentioned below, as well as many others, are clearly reviewed by Beck (1980). Certain crabs pick up anemones (which give off stinging nematocysts when disturbed) and either hold them in their claws to ward off attackers (Duerden 1905; Thorpe 1963) or hold them close to parts of their exoskeletons so that the anemones attach themselves and form a protective outer covering (Ross 1971). Ant lions and worm lions were mentioned in chapter 5, because the pits they construct are functional artifacts. But their prey-catching behavior also includes the throwing of small grains of sand at the ants or other small insects that fall into their pits, and this constitutes a simple form of tool use, as described by Wheeler (1930) and Lucas (1989). But the ant lion does nothing to modify or improve the grains of sand or other small particles that it throws. It is not even clear that the throwing is particularly directed at the prey.

Other insects, however, carry out more specialized tool-using behavior. Ants of the genus *Aphaenogaster* use crude sponges to carry back to their nests semiliquid foods such as fruit pulp or the body fluids of prey, as described by Fellers and Fellers (1976). They pick up bits of leaf, wood, or even mud and hold them in the liquid long enough for appreciable amounts to be absorbed before carrying back the wetted spongelike object to the colony. This enables them to transport as much as ten times as much of the liquid food as they could otherwise carry. Wasps of the genera *Ammophila* and *Sphex* hold in their jaws a small pebble, a piece of wood, or other small object and use it to tamp down soil used to close a burrow where they have laid an egg, as reviewed by Evans and West Eberhard (1970) and Beck (1980). An example of this type of behavior has been described in detail by Haeseler (1985). Finally, the use of silk-secreting larvae by the weaver ants, described in chapter 5, involves manipulating these living tools during the construction of the leaf nests.

An especially striking case of complex tool use is presented by McMahan, who discovered that certain neotropical assassin bugs "fish" for termites with the corpses of previous victims. These insects actually employ two different types of tool when capturing termites. First, they pick up particles from the outer surface of the termite nest and apply them liberally to the outer surface of their own bodies. These particles apparently provide an effective sort of camouflage, whether tactile, olfactory, or both. Then, having caught one termite at an opening into the nest and sucking its internal juices, the assassin bug dangles the corpse into the opening, where it attracts other termites that are captured in turn. One assassin bug was observed to consume 31 termite workers by means of this combination of camouflage and baiting (McMahan 1982, 1983).

There are scattered cases of tool use among birds, many of which have been reviewed by Chisholm (1954, 1971, 1972) and Boswall (1977, 1978). Usually these involve using some small object such as a twig or piece of bark to remove an edible insect from a crevice where the bird cannot otherwise reach it. The bird may simply pick up a suitable probe or it may break off a twig and bring it to the crevice from which it evidently wants to remove an item of food. But Boswall also reviewed cases in which wild or captive birds use some simple tool as a means to groom the feathers, as a hammer, or as a thrown missile. According to Boswall (1978) only 26 of the approximately 8,600 known species of birds have been observed to use any sort of tool, and only a small fraction of individuals of these species do so. Thus tool use is not a stereotyped pattern but a rare type of behavior, performed only by some individuals of a few species on relatively infrequent occasions.

One of the best-known examples of such tool-assisted food-gathering is practiced by two species of Darwin's finches, discussed in chapter 3 with reference to the specialized habit of feeding on the blood of boobies. *Cactospiza pallida* pries insect larvae, pupae, and termites from cavities in dead branches with the aid of cactus spines or twigs held in its beak, and *C. heliobates* uses similar tools to remove prey from crevices in mangroves (Curio 1976, 164–65; Grant 1986, 3, 372). Bowman (1961) and Millikan and Bowman (1967) studied the learning of tool use by captive Galapagos finches.

Jones and Kamil (1973) observed that a captive blue jay learned to use probes and pieces of paper to obtain otherwise inaccessible food. Somewhat similar tool use by green jays in Texas has also been reported by Gayou (1982). Some of the hungry blue jays tore roughly appropriately sized and shaped pieces from the sheets of newspaper lining the bottom of their cages and used the pieces of paper as crude tools to rake into the cage food pellets that they could not otherwise reach. The behavior varied considerably among individuals; some did not learn it even when given abundant opportunity to observe other birds obtaining food in this way. It is thus anything but a stereotyped, fixed action pattern. It is possible that after failing to reach a food item with its bill a bird intentionally picks up and uses a twig or other object to aid in this endeavor. But many random movements seem to precede even the crude and occasional use of tool-like objects by captive jays.

Reid (1982) reported that a young captive rook used a standard drain plug to close the drain hole in a depression in its aviary cage and retain rainwater that would otherwise have escaped. Janes (1976) describes how a pair of nesting ravens dislodged small stones that fell in the general direction of two people who had climbed close to their nest and young. But Heinrich (1988, 1989) observed similar behavior by wild and captive ravens and believes that it may be "displacement behavior when they are angry or frustrated." When ravens are disturbed near their nest and young they often strike at all sorts of available objects, so that twigs fall from branches on which they are perched; but, as Heinrich points out, the dropped objects have never been observed to strike the intruder.

Clayton and Jolliffe (1995) observed an enterprising case of tool use by hand-raised captive marsh tits *(Parus palustris)* that were being used to study food storing. Two groups of 11 birds were housed in separate cages located in close proximity so that they could see each other. They were provided with equal amounts of food, but for one group this consisted of intact seeds that they often stored, while the other group had seeds ground to a powder, a form of food that according to the plan of the experiment they could not carry away and store. Some of the marsh tits that could not store any food began to remove colored gummed tape used to mark the feeders in each cage, folded the pieces of tape around the powder, and stored these as they stored the seeds. One bird dipped the wet piece of plastic tape into the powdered food, then folded the piece and stored it, having thus devised a way to store the "unstorable" powdered food. A second bird in an adjacent cage began to perform this very unusual form of food "packaging" and storing. Other captive birds have been observed to store inedible objects, but in this case the inedible tape was folded and used to store small amounts of the powdered food that could not previously be stored.

Crows and their relatives are versatile and ingenious birds that sometimes use tools to a limited degree, as reviewed by Angell (1978). An intriguing new case of fairly complex tool manufacture and use by New Caledonian crows *(Corvus moneduloides)* was reported by Ornstein (1972) and has recently been described in greater detail by Hunt (1996) during extensive observations in rain forests at altitudes between 950 and 1,300 m. Hunt observed 68 crows using or carrying tools, and he also saw four crows manufacturing their tools. He collected more than 300 tools dropped by the crows and found that they were of two types. The first were twigs stripped of leaves and often of bark and cut so that a short projection of a branch formed a hook at one end, and this end was inserted into cavities to pull out prey animals. The second kind were long thin strips torn from pandanus leaves but shaped to provide a rough hook at the end that was inserted into a cavity.

The crows held these tools in the bill and used them to extract small animals such as adult insects and insect larvae, spiders, and millipedes from cavities in trees and leaves. The tools were clearly useful to the crows, and Hunt describes their behavior: "Between foraging episodes, crows often transferred their tools to their feet or placed them in a secure position on their perch, retrieving them with their bills before departing. . . . Two birds changed trees after putting down their tools, returning within minutes to retrieve them. When feeding on prey, birds left tools in search sites or transferred them to their feet" (249). These crows thus devoted considerable effort to the preparation of their tools, and after extracting prey they took care to keep the tools available for future use.

Another type of tool use that has been observed in a few species of birds is throwing stones or other hard objects at eggs that are too strong to be broken by pecking or other direct attack. The best-known example is the dropping of stones on ostrich eggs by Egyptian vultures, as described by van Lawick-Goodall (1970). Just how this behavior is acquired is not clear, although Thouless, Fabshawe, and Bertram (1989) suggest that vultures first learn to associate small eggs that they can break with the food they obtain in this way, and then, recognizing large ostrich eggs as potential food, learn that they can be broken by dropping stones on them. However this sort of tool use develops, it seems likely that the vultures are thinking about the edible contents of the ostrich egg as they pick up stones and drop them.

Heinrich (1995, 1999) has recently studied a specialized form of tool use by ravens that suggests insight into novel and somewhat complex relationships. He had captured five ravens as nestlings and held them in a large outdoor cage until they were well grown. The cage was equipped with two horizontal poles well above the ground. The ravens had been fed for many months with road-killed animals and other pieces of meat placed on the ground within the cage. During the experiments Heinrich provided the hungry ravens with only a small piece of hard salami suspended by a piece of string from one of the horizontal poles. At first the ravens flew to the suspended food but were unable to detach anything edible from it; they also seized the string while perched on the horizontal pole and pulled at it from time to time. But the string was too long to allow a single pull to lift the meat within reach. After six hours one raven suddenly carried out a complex series of actions that did bring the suspended meat within reach. This entailed grasping the string in the bill, pulling it up, holding the string with one foot, releasing it from the bill, reaching down again to grasp the string below the pole, and repeating the sequence four or five times.

For a few days only this raven obtained suspended meat in this way, but in time all but one of its companions began to pull up the string, hold it with one foot, and repeat these actions until the food could be reached. They performed this feat in slightly different ways; two birds moved sideways during successive stages of holding the string with the foot so that the string was held at different points along the horizontal perch. The other two piled the string in loops, standing in roughly the same spot while holding the string. All but the first raven to perform this string-pulling action could have learned it by observing the successful bird, but Heinrich's impression was that each bird solved the problem for itself, using slightly different maneuvers, although its efforts to do so may well have been encouraged by watching its companion obtain food in this unusual manner.

Heinrich also observed a further indication of real understanding by the ravens that had learned this specialized form of food-gathering. When a hungry raven is startled while holding a small piece of meat in its bill it almost always flies off without dropping the food. But all four of the ravens that had obtained their food by the string-pulling maneuver always (in more than 100 trials) dropped the meat before flying to another part of the cage. They apparently realized that the string would prevent the meat from being carried away. The fifth raven never did learn to obtain food by string-pulling, although she had obtained food from her string-pulling companions. When this less talented bird was startled while holding a piece of meat still attached to a string she did fly off without releasing the meat—which was jerked from her bill when the string became taut. Two crows failed to obtain suspended meat by pulling up and holding the string, and they initially flew off with meat attached to a string, although they learned not to do so after five and nine trials, respectively.

Small caged birds have been trained to obtain food by pulling on strings, as reviewed in detail by Vince (1961). But in all cases a considerable time of trial and error was necessary before the trick was learned. In Heinrich's experiment the four successful ravens seem to have had the insight that pulling and then holding the string would bring the food within reach. Each bird carried out the whole sequence the first time it was attempted, holding the string with the foot and repeating the process at least four times. Heinrich had watched the ravens at all times when food was suspended from a string in their cage, and they did not engage in partial or incomplete actions that gradually developed into a successful food-gathering procedure. Prior to the experiments they had never had access to strings or stringlike objects such as vines, and in his extensive observations of wild and captive ravens Heinrich has never seen them pulling on vines or other stringlike objects to obtain food.

Tool use by mammals is not widespread, and except among primates and elephants it is scarcely more prevalent than with birds. Beck (1980) described simple cases of tool use by three species of rodents: tool-assisted digging by a pocket gopher that held small stones or other hard objects between the forepaws, leaning of an oat stalk against the wall of a glass aquarium by a captive harvest mouse that used it to climb to the top of the enclosure, and throwing of sand at snakes by ground squirrels. Rasa (1973) has described how the dwarf mongoose breaks eggs by throwing them backwards between its hind legs against some hard object.

Shuster and Sherman (1998) have recently reported that captive naked mole-rats, *Heterocephalus glaber,* often place a wood shaving or tuber husk behind their protruding incisor teeth when gnawing on substances that yield fine particulate debris. This oral barrier may prevent choking or aspiration of foreign material. Elephants are routinely trained to hold various objects with the trunk, and wild elephants have been observed to use sticks to scratch at parts of the body that are difficult or impossible to reach directly. They also throw things held with the trunk, although this is apparently not a common occurrence.

One of the clearest cases of relatively common tool use by a mammal under natural conditions is the use of stones by sea otters, as described by Kenyon (1969), Calkins (1978), Houk and Geibel (1974), Riedman and Estes (1990), and Riedman (1990). Stones are carried to clams, abalones, or other mollusks at the bottom and used to hammer them loose from the substrate. Stones are also carried to the surface and used as anvils resting on the otter's chest as it floats on its back and hammers shellfish or sea urchins against the hard surface. Riedman has observed much individual variability in the ways in which sea otters use stones or other objects such as shells as anvils against which to hammer recalcitrant shellfish. They sometimes keep a favorite stone tucked under the armpit for repeated use. Some otters use discarded bottles instead of stones as anvils (Woolfenden 1985). Others have learned that small octopuses retreat into empty aluminum cans; these otters have been observed to bring the occupied can to the surface, where they tear it open with the teeth to extract and eat the octopus (McCleneghan and Ames 1976). It would seem helpful to think in simple terms about the food they hope to obtain by these specialized procedures.

Some species of primates engage in tool use quite often, but others do so rarely, if at all, as reviewed in detail by Beck (1980). Most of the objects used as tools are stones or pieces of wood, and they are used to break open nuts or to throw at other animals or at people. Parker and Gibson (1990), Westergaard and Fragaszy (1987), and Chevalier-Skolnikoff (1989) have provided extensive reviews of tool use and manufacture of simple tools by monkeys and apes. Chimpanzees and capuchin monkeys (genus *Cebus*) are clearly more inclined to make and use tools than other species. Some but not all captive capuchin monkeys display considerable inventiveness in fashioning probes used to obtain otherwise inaccessible food and also spongelike tools to take up liquids.

Ritchie and Fragasky (1988) describe how a mother capuchin "manufactured, modified, and used simple tools to manipulate her infant's head wound, and applied modified plant materials to the wound" (345). Westergaard and Fragaszy conclude that they are almost as versatile in this type of behavior as chimpanzees. Chevalier-Skolnikoff (1989) concludes from her review of tool use by capuchins that in some individuals this behavior conforms to all six of Piaget's stages of sensorimotor intelligence. The many commentators on her review express a wide range of mutually contradictory interpretations, but it seems clear, on balance, that regardless of theoretical debates, these monkeys, as well as the Great Apes, must think about the objectives they are achieving by the use of tools.

It is important to emphasize that although some individual monkeys and apes become very proficient makers and users of tools, others never do anything of the kind, even under virtually the same environmental conditions. Furthermore, the details of tool-making and -using vary widely among individuals and across populations of the same species. For example, the well-known manufacture and use of sticks to "fish" for termites varies in detail among populations of chimpanzees in different regions of Africa, as reviewed by Sugiyama and Koman (1979), McGrew,

Tutin, and Baldwin (1979), Goodall (1986), and McGrew (1994). Captive apes have been trained to use a wide variety of human devices from socket wrenches to bicycles. An orangutan has even been taught to make stone tools (Wright 1972), although of course this is not something that they do under natural conditions.

Boesch and Boesch-Ackermann (1984, 1990) have observed that chimpanzees use different kinds of stones for cracking different sorts of nuts, and remember where both stones and nuts are to be found, so that when a particularly tough species of nut is gathered they know where to find an appropriate type and size of stone with which to open it. They use stones in a variety of well-coordinated ways to open different types of nuts. Indeed their manual dexterity in this activity has led to likening them to early hominids. Boesch (1991) described observations of mothers aiding their young and to a limited degree demonstrating to them some of the procedures they used to crack open nuts. This and other cases that suggest active teaching of one animal by another have been critically reviewed by Caro and Hauser (1992), who concluded that active and presumably intentional teaching is very rare.

Burying and Cooling Young

Egyptian plovers *(Pluvianus aegypticus)* resemble the migratory plovers of North America but nest on sandbars on the shores of African rivers. Herodotus, who called this bird the Trochilos, and more recent writers have stated that it picks leechlike parasites from the teeth of crocodiles. It is no longer found in Egypt because of climatic and ecological changes since ancient times. In 1977, Thomas Howell (1979) studied Egyptian plovers nesting along the Baro River in southwest Ethiopia. Although he and other recent observers have not seen them picking the teeth of crocodiles, the behavior patterns they use to conceal their eggs and young and to protect them from overheating are in many ways more striking and significant.

Egyptian plovers nest only on sandbars exposed during the dry season. When nesting they are very aggressive toward predators of eggs or young and toward other Egyptian plovers as well as other birds that eat the same types of food. Their nest is a simple depression in the sand, known as a scrape. Only island sandbars are selected for nesting, and very similar sandy areas connected to the mainland are avoided. This probably reduces predation on eggs and young by mongooses, which apparently do not venture into even shallow water: some nests were on islands separated from the shore by only about a meter with the intervening water depth only about 10 cm. Both male and female

Egyptian plovers make several scrapes before eggs are laid in one of them. As Howell describes this behavior, "a pair may make dozens of scrapes before finally settling on a nest site, and a small islet may resemble a miniature battlefield pocked with bomb craters" (24).

As soon as an egg is laid it is covered with sand by use of the bird's bill. Both parents take turns sitting on the nest during most of the daylight hours; while incubating eggs they sometimes remove the sand and bring the egg in contact with the incubation patch (an area on the ventral surface where the feathers are much thinner than elsewhere so that the egg can come in contact with the skin). Incubation is almost continuous during the night, and the eggs are about two-thirds covered with sand. During the six hottest hours of the day, when the temperature rises to 45°C in the shade and more than 50° in the sun, each adult frequently soaks its ventral feathers in the river and returns to settle on the buried eggs, thus surrounding them by wet sand, which keeps the eggs distinctly cooler than they would otherwise be.

The action of wetting feathers is distinctly different from wading in the river in search of food, which involves wetting only the feet and lower legs:

> To soak, an adult Egyptian plover wades into quiet, shallow water at the river's edge until its body surface is immersed. The bird then rapidly rocks up and down, alternately lowering and raising its fore- and hindparts in an antero-posterior plane. . . . On reaching its nest, the bird fully extends the wet ventral feathers and settles on the substrate, often with widely spread legs. . . . As the ambient temperature rises and nest-soaking behavior commences, the parents quickly shorten the intervals between change-overs (from one to several hours when it is cooler) and may relieve each other every few minutes. (Howell 1979, 31–37)

The chicks are precocial and do not return to the nest after their first day. The parent birds bring food to them and expose food for them by turning stones. As Howell describes the response to approaching predators, "the chicks crouch down and are completely covered with sand by a parent in the same manner that eggs are covered. Buried chicks are also wetted with soaked ventral feathers. . . . Even juveniles up to three weeks of age may be covered with sand by a parent" (37). If an approaching kite or pied crow is spotted at a considerable distance, the parents manage to cover the chick so thoroughly that Howell found it nearly impossible to locate. When human or other predators approach a nest on very hot days the parents continue the soaking behavior even though they seem nervous when doing so. This of course risks disclosing the nest location, but apparently the danger of overheating is so great that this behavior is worth the risk.

Egyptian plovers wet their eggs or young only when the temperature rises above roughly 40°. They need only go a short distance to the river to gather water for this purpose. The male sandgrouse, in contrast, must add to its own drinking behavior the special action of wetting its ventral feathers and then flying many miles to where its chicks are located. The end result of the specialized behavior is far removed in space and time from the situation in which the belly feathers are wetted. Although this displacement is not conclusive evidence that these birds think about keeping their eggs and young cool when they wet their feathers, the whole pattern is at least suggestive.

Other species of waders related to the plovers are known to soak belly feathers and wet their eggs or young when the temperature is very high (MacLean 1975). But this habit reaches an extreme form in the sandgrouse that nest in the Kalahari desert of southern Africa (Cade and MacLean 1967; MacLean 1968). These birds nest far from the few rivers where water is available, and they concentrate there in large flocks even when raising young at nests that are far away—in extreme cases as far as 80 km. The males, but not the females, have specialized feathers on the ventral surface that can hold up to 40 ml of water. Even after flying for miles these feathers still retain 10–18 ml. When it is very hot the chicks go to males, which squat over them and wet them with the water-carrying feathers.

Fishing Techniques of Herons

Herons are specialized for catching fish and other prey by a very rapid stabbing motion of their long, thin bills. This action consists of an extension of their elongated necks, which have been curved back into a compact resting position beforehand. Prey are ordinarily seized in the slightly opened bill, although the motion is so fast that the heron seems to be stabbing, and on rare occasions it may actually impale a fish. Herons have excellent vision and aim quite accurately at fish, crayfish, or other prey even when they are underwater. They usually forage in quite shallow water, so that refraction of light at the surface is probably not a serious problem. The impression one gathers from reading general accounts of heron behavior is that they obtain their food by standing still and watching intently until they see a fish or other prey and then strike at it so fast that the human eye cannot follow details of the motion. But the prey is seldom so obliging as to make itself readily available, and herons often spend long periods waiting or searching before they strike—and sometimes succeed in seizing an edible morsel.

Several specialized fishing techniques of North American herons have been described by Meyerriecks (1960), and recent observations by Higuchi and others demonstrate that one species occasionally uses bait to attract fish. Because they have been observed to use a variety of feeding techniques, I will consider here primarily the green-backed heron (now known as *Ardeola striata,* although other scientific names have been applied to it in the past) and the reddish egret, *Dichromanassa rufescens.* But most of these special techniques are also used by other species on occasion. The simplest procedure adopted by hungry herons is to stand and wait with the neck retracted and to stab at small fish or other prey when they appear within striking distance.

Although herons typically feed in shallow water, they also sometimes hunt for insects on land or snatch flying insects from the air. Often no prey is sighted during long periods of standing and waiting, and the heron begins to walk slowly, usually wading in shallow water. Meyerriecks describes this behavior of green-backed herons: "As the bird stalks closer to its prey, its steps become slower and longer; each foot is brought forward, placed, and then lifted so slowly that the movements are barely perceptible. . . . The bird may retract its head and neck or hold them extended over the water or ground; or rarely the head and neck are held momentarily in an extended 'peering over' attitude" (8–9).

Herons sometimes add "wing flicking" to their slow and careful walking in search of prey. As Meyerriecks describes this in the great blue heron *(Ardea herodius),* the bird "suddenly extends and withdraws its wings about a foot in a short, rapid flick. Such wing-flicks may be repeated as many as five or more times, but usually two or three flicks in rapid succession are made, and then the bird resumes wading. I have seen wing-flicking only on bright days, in open, shallow water, when each wing-flick created an obvious sudden shadow on the surface of the water. The function of wing-flicking is to startle prey" (1960, 89).

The reddish egret spends less of its feeding time standing and waiting or walking slowly, and wing-flicking is quite common. Its most common feeding technique

> is a lurching, weaving type of half-run, half-jump progression. As the bird reels forward, it stabs rapidly to the right and left, attempting to seize any prey disturbed by its activities. This method is used primarily in very *shallow* water. . . . Open Wing Feeding is another characteristic feeding technique of *rufescens.* The feeding bird begins by running slowly, wings partly extended. Then at the sight of the prey it runs rapidly and extends the wings fully. As the prey dash about, the egret turns and twists, wings still extended fully, and the bizarre performance may continue for several minutes. On occasion the bird may halt suddenly in the middle of a run, retract one wing, rapidly extend and retract the other wing, and then resume its forward run. (Meyerriecks 1960, 108)

Reddish egrets frequently change from the sort of open-wing feeding described above to what is called canopy feeding. As Meyerriecks describes it, the bird

> runs forward with the wings extended, then halts and peers into the water, and then brings both wings forward over its head, forming a canopy over the head and neck. This pose is held for a few moments to several minutes. I could clearly see the rapid fish-catching movements the bird made under the canopy of the wings. . . . My observations of *rufescens* at extremely close range (three feet) indicate that the shadow provides a false refuge for the fish startled into motion by the previous dashing activities. Reddish egrets typically hold their wings in the canopy attitude for a minute or two *before* they make a strike. I could clearly see many fish enter the shade of the canopy in the shallow water, and when a number of prey had thus "fallen for the ruse," the egret would stab rapidly under its extended wings. (108–9)

This is also a common behavior pattern of the African black heron *(Melanophoryx ardesiaca),* which employs canopy feeding extensively, and Winkler (1982) found that they attracted more small fish when canopy feeding than would otherwise have come within their reach.

Another food-gathering procedure observed by Meyerriecks in the reddish egrets is foot-stirring:

> As the egret waded forward rather slowly, it would vibrate its feet over the surface of the mud, imparting a scraping motion to the feet; then it would stop and peer at the surface of the water and either strike at prey or move on, usually resuming the scraping motions. Hovering-stirring is an aerial variant of typical stirring or scraping. The feeding bird moves forward slowly, then suddenly launches into flight, hovers over the surface of the water, and very gracefully scrapes the mud or aquatic vegetation with one foot. This behavior is repeated while the bird continues in flight. The strike is made from the hovering position. (109)

Foot-stirring is also used by snowy egrets *(Leucophoryx thula)* and other herons (Meyerreicks 1959; Rand 1956). Another variation on this theme is vibrating the bill while it is in contact with the water. Kushlan (1973) has found that this also attracts small fish, which the heron then captures.

All of these specialized types of feeding behavior suggest that the heron is actively trying to detect or attract small fish or other prey. Nothing is known about the development of these feeding techniques in the individual birds or the degree to which they involve learning or genetic influences. A behavioristic interpretation would be restricted to noting that the obtaining of food would have reinforced immediately preceding actions, and an evolutionary view would emphasize the obvious adaptive value of obtaining food by these techniques. But the versatility of the behavior also suggests that the heron thinks consciously about these uses of its wings and feet to attract or startle prey and thus make them visible and catchable.

The extreme example of specialized, and apparently purposeful, feeding techniques of herons is occasionally exhibited by the green-backed herons *(Ardeola striata)*. Although most of these birds feed in relatively simple ways, they have occasionally been observed to use bait to attract small fish (Lovell 1958; Sisson 1974; Norris 1975; Walsh et al. 1985). Higuchi (1986, 1987, 1988a, 1988b) has studied this behavior intensively, both in Japan and in Florida. The herons pick up some small object, which may or may not be edible, carry it to an appropriate spot on the shore, drop it into the water, and watch it intently. When small fish approach the bait the heron seizes them with its usual rapid neck extension. A wide variety of small objects are used as bait, including twigs, leaves, berries, feathers, insects, earthworms, pieces of bread or crackers dropped by human visitors to the park, and even bits of plastic foam. If the bait drifts away, the heron may pick it up and drop it again within reach.

Adult birds are more successful than juveniles, suggesting that learning plays a role in the acquisition of bait-fishing behavior. Small fish often come to the surface to nibble at such floating objects, and the heron seizes one with its characteristic stabbing motion. In some situations the heron may wait on a branch a meter or so above the water after throwing the bait out from the shore and then fly out to seize fish attracted to the floating object. In the colony studied by Higuchi, only a few out of 20 or 30 green-backed herons nesting near a city park engaged in this type of fishing. He suspected that they had first observed small fish attracted to pieces of bread dropped into the water by children, but he was unable to induce bait fishing by dropping crumbs near other birds of the same species. Further investigation will be necessary to learn just how a few individual herons acquire this specialized and enterprising behavior.

When no suitable small objects are available near the shore, herons sometimes gather small bits of vegetation and bring them to the water's edge. They may even break twigs into smaller lengths for this purpose. It is important to recognize that only a very small fraction of the green-backed herons have been seen to use bait for fishing, but a bird that has acquired the habit does engage in bait fishing repeatedly. As with virtually all cases of enterprising behavior by animals, a behavioristic interpretation is quite possible. But it does seem more parsimonious to infer that the bird thinks about its behavior and the probable results.

CHAPTER SEVEN

Categories and Concepts

The previous chapters have reviewed evidence that on at least some occasions animals are consciously aware of objects and events and that they experience perceptual consciousness roughly corresponding to what Natsoulas called Consciousness 3. But do they grasp any sort of generalizations, or is each perception experienced in total isolation from all others? For example, do animals ever think in terms of categories such as food, predators, or members of their own group? This is a difficult question to answer, like most of those discussed in this book, but it is an important one. For the ability to classify and think about categories as well as specific individual items is a powerful facet of conscious thinking, and if it lies within the capabilities of any animals, this is an important attribute that must be appreciated to understand them adequately. Some evidence does suggest a capacity to think in terms of simple concepts, and this chapter will review a few well-studied examples.

The natural world often presents animals with complex challenges best met by behavior that can be rapidly adapted to changing circumstances. Environmental conditions vary so much that for an animal's brain to have programmed specifications for optimal behavior in all situations would require an impossibly lengthy instruction book. Providing for all likely contingencies would require a wasteful volume of specific directions, regardless of the degree to which these instructions stem from the animal's DNA, from learning and environmental influences within its own lifetime, or any interacting combination of the two. Concepts and generalizations, on the other hand, are compact and efficient. An instructive analogy is provided by the fact that the official rules for a familiar game such as baseball run to a few hundred pages, although once the general principles of the game are understood even a small child knows what each player should do in almost all game situations.

Animals often react not to stereotyped patterns of stimulation but to *objects* that they recognize despite wide variation in the detailed sensations transmitted to the central nervous system. This type of discrimination is somewhat akin to recognizing a category. For example, as described in chapter 4, a Thompson's gazelle recognizes a lion when it sees one. The lion's image may subtend a large or small visual angle on the retina, and it may fall anywhere within a wide visual field; the gazelle may see only a part of the lion from any angle of view. Yet to an alert tommy, a lion is a lion whether seen from the side or head on, whether distant or close, standing still or walking. Furthermore, its perceptions of lions are obviously separated into at least two categories: dangerous lions ready to attack and others judged to be less dangerous on the basis of subtle cues not obvious to a human observer without considerable experience. Comparable behavior is so common and widespread among animals living under natural conditions that it seems not to call for any special scientific analysis. Yet the ability to abstract salient features from a complex pattern of stimulation requires a refined ability to sort and evaluate sensory information so that only particular combinations lead to the appropriate response.

It is of course obvious that *we* can classify stimuli into groups to which an animal gives the same or a very similar response. But such evidence does not suffice to demonstrate that the animal thinks in terms of a category that includes these and only these stimuli. For example, animals approach and eat a wide variety of foods and flee from a variety of dangers, but they might think about each one quite independently of all others and never think about the categories of edible or dangerous things. In social groups individual animals often recognize other members of their group, or at least react differently to kin and nonkin (Fletcher and Michener 1987; Hepper 1990). For instance, honeybees are more likely to deliver food to their full sisters than to the half-sisters that result from the queen having mated with several drones (Oldroyd, Rinderer, and Buco 1991).

Do any animals think in terms of such categories as "one of us"? Differential reaction to group members or close relatives as compared with others of the same species might be based on familiarity versus strangeness and not entail any conscious thinking about the categories of "group member" or "kin." In an especially suggestive case, Porter (1979) observed that in a captive colony of the neotropical fruit bat *Carollia perspicillata,* when the harem male heard distress calls from a baby that had fallen to the floor he would frequently crawl to the mother of that particular baby and stimulate her to retrieve her infant. This male had at least some internal representation of which baby belonged with which mother.

Some of the strongest evidence that animals can think in terms of categories or concepts has become available from the detailed analyses of animal learning by experimental psychologists, as reviewed by Mackintosh (1974), Hilgard and Bower (1975), Dickinson (1980), Bolles and Beecher (1988), and Shettleworth (1998). For example, Davis (1989) arranged an experiment in which hungry rats learned that they could obtain food only when the experimenter was not present. They then refrained from efforts to reach the food when the experimenter was nearby but took it when he was absent. They had thus learned that the presence of an object quite different from the food meant they would be prevented from obtaining it. This is so simple a relation that it scarcely deserves to be called a concept, but it is an example of the sorts of contingencies that animals often learn.

Expectations

One simple type of category is what an animal expects to happen. Of all the innumerable events that might occur in a particular situation, some are more likely than others. This raises the question whether animals think about events that have not yet occurred but are likely to happen in the near future. Rats and other laboratory animals easily learn that a certain light or sound will be followed by an electric shock. They may cringe or show other signs of expecting the unpleasant shock before it is delivered. They can also learn how to prevent the shock by taking some specific action, such as moving to a different part of the cage or pressing a lever. After learning this so-called conditioned avoidance, the animal continues for long periods to react to the warning signal by taking the same avoidance action, even though it no longer receives any shocks (reviewed in Macintosh 1974). This certainly indicates that the rat fears that it will be hurt shortly after the warning signal unless it does what it has learned will prevent this unpleasant experience. But psychologists are so anxious to avoid involvement with subjective experiences that they usually refrain from describing conditioned avoidance as evidence that the animal *expects* a painful shock following the warning signal or *anticipates* that it will be hurt unless it takes the preventive action it has learned to be effective.

Yet students of animal learning have noted that animals often act as though they are expecting something, and if it does not become available they appear surprised or disappointed. Tolman (1932, 1937) emphasized this sort of behavior in rats that were required to learn complex mazes in order to obtain food. In a typical experiment, after the rat had learned a moderately complex maze and was performing almost perfectly, choosing correctly a long series of right or left turns, the experimenter withheld the reward. On reaching the goal box and finding no reward, the rats would appear confused and search about for the food they had reason to expect. Tolman concluded that such rats and many other animals expect certain outcomes when they perform various learned behavior patterns, such as running mazes.

Tolman called his general viewpoint "purposive behaviorism," and he clearly believed that rats and other animals intentionally try to obtain desired things such as food and to avoid unpleasant experiences like receiving electric shocks. Yet the positivistic Zeitgeist of his time was so influential that Tolman refrained from explicit suggestions that animals might consciously think about what they were doing. His ideas were not widely accepted by behavioristic psychologists, but in recent years he is often acknowledged as having anticipated the development of cognitive psychology in the 1950s and 1960s, as reviewed by Burghardt (1985b). Only toward the end of his long and distinguished career did he confess to having been a "cryptophenomenologist" (Tolman 1959, 94).

One of the most dramatic examples of expectation is still one described by Tinklepaugh (1928, 224, also quoted in Tolman 1932, 75). He trained monkeys to watch the experimenter place a favorite item of food, such as a piece of banana, under one of two inverted cups that remained out of reach until a barrier was removed. The purpose of the experiment was to measure how long the monkey could remember which cup hid the piece of banana. When a monkey had learned to select the correct cup almost every time, provided it did not have to remember the situation for too long, the banana was replaced by lettuce during the short waiting period when the monkey could not see the cups or the experimenter.

As Tinklepaugh described the results, the moderately hungry monkey now "rushes to the proper container and picks it up. She extends her hand to seize the food. But her hand drops to the floor without touching it. She looks at the lettuce but (unless very hungry) does not touch it. She looks around the cup . . . stands up and looks under and around her. She picks up the cup and examines it thoroughly inside and out. She has on occasion turned toward the observers present in the room and shrieked at them in apparent anger."

Numerous other experiments have confirmed Tolman's thesis that animals act as though they expect a particular outcome at certain times. For example, Capaldi, Nawrocki, and Verry (1983) demonstrated that rats anticipate the patterns of reinforcement they have experienced in ways that support what they term "a cognitive view of anticipation." Naturalists and ethologists have gathered abundant evidence that anticipation occurs very commonly in the natural lives of animals, and the resulting behavior strongly suggests that they understand in an elementary fashion what problems they face and how their behavior is likely to solve them. Animals appear to think in "if, then" terms: "If I dig here, then I will find food," or "If I dive into my burrow, then that creature won't hurt me." Likewise, in the laboratory, "If I peck at that bright spot, then I can get grain," or "If I press the lever, then the floor won't hurt my feet."

A relatively simple case of expectation is demonstrated by the ability of numerous animals, including many invertebrates, to learn that food is available in a certain place at a certain time of day. They typically return to this place at or shortly before the appropriate hour on subsequent days and may continue to do so, though with decreasing regularity, even after many days when no food has been found there. It seems reasonable to infer that these animals really do expect food at a certain time and place and that they experience disappointment, annoyance, or other subjective emotions when their expectations are not fulfilled.

One powerful experimental approach to studying what animals or young human children expect is the analysis of looking time. As reviewed by Hauser (1996, 552–63), such experiments entail presenting the subject with situations and events that are either very similar to, or strikingly different from, what it has previously seen so often that no appreciable response is elicited. For example, if an object is seen to fall toward a horizontal shelf but to continue through the solid material, both human infants and monkeys look longer at the scene than when the object stops when it strikes the shelf, as one has reason to expect. Many variations on this experimental theme have been employed to test the perceptions and expectations of young children, as reviewed by Baillargeon (1994) and Spelke (1994). In one sense these experiments have demonstrated that monkeys act as though surprised when they see something that was *not expected*. This in turn implies that familiar events *were* expected.

Numerical Competence

Another important type of simple conceptualization is to recognize the number of items in a particular assembly. For instance, all sets of objects that number four can be distinguished by recognizing a category of "fourness," as contrasted with "threeness" or "fiveness." In a few special situations animals give evidence of thinking in terms of numbers. The German ethologist Otto Koehler and his colleagues carried out some of the earliest experiments on the abilities of birds to solve problems that required what he called "wordless thinking," meaning that they thought about objects and relationships but not in terms of words (Koehler 1956a, 1956b, 1969). In one of the most impressive of these experiments birds were trained to select from a number of covered boxes the one having a certain number of spots on the lid. The spots varied in size, shape, and position, but a well-trained raven could reliably select the box with any number from one to seven spots.

From the results of many such experiments Koehler concluded that these birds had the concept of numbers from two to seven, which he called unnamed numbers. This ability may be comparable in some ways to the very earliest stages of understanding of numbers in preverbal children (Gelman and Gallistel 1978). Koehler also believed that animals understand other relatively simple concepts as unnamed thoughts. To be sure, Seibt (1982) argued that since pigeons can learn as easily to peck three times as to peck twice when shown two lighted spots, and vice versa, there is no basis for the claim that birds have an unnamed number concept in the sense claimed by Koehler. But these data can also be interpreted by crediting pigeons with the ability to learn two correlated unnamed numbers, that of the stimulus and that of the required response. Several experiments have subsequently demonstrated that rats and other mammals and birds react differentially to the number of objects (up to about seven) in a set presented to them, as reviewed in a volume edited by Boysen and Capaldi (1993) and by Hauser (1996) and Shettleworth (1998).

Davis and Memmott (1982) have pointed out that although several birds and mammals have been able to learn to respond selectively to different numbers of objects, this is a relatively unnatural sort of behavior that has only been elicited by relatively stressful experimental conditions. Capaldi and Miller (1988a, 1988b) and Davis et al. (1989) demonstrated that rats are capable of discriminating what is sometimes called "numerosity" to distinguish it from the sort of counting by mentally assigning successive numbers that we usually do.

Irene Pepperberg (1999) has trained African grey parrots to use their imitations of human speech in a meaningful way, as discussed in detail in chapter 9. One of her extensively trained parrots can say the correct number from one to six when asked how many objects are presented to him. Chimpanzees have also been trained to indicate how many objects are in a given set, using symbolic communication systems they have learned from experimenters (Matsuzawa et al. 1996; Whiten 1998). These exciting developments in the use of animal communication to investigate their thoughts will be described in detail in chapters 9 and 11. Chimpanzees have also been trained by Matsuzawa (1985) and by Boysen and Berntson (1989) to select the arabic number that corresponds to the number of objects presented to them. A similar ability was also taught to squirrel monkeys by Olthof et al. (1997). The chimpanzees studied by Boysen and Berntson could even choose the printed number corresponding to the sum of two sets of objects, a rudimentary form of arithmetic. They even use "indicating acts" when they appear to be counting.

Shettleworth (1998, 363–78) has reviewed the considerable controversy about whether rats, pigeons, monkeys, and apes actually count in the sense of thinking about successively increasing numbers as they come to decide which number of objects is present. But recent experiments reported by Biro and Matsuzawa (1999) indicate that chimpanzees may be doing something closer to our kind of counting than previous investigations had demonstrated. A female chimpanzee named Ai, who had learned to identify numbers printed on cards and to select the appropriate card to represent the number of objects presented to her, was taught to touch arabic numbers displayed on a computer monitor in ascending numerical order—for instance, 2, 5, 6 or 1, 4, 9. The numbers were randomly located on the monitor surface, so that Ai had to move her arm in different directions on successive trials in order to touch the correct numbers. When she touched the lowest number it turned off, as did the second and third; after touching all three in the correct order a small piece of food was delivered. If she touched the wrong number the screen went blank, a buzzer sounded, and she received no food.

After Ai had learned this procedure, 1 out of 11 trials was randomly changed so that after the correct lowest number was touched the other two correct numbers were shifted in position on the monitor, so that the arm movement required to touch them was different from what would have been needed when they were in their original positions. When this happened Ai hesitated before choosing which second number to touch, and she made many more errors. This indicates that she had already planned the appropriate arm movements on first seeing the array of numbers but before moving her arm to reach the second and third. Had she searched for each successive correct number after touching the previous one, one would not expect this significant difference in performance. Biro and Matsuzawa therefore concluded that "a three unit ordering task is supported in the chimpanzee, much as it is in humans, by planning, executing, and monitoring phases" (178). But regardless of the similarity or differences between human counting and the ability of at least some animals to discriminate numerosity, it seems clear that they can think in at least simple terms about how many objects they perceive.

Thinking about numbers may seem to be a matter that would seldom have been useful enough in the past for natural selection to have favored it. Yet when it becomes important to think in this way in order to get food, ravens and a few other birds, as well as rats, monkeys, and apes, have learned to do so, apparently employing a general ability to learn simple concepts. Furthermore, there is one known case where counting has been shown to be important to the animals concerned. In field experiments by McComb, Packer, and Pusey (1994) tape-recorded calls of varying numbers of female lions led to attempts by resident lionesses to repel the simulated intruders, but only if the defending pride outnumbered them by at least two. To quote Packer and Pusey (1997, 59), "Females can count, and they prefer a margin of safety. Numbers are a matter of life and death."

Simple Abstract Rules

Animals can learn simple but relatively abstract rules, such as oddity, or the difference between a regular and an irregular pattern, as reviewed by Mackintosh (1974) and Walker (1983). Here too the animal can be considered to have recognized a type of category. In oddity experiments, the animal is presented with a number of stimuli or objects, one of which differs from the others in some way, and it must learn to distinguish this "oddball" from the other members of the set. For many animals, learning a single case of this sort is not difficult. Chimpanzees, however, have learned to generalize oddity as such, and having learned to select a red disk placed with two blue disks, and a blue disk accompanied by two reds, they also selected the oddball when it was a triangle with two squares. Pigeons have much greater difficulty with comparable problems, but they do better than cats and raccoons.

Variations on this experimental theme have led to other unexpected results. For instance, Zentall and colleagues (1980) compared the performance of pigeons faced with two types of oddity problem. In one case the birds saw a 5 × 5 array of 25 disks, 1 of which differed in color from the remaining 24. In the other problem there were only 3 disks in a row, 2 alike and the third a different color. In that case, if the position of the odd colored disk was varied randomly, or if the actual colors were changed (for instance, from 1 green and 2 reds to 2 greens and 1 red), the pigeons failed to solve the problem. But with the array of 25 they quickly learned to peck at the disk that differed from the rest in color, even when the colors were shifted randomly from 24 reds plus 1 green to 24 greens and 1 red. This is an example of an ability that animals demonstrate under one set of conditions even though they fail to do so in other circumstances that seem to us relatively similar. This in turn means that failure to show a particular cognitive ability in one situation may not always mean that the animal is totally incapable of making the distinction in question. Of course, this does not mean that we should always assume that by changing conditions any failure on an animal's part can be overcome.

In a related type of experiment, Delius and Habers (1978) trained pigeons to distinguish pairs of visual patterns according to their relative symmetry or asymmetry. Having learned this task, the pigeons were also able to make the correct distinction on the first try when given new pairs of shapes, some of which were symmetrical and others not. Furthermore, Bowman and Sutherland (1970) trained goldfish to distinguish between a perfect square and one with a bump in the top edge. In one of many variations, goldfish that had been trained to swim toward a square having a small triangular extension from its top also selected a circle with a small semicircular indentation in the upper edge in preference to a plain circle. They seemed to have learned to distinguish simple shapes from the same shape complicated by either an indentation or an outward bulge.

Walker (1983, 266) expressed surprise that "even a vertebrate as small and psychologically insignificant as a goldfish appears to subject visual information to such varied levels of analysis." Why should this be surprising, when it is well known that fish can discriminate many types of patterns that signal food or danger? We should be on guard against the feeling that only primates, or only mammals and birds, have the capacity for learning moderately complex discriminations, for the natural lives of almost all active animals require an ability to discriminate among a wide variety of objects and to decide that some are edible, others dangerous, and so forth.

Life in the Skinner Box

Many ingenious experiments on animal learning provide strong evidence of simple perceptual consciousness. But to understand the situation in which the animal finds itself in these experiments, it is helpful to review the general methods customarily employed in laboratory studies of animal learning. Many such investigations of learning and discrimination between stimuli have employed the Skinner box, in which a very hungry animal, typically a rat or a pigeon, is isolated from almost all stimuli except those under study. To obtain food the animal must manipulate something in the box when a particular stimulus is presented to it. Opaque walls prevent it from seeing anything outside the box, and a broadband hissing noise is often provided to mask any outside sounds.

The animal in a Skinner box has almost nothing to do but operate devices within the box that were originally selected because they were things that members of its species easily learn to manipulate. For rats this is a lever close to the floor that they can depress with one forepaw, and for pigeons it is a key or small backlit piece of translucent material flush with the wall at a height easily reached by the bird's beak. A third fixture of the Skinner box is a mechanically operated food hopper or magazine that provides access to food, or sometimes water, but ordinarily for only a few seconds at a time. Both levers and backlit keys are attached to microswitches connected so as to control the food hopper or produce changes in the stimuli presented to the animal. General illumination in the box is also provided, and since pigeons rely heavily on vision, turning off this "house light" tends to inhibit most activities, including key-pecking.

Studies of learning in Skinner boxes have ordinarily been conducted and analyzed in strictly behavioristic terms, but some of the results provide significant though limited evidence about what the rats or pigeons may be thinking and feeling as they work for food or water. The most revealing evidence of this type has been obtained with pigeons, which rely much more than rats do on vision. This facilitates complex types of experimental stimulation that we, as equally visual animals, can more easily appreciate. Psychologists who study this sort of learning emphasize what they call contingencies of reinforcement, that is, the rules relating what the apparatus does in response to the animal's bar-pressing or key-pecking. A very simple rule would be for the food hopper to make food accessible for a few seconds whenever a pigeon pecks the key. Or the bird may be obliged to deliver two, ten, or some other number of pecks to obtain food. An early discovery in this type of investigation was that animals work harder if the food hopper opens and makes food available only occasionally, after a variable and unpredictable number of pecks or bar presses. This variable ratio reinforcement elicits a very high and sustained rate of response. In this situation the hungry pigeon might think something like "Pecking that bright spot gets me food, but not always. It's almost like picking up seeds—so I'll keep trying until that box clanks and I can get some seeds."

In these experiments pigeons are ordinarily deprived of food or water long enough to make them very hungry or thirsty; a standard procedure is to hold the bird's weight at 80 percent or even 70 percent of what it would be with food available at all times. When first put into a Skinner box most hungry or thirsty pigeons soon peck the key; after all, there is nothing else to do, and pecking is a natural action for hungry pigeons, which pick up seeds or other small objects as a routine part of their daily lives. The new, learned behavior is to peck at the bright spot on the wall instead of an actual bit of grain. Since this causes the food hopper to open and gives the bird a chance to pick up edible seeds for a few seconds, it is not surprising that most pigeons soon learn the basic rules of the game.

After this stage has been attained, the experimenter may change the rules so that the food hopper operates only some of the time or only when some other information is supplied. One or more additional keys may be provided and their microswitches connected to circuits that operate the food hopper only when this new key is lit. The pigeons then learn to peck much more often when this second light is on, though they usually try occasional pecks at the food-getting key when this light is off, perhaps thinking that it might still work. Or the apparatus may provide food only when a certain color or a specific pattern is displayed, and endless variations on this theme have been used to measure sensory capacities and the ability to discriminate between similar stimuli.

These procedures were originally developed by psychologists who denied any interest in whatever subjective, conscious thoughts or feelings their experimental animals might experience. They had been conditioned by the intellectual Zeitgeist of behaviorism to restrict their concern to overtly observable behavior and how it could be altered by learning. Indeed one of Skinner's reasons for developing the Skinner box was that the animal's behavior could be recorded mechanically and objectively as numbers and rates of bar-pressing or key-pecking. Any other behavior was ordinarily ignored, and the opaque walls of the Skinner box prevented the experimenter from seeing what else the animal might be doing. The psychologists who conducted these experiments almost never spoke of the animals as hungry: they were food-deprived, or maintained at 80 percent of free feeding weight. And they never let themselves be caught saying, in print, that such a pigeon might *want* the food-hopper to open or *believed* that pecking the key would get the food it must have craved. When the animal learns to do what gets it food or water, such behavior is said to be "reinforced," rather than rewarded. This choice of terms reflects the behavioristic insistence on ignoring any mental experiences of the experimental subjects, as recently analyzed in detail by Crist (1999).

Once we allow ourselves to escape from what the philosopher Daniel Dennett (1983) called "the straitjacket of behaviorism" we can ask ourselves what it may be like to find yourself famished or very thirsty in a closed box where you can get a little food or water by playing the Skinner box game. Consider a typical experiment where the box contains two backlit panels in addition to the white key that sometimes, but not always, activates the food hopper. One panel is red when turned on, the other green, and the food hopper operates occasionally when the red light is on, but never with the green panel illuminated. A plausible inference is that the pigeon might think something like: "When that spot is red I can sometimes get food by pecking the white spot." But how can we test such an inference? Perhaps it is quite wrong; maybe the bird is thinking about something entirely different, such as the perch where he spent last night, or the hen he was courting when last given an opportunity to do so. Or perhaps he does not think about anything at all.

Straub and Terrace (1981) trained pigeons to peck at colored keys in the wall of the Skinner box and to do so while following a particular sequence of colors. To get its food, one pigeons might have to peck first red, then blue, yellow, and green, while another was required to peck in the sequence yellow, red, green, blue. These pigeons were faced with two rows of three spots that could be illuminated with different colors. In the most significant experiments four of the six spots were illuminated simultaneously, each one a different color, but the positions of the colors varied from trial to trial; the pigeon had to ignore the position of the spots and select the appropriate colors in the correct sequence in order to obtain food. Several pigeons solved this problem and performed at a level far above chance, indicating that they had learned a sequential rule that guided their decisions about which spot to peck. It seems plausible that they thought something like "I must peck first at red, wherever it is, then blue, next yellow and then green."

One set of experiments with pigeons in Skinner boxes has provided a significant though incomplete indication of what the birds were thinking about. Jenkins and Moore (1973) departed from custom by actually watching what their pigeons were doing. Close observation and photography showed that when pecking a food-getting key pigeons held their bills in a position closely resembling that used in picking up actual seeds. But thirsty pigeons held and moved their bills in ways that were much like drinking. (Pigeons swallow water with a distinct set of movements that differ from those used to swallow seeds.) To avoid observer bias and expectations, ten judges who did not know whether the pigeons were hungry or thirsty, or whether they were rewarded with food or water, were shown motion pictures or videotapes of pigeons pecking keys in a Skinner box. Eighty-seven percent of their judgments were correct, and two of the judges made no errors at all. As summarized by Jenkins and Moore, "The basis for judgment most commonly mentioned was that eating-like movements were sharp, vigorous pecks at the key. In contrast, the drinking-like movements, it was said, involved slower, more sustained contacts with the key (or other object) and were often accompanied by swallowing movements" (165).

Later investigations have confirmed and extended these results. La-Mon and Zeigler (1988) reported detailed measurements of the pigeon's behavior when pecking at a key located on the floor of the Skinner box and when actually taking grain or water. The most obvious difference between keypecks for food and water was that the beak was opened much farther for food-reinforced keypecks (about 5.5 mm compared to 0.4 mm). Whether pecking the key or actually eating and drinking, the hungry pigeons used brief pecks with a relatively high force, while thirsty pigeons employed sustained contact movements of the head and beak. Although the scientific papers describing these experiments are constrained by orthodox behavioristic terminology, it is clear that hungry pigeons peck the key as though eating, thirsty birds as though drinking. Since food and water were of the utmost importance to these birds, it seems quite reasonable to infer that they were thinking about eating or drinking when operating the Skinner box mechanism to satisfy their severe hunger or thirst.

In another series of experiments Richard Herrnstein and several other psychologists have presented pigeons with truly challenging problems of memory and perception. In a pioneering experiment by Herrnstein and Loveland (1964), the standard Skinner box was modified so that in addition to the customary food-getting key there was a small screen flush with the wall on which colored slides could be projected. The screen also served as a key that closed an electrical switch when the pigeon pecked it. A wide variety of photographs were projected on this miniature screen: indoor and outdoor scenes, pictures of people, animals, buildings, trees, flowers, and street scenes. In all these experiments some slides, designated positive, signaled the availability of food; pecking them sometimes caused the food hopper to open for a few seconds. Other pictures, termed negative, were not rewarded; pecking them had no effect, or it might turn off the "house lights," leaving the pigeon in darkness.

When the pigeons had learned to peck much more frequently at the positive pictures, the rules of the game were made even more challenging. A large number of miscellaneous scenes were displayed, and key-pecking obtained food only when the picture included a person or part of a human figure. All other pictures were unrewarded, or negative. The positive pictures might show men, women, or children, and the human figure might be large or small, dressed in different sorts of clothing or engaged in a variety of activities (sitting, standing, walking), with or without other people or animals present. In some pictures only part of a human figure such as the face was included. The negative pictures varied just as widely.

Once the pigeons were pecking significantly more often when shown pictures containing people, a series of wholly new positive and negative pictures was projected, ones the birds had never seen, that did or did not include a person and that varied as much as the original set. Surprisingly, some of the pigeons mastered this task and pecked significantly more at the new pictures containing people. It is important to appreciate that the pigeons did not perform perfectly in these tests; typically they pecked at 70–80 percent of the positive pictures and only 20–30 percent of the negatives. But the numbers of pictures used in such experiments are so great that these differences are extremely unlikely to occur by chance.

Herrnstein originally termed this concept learning, for the pigeons had learned not specific pictures or patterns but categories. In other experiments of the same general type pigeons learned to distinguish (1) oak leaves from leaves of other trees (Cerella 1979), (2) scenes with trees from scenes without them, (3) scenes with bodies of water from scenes without them, (4) pictures showing a particular person from others with no people or different individuals (Herrnstein, Loveland, and Cable 1976), and (5) underwater scenes containing fish from similar underwater pictures containing no fish (Herrnstein and de Villiers 1980). This last task was selected because pigeons would never in their individual or species experience have been obliged to discriminate among underwater scenes.

In other experiments by Poole and Lander (1971) pigeons learned to distinguish pictures of pigeons from other animals and birds. After having been trained with positive photographs of normal pigeons, they treated as "pigeon" some pictures of "weird" pigeons described as "fancy varieties having heavily feathered feet, abnormal head, body, or tail structures" (157). This suggests recognition of a variety of pigeons as something equivalent to "one of us," although it has not been demonstrated that pigeons can learn more easily to recognize pictures of pigeons than pictures of other animals such as dogs or hawks. It is important to bear in mind that in all cases the crucial tests were carried out with brand new pictures, ones that had never been shown to the pigeons before. Herrnstein et al. (1989) also trained pigeons to peck at patterns in which one object was inside rather than outside a closed linear figure. Herrnstein (1984, 1990) did not claim that these concepts were equivalent to the rich meaning conveyed by words in human language, but some elementary classification of very diverse scenes is clearly accomplished by the pigeons.

This type of experiment was further elaborated by Wright, Cook, Rivera, Sands, and Delius (1988) to test whether pigeons could master the concept of same versus different. In several previous experiments pigeons had failed to make this distinction in a reliable and convincing fashion. Wright et al. developed a modified Skinner box in which three pictures 5 × 6 cm in size were projected side by side on the floor instead of the wall of the Skinner box. The pictures were cartoons produced by a computer graphics program. When the pigeon pecked at the correct picture a simple mechanism dispensed seeds directly on the screen where the picture was projected; thus the association of a given picture with food was closer than in the standard Skinner box. In each trial three pictures were projected on the three screens, and either the right-hand or the left-hand picture was identical to the central one. If the pigeon had grasped the rule that the correct picture to peck was the one that was the same as the middle picture, it should perform correctly when presented with wholly new sets of pictures.

Two pigeons were trained by repeated trials with two cartoons, a duck and an apple. Sometimes the duck was the central picture and at other times the apple, and the matching picture varied irregularly in position, left or right. After 1,216 presentations during 76 trials, they were making the correct choice 75 percent of the time. But when tested with new sets of pictures they performed only at the chance level. It proved very difficult to train these birds to make the correct choices with a third and fourth set of pictures even after many sessions. Two other pigeons were shown 152 different sets of pictures in each day's testing session, and each picture was shown only once in the 76 trials. Learning was slow and uncertain, but after 360 training sessions over 18 months they were making about 75 percent correct choices. When these pigeons were presented with new sets of pictures, however, they did slightly better, 83 percent correct. Evidently the lengthy training with 152 sets of pictures had enabled the pigeons to recognize that the correct picture, whatever its nature, was the one that matched the central image. Wright et al. express their conclusions as follows: "The ability to learn a concept (same-as-the-center-picture), however, does not mean that this is the pigeon's preferred learning strategy. Quite the contrary, it is clear from the vast amount of research with pigeons that they prefer to attend to absolute stimulus properties and to form item-specific associations" (442).

It is interesting that this problem of matching to a sample seems much more difficult for a pigeon than selecting pictures containing a certain feature such as a tree, person, or fish when the actual appearance of these objects varies enormously in prominence and other attributes. Distinctions that seem elementary to us may not be at all obvious to another species. And conversely, tasks such as recognizing an important class of object even when it differs widely in size and other attributes is probably of crucial importance in identifying edible foods or detecting dangerous predators at a distance sufficient to permit successful escape. Noticing which two of three objects are the same does not have any particular salience in the real world where animals live under natural conditions. But learning to recognize some category of important objects such as food or predators is often a matter of life and death.

When a new type of food becomes available, it is important for many animals to recognize it whenever at all possible without expending enormous time and effort trying and rejecting a huge range of objects. The same consideration is even more important with respect to dangers. A similar situation is indicated by the experiments of Roberts and Mazmanian (1988), who found that pigeons could more easily distinguish one kind of bird from another than make more general distinctions such as birds versus other kinds of animal. The results of these experiments make excellent sense from the perspective of a naturalist. Experiments on animal learning will probably become more significant as they employ stimuli and discriminations comparable to those that are important in the natural lives of the animals concerned.

It seems reasonable to suppose that when the pigeons are working hard in Skinner boxes to solve these challenging problems, they are thinking something like "Pecking that thing gets me food," that is, it seems most plausible to suppose that they classify the items simply as those that do and don't produce food. In any event it seems difficult to imagine learning to solve such problems without at least basic and elementary perceptual consciousness. To be sure, behaviorists reject any such interpretation and insist that it is unscientific to speculate about even the simplest sort of subjective thoughts of other species. Although we cannot prove conclusively whether a pigeon making these categorical discriminations thinks consciously about the pictures or the features that lead it to peck or not, its brain must, at the very least, classify complex visual stimuli into one of two categories.

These experiments stimulated a number of further investigations, because they demonstrated that pigeons could learn not only specific stimuli but general categories. The basic finding has been replicated in other laboratories using somewhat different procedures. For example, Siegel and Honig (1970) repeated the Herrnstein and Loveland experiment using a Skinner box but varying the procedure by showing the positive and negative pictures (that is, scenes with and without people) either simultaneously on adjacent parts of the screen or sequentially, as in the original experiments. Again the pigeons transferred their discrimination to brand-new examples, and they performed above chance levels when the pictures were upside down and when the negative pictures were photographs of the same scene as the positives but without a human figure.

In another replication Malott and Siddall (1972) used a modification of the Wisconsin General Test Apparatus, developed for training monkeys and other animals to discriminate among various objects. The pigeons poked their heads through an opening in the wooden box where they were confined and looked at two wooden cubes. On the faces of the cubes visible to the bird were glued colored photographs clipped from an illustrated magazine. When the bird pecked at a positive picture containing people, the cube was pulled away, uncovering a shallow well containing a kernel of corn; pecking negative cubes with miscellaneous pictures of geometric shapes, machinery, landscape, furniture, or animals yielded no food. Each positive picture was presented until the pigeon had made five consecutive correct choices, and then two new picture cubes were presented. After somewhere between 3 and 17 such problems had been solved, the pigeons very seldom pecked at cubes with negative pictures that contained no human figures. Then, in the critical tests, wholly novel pairs of pictures of the same general sort were presented, and the pigeons performed almost perfectly.

These results were so surprising to those who had tended to view pigeons as stupid "learning machines" that various alternative interpretations have been advanced to explain these findings without inferring any conscious thinking. Many psychologists have been reluctant to agree with Herrnstein that pigeons can learn a concept. Yet something in the pigeon's brain must correspond at least roughly to what we call the concept of person, tree, or fish. In a review of animal cognition Premack (1983a, 358–59) expressed frustration because of "the inability of the reader [of these papers]—who is shown only a few of the test photographs—to judge for himself the author's claim that the concepts could not be formed on the basis of 'simple features.' I have never found this claim entirely convincing. . . . Pigeons have never been shown to have functional classes—furniture, toys, candy, sports equipment—where class members do not look alike; they only recognize physical classes—trees, humans, birds—where class members do look alike." But in many of these experiments the examples of positive and negative categories did not look very much alike, and from the pigeon's point of view they did fall into two functional categories—those that yielded food and those that did not.

Lea (1984) doubtless spoke for many psychologists when he worried about just what we mean by acquisition of a concept and whether the experiments of Herrnstein and others suffice to show that these birds have the concept of person, tree, or whatever object was present in a variety of forms in the positive pictures. He and others seem to agree that the best definition of concept recognition, in contrast to learning to respond to a stimulus attribute such as color or shape, is that recognition of a concept can only be inferred "when there is no simple single perceptual feature on which a discrimination could be based" (274). This definition is unsatisfying because it rests on a negative criterion, and proving negatives is notoriously difficult; strictly speaking, it is impossible. Thus one can always postulate that some simple feature has escaped the notice of the investigators but has been recognized by the pigeons as a signal meaning "pecking that gets me food."

Another complication that arises in interpreting these experiments stems from the remarkable ability of pigeons to learn and remember hundreds of individual pictures and to respond appropriately to most of those that yielded food even weeks or months after they were last seen (Skinner 1960; Greene 1983; Vaughan and Greene 1984). For instance, Wilkie, Wilson, and Kardal (1989) trained pigeons to recognize airplane views of a particular geographical location, and similar experiments are described by Wasserman (1995). Other birds have comparable abilities, as demonstrated, for example, by the experiments reviewed by Balda and Kamil (1998) showing that Clark's nutcrackers remember where they have stored hundreds of seeds. But the ability of pigeons to respond correctly to most if not all of *new* positive or negative pictures they have never seen before rules out an explanation based on a simple memory of specific pictures.

As might be expected, behavioristic psychologists have been very reluctant to interpret these experiments as evidence that pigeons might think consciously about the categories that they learn to distinguish. For example, Premack (1983a) described experiments by Epstein, Lanza, and Skinner (1980) in the following terms: "The basic approach has been to find performances in apes or monkeys that are recommended as proofs of mind and then demonstrate the same performance in the pigeon. Although this could backfire (and be taken as showing mind in the pigeon) . . . the opposite conclusion is drawn: what need is there for mind when there are contingencies and reinforcement?" (359). In this exchange Epstein et al. were expressing doubt that experiments on self-recognition by chimpanzees (discussed in chapter 12) demonstrate that they have minds, using as an argument the fact that somewhat similar performances could be elicited from pigeons by means of operant conditioning, apparently taking it for granted that pigeons are mindless. Premack's use of the term "backfire" captures nicely the widespread reluctance of psychologists to credit even as complex an animal as the pigeon with any sort of mental experience.

Important extensions of the pioneering experiments of Herrnstein and Loveland have strengthened the case for something approaching conscious awareness of simple concepts or categories, although, judging by some of their other publications, the psychologists who conducted the experiments would vigorously dispute this interpretation of their findings. Wasserman and several colleagues trained pigeons to recognize and distinguish four categories simultaneously (Bhatt, Wasserman, Reynolds, and Knauss 1988; Wasserman, Kiedinger, and Bhatt 1988; Wasserman 1995; Astley and Wasserman 1998), also reviewed with helpful illustrations by Shettleworth (1998). Their apparatus and procedures are of interest, because they provide additional hints about how the situation may appear to the pigeons. The Skinner box was provided with a 7 × 7 cm viewing screen and four circular keys 1.9 cm in diameter and located 2.3 cm diagonally from the four corners of the picture screen. When activated, these keys differed in color. In preliminary training the pigeons learned to get access to food by pecking first the picture screen and then whichever one of the colored corner keys was illuminated. At this stage the picture screen was a uniform white.

After this task had been mastered, the pigeons were presented every day with a series of 40 pictures, 10 of which included a cat, 10 a flower, 10 an automobile, and 10 a chair. As in earlier experiments of this type, the pictures varied widely in content, and the cats, flowers, autos, and chairs also varied widely in size, color, and position in the picture. Along with these pictures all four of the corner lights were turned on, and the pigeon was required to peck a different corner light if the slide on the central screen contained a cat, a flower, an automobile, or a chair. The pigeons obtained food only if they first pecked at the picture about 30 times and then pecked at the correct one of the four colored keys. Only this key opened the food hopper and provided something for the hungry bird to eat.

Initially the pigeons had no way of knowing the rather complicated rules of this particular variant of the Skinner box game, and they were equally likely to peck at any of the four corner keys, so their choices were correct only about one-quarter of the time. Every day the pigeons were given 40 trials, 10 with each type of picture, and after 10 days they were performing better than chance. By 30 days they were making, on average, 76 percent correct choices. By this time they had had ample opportunity to learn all 40 pictures, each of which they had seen 30 times. Since pigeons can learn and remember dozens or hundreds of pictures and identify at far better than chance levels those that get them food, at this point the experiment had only shown that pigeons can learn four sets of pictures at the same time and respond to them by pecking the correct one of the four colored keys.

The critical stage of the experiment consisted of mixing in among the 10 familiar pictures of each type entirely new and different pictures containing one of the four key features. The pigeons still made primarily the right choices, although they were correct a somewhat smaller fraction of the time. But they did generalize to new examples of these four types of picture at much better than chance levels. In another experiment of the same type the pigeons learned to classify stimuli into four types without ever seeing the same slide twice.

In experiments described at a scientific meeting, Wasserman and his colleagues trained pigeons to discriminate between pictures of human faces expressing strong emotions such as anger or sadness. The pigeons responded correctly at far above chance levels to new pictures of different persons displaying the same emotions. Of course, the training could not teach the birds anything about the feelings of the people photographed when sad or angry. But these experiments do show that fairly subtle categories of visual patterns can be learned by birds. This ability probably stems from the very widespread need to evaluate the likelihood that a predator will attack, that a given object is or is not something edible, or what the behavior of another bird is likely to be.

Wasserman and his colleagues concluded from these and many related findings that "the conceptual abilities of pigeons are more advanced than hitherto suspected" (Bhatt et al. 1988, 219) and that "these results suggest that many words in our language denote clusters of related visual stimuli which pigeons also see as highly similar. To the degree that reinforcement contingencies correlate with these human language groupings, pigeons' discrimination learning is hastened and generalization to new and altered examples is enhanced" (Wasserman et al. 1988, 235). But in keeping with the behavioristic tradition, the papers describing these impressive achievements of pigeons are titled "Conceptual Behavior in Pigeons." Presumably this wording was chosen to reinforce the customary insistence that any mental terms be scrupulously avoided. Animals may behave as though they use simple concepts, but behaviorists are constrained to deny that they might consciously think about the categories or concepts that must be postulated in order to explain their behavior.

It is significant that Wasserman (1981, 1982, 1983, 1984, 1985) argued vigorously in favor of the taboo against any implication of consciousness, despite the recent revival of research on animal cognition to which he has made important contributions. For example: "I, for one, have tried to steer clear of the possibility of subjective experience in my animal subjects; the more prudent of my professional colleagues have as well . . . cognitive psychology need not be construed as mentalistic. Those cognitive processes that are said to mediate behavioral relationships are the public behaviors of scientists, not the private experiences of their subjects" (1983, 10–11). And: "No statement concerning consciousness in animals is open to verification and experiment. Isn't it time we set aside such tantalizing, but unanswerable, questions and direct our energies to more productive pursuits?" (1985, 6). Blumberg and Wasserman (1995) liken inferring animal consciousness to explaining biological diversity on the basis of divine intervention. But as psychologists come to recognize such similarities between human and animal cognition they may gradually begin to suspect that human and nonhuman mental experiences may also have much in common. Thus Wasserman (1995, 235), although still clinging tenaciously to the behavioristic taboo against considering subjective experiences, recognizes that "conceptualization is not unique to the human brain. . . . Cognitive ethology is still with us. . . . And, based on the growing number of advocates who enthusiastically espouse its mentalistic premises . . . can even be said to be flourishing" (Wasserman 1997, 129).

CHAPTER EIGHT

Physiological Indices of Thinking

The basic structure and functioning of neurons, synapses, and glia is quite similar, as far as we know, in all animals with organized central nervous systems. I will assume that consciousness results from physiological processes occurring in our brains, and therefore there must be some neural structures or processes that give rise to it. There is considerable evidence that certain parts of the human brain are most important for consciousness, as reviewed by Baars (1988, 1997) and Milner (1998). But as Mountcastle (1998) put it recently, these data are "geographical." The effects of localized damage to the brain, recording of electrical potentials, and the exciting new methods of visualizing which areas are most active under varying conditions show *where* in the brain some activity is taking place. But none of this evidence has yet revealed just *what* is going on when the activity of a particular area of the brain renders us conscious. Areas most important for one type of conscious experience are active to some extent when our conscious attention shifts to something else. But localization is a first step toward finding what neural events lead to consciousness.

This raises the intriguing question whether conscious experience requires the specific structure of human or primate brains. *Something* goes on when we are conscious that does not occur when we are not. Many talented neuroscientists are hard at work seeking to answer this question, as reviewed by Taylor (1999). But does this unknown sort of neural activity depend on particular arrangements of gross neuroanatomy? Or could other anatomical arrangements of interacting neurons and synapses accomplish the same basic function of producing conscious experience? Since we do not know just *what* neural activity is needed for consciousness this remains an unanswerable question, and therefore it is best to keep an open mind about the possibility of consciousness in at least the animals that exhibit versatile behavior or communicate in ways that indicate they may be expressing thoughts or feelings.

During the 1990s neuroscientists made substantial progress toward identifying structures and functions in human brains whose activity is correlated with particular kinds of consciousness. Have they discovered any specific set of neurons or any particular neural process that is uniquely responsible for producing conscious experience rather than processing information unconsciously? If so, it would seem to be a simple matter, at least in principle, to look for such structures or functions in other species. Can we specify just what we should look for in an animal brain in order to determine whether the animal is conscious? The short answer to this question is "Not yet."

Although a thorough review of cognitive neuroscience is far beyond the scope of this book, a few especially significant recent discoveries are important because they indicate that conscious experience is probably not a human monopoly. Neurons, synapses, glial cells, and neurotransmitters appear to be basically similar in all animals. Thus if one were to rely exclusively on anatomical evidence at the cellular level, there would be no strong reason to deny that any animal with a central nervous system could be conscious. Yet we tend to assume that only the largest and most complex brains, such as ours, are capable of producing conscious experience. Koch and Laurent (1999), Tononi et al. (1996, 1998), and Crick and Koch (1998) explore this assumption cautiously, for in the absence of clear criteria for detecting consciousness in animals it is difficult to be certain that it is absent. After considering the accomplishments of insect brains, Koch and Laurent conclude that "no brain, however small, is structurally simple" (97).

In human brains the reticular system and its massive connections to the thalamus and cerebral cortex, especially the prefrontal cortex, are obviously important for consciousness. But some neurons in these areas do not cease all activity when we are not conscious, although the level and pattern of activity are very different. Perhaps the cellular components of all central nervous systems are similar but it is something about the pattern of organization of our brains that leads to consciousness. But no specific part of the human brain, nor any specific process, has yet been shown to be active when *and only when* a person is conscious.

Numerous aspects of brain function have in the past been suggested as neural correlates of consciousness. For example, John (in Thatcher and John 1977) equated consciousness with internal feedback whereby information about one part of a pattern of information flow acts on another part. But, although such feedback may be necessary, it is not a sufficient condition since it is also a feature of many physiological processes, such as postural reflexes, that operate without any conscious awareness.

One type of experiment that suggests a possible approach to finding neural correlates of consciousness is typified by the work of Georgopoulis et al. (1989), who recorded action potentials from neurons in the motor cortex of a monkey while it moved a lever to follow the motion of a spot of light. The activity of these neurons took place with the same spatial and temporal pattern as the activity during the actual hand movement a fraction of a second later. This suggests that perhaps the monkey was thinking consciously about the anticipated movement. But the time interval between the "anticipatory" neural activity and the actual movement was so short that one can interpret these data as simply reflecting an early stage in the neural activity leading up to the movement of the hand.

Frith, Perry, and Lumer (1999, 105) have recently reviewed several lines of evidence that point to promising "areas where progress is likely to be made" in the search for neural correlates of consciousness. They place special emphasis on types of visual imagery that appear to be very similar in humans and monkeys. But it has not yet been possible to pinpoint any specific set of neurons or any specific pattern of activity that provides a "robust neural signature . . . that consistently correlates with conscious experience." They consider the contents of consciousness to be "mental representations; mental entities that stand for things in the outside world," but they do not appear to mean *any* internal representation that is closely linked to the outside world. For example, they would almost certainly not include the optical images formed on the retina, even though these are highly correlated with things in the outside world. Thus *mental* representations are apparently considered to be those of which one is conscious, as demonstrated by the ability to report them.

Frith et al. emphasize the importance of reports about the content of consciousness as data about mental representations: "To discover what someone is conscious of we need them to give us some form of report about their subjective experience. Such reports are qualitatively different from behaviour: reports, like consciousness, have content. They are about something. Behaviour simply occurs. . . . Such reports depend on a shared communication system, such as language. . . . However, we do not need to use language to report our mental experiences. Gestures and movements can be made with a deliberate communicative intent. . . . The same procedure can be used in studies of animals" (107). Of course, a skeptic could argue that in the case of nonhuman animals the representations are reported unconsciously. This question will be discussed further in chapter 14. But this recognition by neuroscientists that animal communication provides objective evidence about conscious experiences serves to focus attention on the neural mechanisms of communication as especially significant in the search for neural correlates of consciousness.

Lateralization of Brain Mechanisms of Communication

Human language and conscious thinking are closely linked; many have argued that they are inseparable. Therefore it has seemed logical to search for neural correlates of consciousness by investigating the neural control of speech, which depends heavily on Broca's and Wernicke's areas of the temporal cortex, ordinarily on the left side. This localization is not precise and absolute, for damage to, or electrical stimulation of, particular parts of the speech control areas does not always produce the same effects, as reviewed by Lecours et al. (1989), Feindel (1994), and Ghazanfar and Hauser (1999). Lateralization of speech control used to be considered a qualitatively unique human attribute that provides the physiological basis for our vastly superior mental abilities. But these areas are not anatomically unique to our species. Homologs of Broca's and Wernicke's areas are present in other mammals, but they are not called by the same names because the animals are not capable of human speech.

For many animals, specific sounds or other communicative signals are especially important, and often control of these salient signals is concentrated on one side of the brain. One of the clearest examples is the lateralized control of singing in songbirds (Nottebohm 1979; Konishi 1985; Arnold and Bolger 1985; McCasland 1987).

An example of lateralization of communicative function was provided by Hamilton and Vermeire (1988), who studied twenty-five monkeys *(Macaca mulatta)* in which the corpus callosum connecting the two cerebral hemispheres had been surgically cut. In such animals, and in human patients whose corpus callosum has been cut in order to control severe epilepsy, the two halves of the brain operate more or less separately. In these monkeys the hippocampal and anterior commissures and the optic chiasm were also cut at the midline, thus separating the two sides of the cerebral cortex more completely than in many other "split brain" experiments of this general type.

These monkeys behaved quite normally, although when allowed to see things with only one eye, only one cerebral cortex received visual input. This allowed the experimenters to train them to make various visual discriminations separately with either the right or the left cortex. When required to discriminate between straight lines differing in slope by as little as 15 degrees, the monkeys performed significantly better when using the left hemisphere. But when the problem was to discriminate between pictures of the faces of individual monkeys, the right cortex was superior. These differences were quantitative and not absolute; both hemispheres could learn to perform both discriminations, but the superiority was clear and statistically significant.

Some primate vocalizations convey specific information in addition to levels of emotional arousal, as described in chapter 9. When Green (1975) analyzed the sounds exchanged by Japanese macaques *(Macaca fuscata)* in relatively relaxed social situations, he found that a group of sounds most readily described as "coos" differed in acoustical details, even though they had at first seemed much the same to human listeners. Furthermore, certain types of coo-like sounds were used most often in specific situations. For example, "smooth early highs," which began at a fairly high frequency, rose slightly, and then declined in frequency, were usually emitted by infants sitting apart from their mothers. On the other hand, "smooth late highs," in which the frequency rose steadily to peak near the end of the call before dropping slightly, were most often used by sexually receptive females. This and other observations indicated that these cooing sounds were especially important in social communication. Different types of cooing are used in different situations and probably convey different messages. This led Zoloth and Green (1979) to suggest similarities to human speech.

The coos are not the only vocalizations used by Japanese macaques to communicate with their social companions when they are close together and interacting amiably. Another type of sound, called "girneys" by Green, has been studied in detail by Masataka (1989). These sounds have multiple harmonics that rise in frequency and fall toward the end of each vocalization. Masataka distinguished two general sorts of girney: those in which the peak frequency occurred during the first third of the sound and others that peaked in the final third. After the first type the caller often groomed the receiver, whereas the latter type usually resulted in the receiver grooming the caller. Thus these two types of sound seem to mean something like "I'll groom you" and "You groom me." This interpretation was supported by the results of playbacks of tape-recorded sounds, which were followed by motions and gestures that strongly indicated that the hearer expected to be groomed or to groom the companion nearby. This is an additional example of the subtle differences in animal signals that once seemed meaningless but have been shown by careful experiments to convey different messages.

Building on these findings, Zoloth et al. (1979) trained Japanese macaques and three other species of monkeys to discriminate between tape-recorded coos on the basis of two acoustic features. The first was the position of the frequency peak, early or late in the sound, which seems to convey different meanings to the Japanese macaques. The second was the frequency at which the coo began. One of the other species, the vervet monkey *Cercopithecus aethiops,* does not use coos in its social communication; the other two do use coo-like sounds, but it is not known whether they have any special significance to the animals themselves. The Japanese macaques consistently learned the discrimination of peak position more easily than the other three species, but they did not do as well at learning to discriminate on the basis of the initial frequency. Thus the perceptual capabilities of these species of monkeys appear to be correlated with the acoustic features of the sounds they use for social communication.

In related experiments by Petersen et al. (1978, 1984) these coo-like sounds were presented to Japanese macaques and to five other species of monkeys through earphones that allowed the experimenters to present the stimuli to either the right or the left ear. Since most of the auditory neurons of mammals cross the midline before reaching the cerebral cortex, a right ear advantage means that the left auditory cortex is playing a larger role than the right in processing such signals. When sounds are presented to the human right ear, we can usually discriminate small differences better than with sounds arriving at the left ear. All five Japanese macaques showed a significant right ear advantage for detecting the position of the frequency peak.

This evidence that Japanese macaques, but not the other species tested, process coos primarily with their left auditory cortex was further supported by the experiments of Heffner and Heffner (1984), who trained animals to make the same discrimination between coos on the basis of the position of the peak frequency. Then parts of the brain were surgically removed, and after recovery the monkeys were tested again in the same way. When the auditory cortex on both sides of the brain was removed, the monkeys could no longer make this discrimination even after additional training. But they performed normally when only the right auditory cortex was removed. When only the left auditory cortex was destroyed, the monkeys' performance was initially poor, but with further training it improved and reached its former level.

Evidently these monkeys had been using the left auditory cortex before the operation but could relearn the discrimination with the right cortex when necessary. This result is intriguingly similar to the localization of human speech perception in the left auditory cortex, more specifically in Wernicke's area. The finding of similar lateralization in songbirds and in Japanese macaques is reminiscent of the nineteenth-century controversy between Owen and Huxley over the former's mistaken claim that there was no hippocampus in the brains of nonhuman primates. Human mental superiority, enormous as it is, does not seem to be based on any single, unique feature of neuroanatomy.

These and other relevant experiments have been reviewed by Hauser (1996). One important finding is that the right cerebral hemisphere is more important than the left in emotional expression in chickens, rats, monkeys, and humans (Bradshaw and Rogers 1993). And recent experiments by Hauser (1993) have shown that in human subjects and rhesus monkeys the left side of the face, which is controlled by the right hemisphere, is more effective in expressing emotions. Furthermore, Hauser and Anderson (1994) found that when rhesus monkeys were stimulated by sounds from a loudspeaker located directly behind them, they turned more often to the right when the acoustic stimulus was a monkey call having a clear emotional salience (aggressive, fearful, or affiliative). But when the alarm call of a common local shorebird was played from behind, they turned to the left. The right ear sends stronger sensory input to the left cerebral cortex, so turning right brings the right ear into a more favorable position for listening to a source behind the animal. In short, cerebral lateralization is by no means a unique human attribute.

Electrical Correlates of Consciousness

Neurophysiologists attempting to understand brain function seek objective data that can be measured when brains are engaged in their complex and enormously significant activity. Until recently the principal data available were electrical potentials. These clearly accompany not only the conduction of impulses along neurons but also the all-important modulating processes that occur at synapses. Ideally, a neurophysiologist prefers to record with microelectrodes from individual neurons or synapses, or at most a few at a time. But when dealing with the more complex and significant processes of cognition, monitoring the activities of single cells is clearly of limited use, because whatever the activities of central nervous systems that lead to cognition and conscious thinking, they clearly entail complex interactions of large numbers of cells.

Relatively weak electrical signals, generally known as electroencephalograph (EEG) potentials, or, more popularly, as brain waves, can be recorded both from electrodes inside brain tissue and from outside an animal or human skull. They have been recorded extensively from the human scalp and often give useful signs of clinical abnormalities. They are ordinarily only a few microvolts when recorded from the human scalp. The most prominent are the alpha waves with a frequency on the order of 5 to 8 Hz that are most evident when the human subject is lying quietly with closed eyes and is not engaged in any particular mental activity. If the subject performs some mental task such as solving arithmetic problems, these alpha waves diminish in amplitude and merge into more irregular, noisy signals covering a broad frequency band. But EEG potentials seem too coarse an index to reveal more than the most general sorts of activity.

Nevertheless, the analysis of electrical potentials recorded from both human and animal brains offers a tantalizing hope of identifying neural correlates of some types of consciousness. There are small quantitative differences in the EEG waves when the subject engages in thinking about different topics. For example, verbal problems, such as selecting synonyms from a series of words, produce slightly larger potentials over the left cerebral cortex, as one would expect from the well-known fact that the left side of the cortex is much more heavily involved in processing and recognizing speech.

An especially significant type of electrical response called evoked potentials can be recorded from the human scalp after discrete sensory stimulation with flashes of light or brief sounds. Ordinarily these are too low in voltage to be detected reliably against the background of other EEG signals, but this difficulty has been surmounted by repeating such discrete stimuli and averaging the EEG potentials. Several hundred or even a few thousand responses must be averaged to obtain a clear graphic display, and this sets obvious limits to the kinds of stimuli that can be studied. They must be brief enough that excitation from different parts of the stimulus do not overlap in time and interfere with each other, and they must be repeated many times.

The potentials resulting from discrete sensory stimulation are ordinarily described in terms of electrical polarity and time or latency after the stimulus. The portions of the evoked potentials occurring within about one-tenth of a second are primarily reflections of the sensory impulses traveling from the peripheral sense organs to successively more anterior portions of the brain. A subset of evoked potentials usually having longer latencies are clearly correlated with moderately complex information processing. They are called event-related potentials (ERPs), because they are not a direct function of the sensory input but are also affected by internal processes within the brain, including previous events.

This discussion will concentrate on only a few of the more clear-cut types of ERPs that provide significant suggestions about the occurrence of conscious thinking. Donchin et al. (1983), Picton and Stuss (1984), Stuss, Picton, and Cerri (1986), and Sommer, Matt, and Leuthold (1990) reviewed the evidence relating these potentials to human consciousness. The whole subject was also discussed in great detail by Verleger (1988) and Donchin and Coles (1988) in the journal *Behavioral and Brain Sciences,* which publishes numerous comments by other interested scientists and responses to these comments by the authors.

Of a wide variety of components of ERPs, much of the experimental attention, and the most interesting implications, involve what is usually called the P300 wave. This is a positive potential occurring about 300 milliseconds after a stimulus. Actually it lasts in many cases 100–200 msec, and the peak varies somewhat but is ordinarily in the range of 300–400 msec. The defining characteristic of the P300 wave is not so much its electrical or temporal properties as its relation to at least simple types of cognition.

One type of event-related potential that seemed at one time to be a reasonable candidate for a neural correlate of consciousness is recorded by presenting a long series of uniform stimuli, usually sounds, one of which is occasionally omitted. The P300 waves occur after all of these sounds, but the one following the omission of an expectable signal is often as large as or larger than those following actual stimulation. Since there was no stimulus at all preceding the P300 waves for an omitted stimulus, these waves must reflect some sort of activity in the brain related to the general pattern that had been established by the repeated stimuli. When such data were available only from human subjects it was tempting to suspect that they might be neural correlates of consciousness.

This conclusion has been cast in a quite different light, however, by the experiments of Bullock, Karamürsel, and Hofmann (1993) showing that so-called omitted stimulus potentials can be recorded from the primary sensory nucleus of the medulla of sharks and rays, and even in the retina and some afferent nerves. Responses to stimuli omitted in the midst of a long series are thus widespread among nervous systems and not likely to be correlated with complex neural processing. This case is a cautionary example of the subtleties and complexities of neuroscience; it can warn us against premature conclusions based on limited evidence.

A somewhat similar experiment is to present to human subjects a train of stimuli including two types, one much more common than the other. The relatively rare stimulus has come to be called the "oddball" stimulus. Under many conditions all stimuli produce P300 waves in human subjects, but the oddball stimuli generate larger ones, as reviewed by Galambos and Hillyard (1981). Perhaps more relevant for conscious thinking is a variation on this experiment in which the common and the rare stimuli differed in semantic meaning. In one set of experiments the subject heard the spoken name *David* 80 percent of the time and *Nancy* the remaining 20 percent (Donchin 1981). Or 80 percent of the names were masculine and 20 percent feminine. In another variation on this experimental theme, 20 percent of the words rhymed with *cake* and 80 percent did not. In the final experiment the oddballs were synonyms of *prod* and the other words were not. The subject's task was to count the number of times the rare stimuli were heard. All the words elicited prominent P300 potentials, but those following the rare stimuli were larger and showed longer latency. The difference in latency was greater when the subject's task concerned semantic meanings. It apparently takes the brain longer to deal with the problem of deciding whether the sound was a word synonymous with *prod* or rhyming with *cake* than simply to distinguish between the common and the less common of two words.

Do animal brains show potentials comparable to human ERPs? Those of cats and monkeys certainly do, although the detailed form of the ERPs may differ from the human P300 waves to some extent. Wilder, Farley, and Starr (1981), Buchwald and Squires (1982), and Harrison, Buchwald, and Kaga (1986) stimulated cats with brief tones or light flashes and recorded evoked potentials very similar to the human P300. These electrical responses were prominent, however, only after the stimulus had been associated with the delivery of an electrical shock to the cat's tail. In other words, the stimulus did not produce a significant P300 initially, but did so after the cat had learned that it signaled an unpleasant event. In one experiment a cat learned that a light flickering 7.7 times per second signaled that it would receive an electrical shock unless it made a simple response, but that 3.1 flashes per second meant it could obtain food. The ERPs increased in amplitude and changed their waveform after the cat had learned what to expect.

In a similar experiment with monkeys *(Macaca fascicularis)*, two tones of 500 and 4,000 Hz were presented, the latter occurring less often than the former. The rare stimulus was accompanied by an electric shock. After the monkey learned that the rare tone would be followed by a shock, its brain showed a clear P300 wave in response to it but not to the other sound. Neville and Foote (1984), Glover et al. (1986), and Pineda et al. (1988) showed that in squirrel monkeys *(Saimiri sciures)* an oddball tone elicited a larger P300 wave. These experiments can be interpreted conservatively as showing only that meaningful stimuli activate larger numbers of neurons and synapses in both human and animal brains, which is scarcely surprising since the animals clearly learn to respond appropriately to stimuli they have learned have a particular significance. As the functioning of some animal brains is found to resemble to a greater degree the comparable functions in human brains, the possibility of conscious awareness certainly does not diminish. It is therefore appropriate to consider what is known about the relation between the human P300 wave and the conscious thinking that can be reported by human subjects.

Sommer, Matt, and Leuthold (1990) found that conscious expectations modified the human P300 to some extent, although the subjects were not conscious of most of the factors affecting these event-related potentials. Donchin et al. (1983) addressed this question directly, accepting verbal reports as relevant objective data about the conscious experiences of human subjects. Although there was a rough correlation between the presence of a P300 wave and the subject's awareness of the stimulus, the two did not always occur together. Some P300 waves have been recorded following stimuli that the subjects did not consciously notice, and the absence of P300 waves has been recorded following stimuli the subjects did notice. Given the complications of recording P300 waves and the numerous other electrical events that often obscure them, the occasional mismatch between their occurrence and conscious awareness of the stimulus is not altogether surprising.

The general conclusion, after many detailed studies of human P300 waves, is that they are endogenous in the sense that their properties are not completely determined by the stimulation. They are elicited by stimuli that are unexpected and yet are relevant, stimuli that signal something important, whether pleasant or unpleasant, or signals informing the subject of something he or she is expected to do. This has led to the general interpretation that the P300 results from an updating of the internal representation of some important aspect of the subject's situation. When the process of updating is more complex, there tends to be an increase in the P300 latency, but this is not invariable. Donchin et al. (1983, 112) concluded that "whenever P300 occurs, the subject is conscious of the task-relevant information carried by the eliciting stimulus. In this sense, P300 can be used to index the occurrence of conscious processing." Nevertheless, there are exceptions, and the presence of a P300 wave does not demonstrate with absolute certainty that the human subject is aware of the stimuli. Verleger (1989), Donchin and Coles (1988), and numerous commentators on these two papers debated at length and in almost excruciating detail just what these potentials reveal about the activities of the human brain.

Despite such uncertainties, the presence of an electrical potential that correlates with task relevance and the unexpectedness of stimuli is at least suggestive evidence that the subject is consciously thinking about the meaning of the stimuli. The requirement that stimuli be repeated many times in order to measure ERPs means that only some types of stimulation can be studied in this way. But it does seem that ERPs from animal brains deserve much more intensive study than has yet been reported. It would be of great interest to arrange experiments in which the stimuli had clear semantic meanings (such as alarm calls) or required that animals make important decisions on the basis of information thereby conveyed. Certain types of ERPs may be necessary though not sufficient for conscious thinking, and their occurrence, magnitude, latency, and correlation with relevance to the animal might provide very helpful indications of the likelihood that it was indeed thinking consciously about the information conveyed by the experimental stimuli.

The reluctance of behavioral scientists to become enmeshed in the problems of consciousness may have discouraged attempts to study experimentally the correlation between human conscious awareness and ERPs recorded from the human scalp. It would seem possible to arrange conditions under which human subjects made the same or very similar discriminative responses to stimuli that produce ERPs but did so under two sorts of conditions: one in which they were clearly aware, consciously, of their responses and another in which they were not. The avoidance of conscious awareness might be achieved through long repetition and overlearning or by the distraction of competing stimuli. But experiments of this type do not seem to have been carried out with the care and ingenuity required. Thus we do not yet have more than rather general and uncertain correlations between human ERPs and conscious awareness.

One exciting recent development, reviewed by Rizolatti and Arbib (1998) and by Gallese and Goldman (1998), has been the identification in an area of the monkey cortex (F5) of what are called "mirror neurons." These are activated both when the monkey performs a specific movement and also when it sees another monkey or human doing the same thing. The existence of such neurons does not of course prove that the monkey is conscious, but it edges suggestively closer to that possibility.

New Advances in Brain Imaging

In the 1990s neuroscientists developed greatly improved noninvasive procedures that yield detailed three-dimensional images of any portion of a human brain showing which areas are most active under particular conditions, as concisely described by Kosslyn (1994, 45–49). The most useful methods are positron emission tomography (PET) and functional magnetic resonance imagery (fMRI). The activity displayed by both kinds of imaging is primarily local blood flow, or metabolic rate, which is generally assumed to reflect information processing by neurons and synapses. In a broad and thoughtful history of these developments, Raichle (1998, 771) explains that "local increases and decreases in brain activity are reliably accompanied by changes in blood flow. . . . While paired data on glucose metabolism and blood flow are limited, they suggest that blood flow changes are accompanied by changes in glucose metabolism of approximately equal magnitude and spatial extent." Rosen, Buckner, and Dale (1998) have summarized the most recent advances in event-related fMRI by which it is now possible to obtain rapid sequential imaging at rates up to 20 images per second.

What can these powerful new methods contribute to the search for neural correlates of consciousness? Crick and Koch (1998) review recent studies of consciousness of visual perceptions, or visual imagery, which they consider to be a strategic area of investigation. One of the most important recent developments has been investigation of "blindsight." In rare cases a human patient who has a large lesion in the primary visual cortex on one side of his brain, and says he cannot see anything at all in the corresponding half of his visual field, can nevertheless give mostly correct answers when forced to guess about visual stimuli flashed briefly in his blind area.

Philosophers have been very interested in blindsight, as reviewed by Siewert (1998) and by Kentridge and Heywood (1999), along with four other papers in the same issue of the *Journal of Consciousness Studies*. This is because blindsight provides a clear-cut case where visual stimuli are detected and moderately complex discriminations about them are achieved without conscious awareness on the subject's part that he sees them at all.

One such patient studied extensively by Weiskrantz (1997) could distinguish parallel bars oriented horizontally from identical bars that were tilted by as little as 10 degrees, even though he insisted that he could see nothing and was amazed when told that most of his guesses were correct. Humphrey and Weiskrantz (1967) had found that a monkey whose primary visual cortex had been destroyed on both side of the brain and that acted initially as though completely blind nevertheless "came with time to have the ability spontaneously to avoid obstacles and to retrieve tiny visual objects, even specks of dust (although she was not able to identify them before touching or tasting)" (Weiskrantz 1997, 77).

Recent fMRI imaging of a patient with blindsight reported by Sahraie et al. (1997) indicates that the superior colliculus of the midbrain is part of the alternative pathway from eye to brain by which blindsight is achieved in the absence of the relevant part of the visual cortex. These cases confirm what has long been known from other types of evidence: that at least simple behavior can be based on visual stimulation of which the subject is unaware. The studies of blindsight have tempted some to argue that since a human subject can respond appropriately to visual stimuli that he does not see consciously, perhaps all nonhuman animals are like blindsight patients. This conclusion seems unlikely to be correct, if only because the actual visual ability of blindsight patients is very inferior to what they can do with the undamaged half of their brains.

Further experiments with monkeys have revealed something very similar to the blindsight of brain-damaged human patients. Cowey and Stoerig (1995) studied three monkeys that years previously had the primary visual cortex removed surgically on one side of the brain. These monkeys were trained to obtain a food reward by touching the position on a screen where a small bright square was flashed for a fraction of a second. The conditions were arranged so that some of the squares fell in the normal visual field and others in the area where the monkey was blind owing to loss of half its primary visual cortex. They learned to perform this task even with squares in the "blind" field, provided the squares were appreciably brighter than the level needed in the normal field.

These monkeys were then trained to touch a different spot on the screen on "blank" trials when no bright square were presented. The intriguing result was that they signaled "blank," that is, no stimulus, when the squares fell in their blind field—even though in other experiments they responded correctly by touching the appropriate position. In other words, these monkeys signaled that they did not see the stimuli to which they could nevertheless point in order to obtain some food. This seems clearly to demonstrate that they were not conscious of these visual stimuli even though they could point to their position. Thus these experiments allow investigators to distinguish when a monkey is and when it is not conscious of a particular visual pattern.

Another powerful approach to neural correlates of consciousness has been experimentation with binocular rivalry. This occurs when our two eyes see distinctly different images. Although with some effort we can discern both, ordinarily we see one of the two and are at least momentarily unaware of the other. Often our awareness alternates, and we see first one and then the other. Neural input still reaches the brain from both eyes, but on one side it is suppressed between earlier stages of visual processing in the brain and the parts of the cerebral cortex that produce perceptual consciousness. Such cases of binocular rivalry provide an opportunity for significant experiments because the sensory input from both eyes is obviously reaching the brain, but one pattern of stimulation leads to conscious perception while the other does not.

When our conscious awareness alternates between the two, appropriate measurements can indicate what various parts of the brain are doing when we become consciously aware of first one and then the other pattern.

A related area of exciting progress has been the identification of neurons that are selectively responsive to particular classes of stimuli. Dittrich (1990) showed that monkeys can be trained to respond selectively to faces of known companions and even to sketches as well as photographs. As reviewed by Maunsell and Newsome (1987) and Perrett, Rolls, and Caan (1982), some neurons in the temporal cortex of monkeys also respond selectively to pictures of faces, whether these be actual simian or human faces or pictures of them. Baylis, Rolls, and Leonard (1985) even found that some neurons respond selectively to the faces of particular individual monkeys. It would be of great interest to learn whether specialized neurons in the brains of other animals respond selectively to visual, acoustic, or olfactory representations of other individuals.

A significant experiment of this type was recently reported by Tong et al. (1998), who studied fMRI images of areas in the human cortex known to contain cells that respond selectively to faces or to images that relate to specific places, such as pictures of a house. Images of a face or of a house were presented to the two eyes in a way that produced binocular rivalry between the two images. In parallel experiments the fMRI images were obtained when the two images, face and house, were presented alternately to both eyes, so that there was no binocular rivalry but instead an alternation of images in both eyes at approximately the same rate as in the situation of rivalry. The result was that "these responses during rivalry were equal in magnitude to those evoked by nonrivalrous stimulus alternation, suggesting that the activity . . . reflects the perceived rather than the retinal stimulus" (753).

Similar experiments with monkeys demonstrate quite similar effects of binocular rivalry. Sheinberg and Logothetis (1997) studied two *Macaca mulatta* with implanted electrodes that allowed observers to record from cortical neurons. Different visual patterns were presented to the two eyes, and the monkeys were trained to pull and hold a lever on the left for a sunburst-like pattern, to pull and hold a lever on the right for other figures, and to pull neither—or to release an already pulled lever—when they were shown a physical blend of two pictures. Of the neurons that the authors could monitor in the striate cortex, a few discharged only when one of these stimuli was presented. But in more anterior areas of the temporal cortex almost all the neurons studied discharged only when the monkey pulled either the left or the right lever, demonstrating that it was seeing one or the other of the two stimuli.

Thus, in the special conditions of these experiments, specific neurons respond when the animal is apparently aware of the image formed on the right retina, and other neurons respond to the image on the left retina. Meanwhile, the monkey is showing, by its choice of lever, that it is aware of one image but not the other. This type of experiment comes close to localizing the site of conscious visual perception on the part of a monkey, although determined skeptics could argue that the responses of the monkeys *might* occur unconsciously even though in comparable experiments a human observer reports that he is conscious at any one time of only one of the two stimuli presented to his right or his left eye.

Although neuropsychologists have not yet found cells or patterns of neural activity that correlate with consciousness in an absolute 1:1 fashion, tantalizing hints have emerged. Both human blindsight patients and monkeys with experimental lesions in the visual cortex can make simple discriminations between visual stimuli even when not aware that they are seeing anything. In experiments on binocular rivalry, both human and nonhuman primates perceive scenes presented to one eye while remaining unaware of other patterns presented to the opposite eye. Furthermore, the same cortical areas are active when human subjects are, and when monkeys appear to be, conscious of what they see. Yet it is important to recognize that even these elegant experiments do not reveal just *what* is needed for part of a brain to produce consciousness.

One distinct possibility is that there are no specific groups of neurons and no specific patterns of neural activity that are uniquely correlated with consciousness in an "if and only if" sense. Perhaps the interactive functioning of neurons, synapses, and glial cells can lead to various types and levels of conscious awareness in any central nervous system. This would explain the frustrating difficulties encountered when neuroscientists try to identify the neural correlates of consciousness. But of course only a small fraction of the neural activity in large and complex brains like ours becomes conscious, so in one sense the problem is to explain why we are *not* conscious of the rest. This consideration reminds us that we can be consciously aware of only one or a very few items at any one time. This limitation of consciousness thus becomes a major question.

CHAPTER NINE

Communication as Evidence of Thinking

It is much more effective for one animal to anticipate another's actions than to wait until they are under way. This is especially obvious in the case of aggressive encounters. When a dominant animal signals its intention to attack, it is much better for a subordinate to perceive this as a threat than to wait until it is actually injured. For threats can be dealt with in several ways, including retreat, counterthreats that may deter the attack, or submissive behavior. Insofar as animals ever experience conscious thoughts and feelings, these are very likely to accompany social behavior and interactions between predators and prey. Many if not most interactions between animals may well involve at least simple feelings and thoughts about the situation. If so, other animals with which signals are exchanged will benefit by correctly understanding what the communicator feels or wants, as emphasized by Krebs and Dawkins (1984). Communication is often a two-way process, a repeated exchange of signals by which two or more animals can evaluate each other's feelings and thoughts as well as their likelihood of behaving in various ways.

Animal communication can therefore provide a useful and significant "window" on animal minds, that is, a source of objective evidence about the thoughts and feelings that have previously seemed so inaccessible to scientific investigation. Experimental playbacks of communicative signals are of crucial importance because they allow a limited but revealing sort of participatory dialog between animal and scientist. Sounds are the most easily simulated signals, but other sensory channels can also be employed in playback experiments, provided only that technical means are available to reproduce the animal signal with adequate fidelity. This has even been effective with electric fish, which use weak electric signals not only for orientation but in social and predator-prey interactions (Bullock and Heiligenberg 1986; Bratton and Kramer 1989; Kramer 1990, 1997; Bradbury and Vehrencamp 1998, 319–50; Hopkins 1999; Crawford and Huang 1999; Stoddard 1999).

The implications of this general proposition—that animal communication provides objective, verifiable data on animal feelings and thoughts—are so far-reaching and so significant for cognitive ethology that they call for thoughtful consideration. Ethologists have seldom inquired whether an animal may want or intend to attack, or whether another may fear injury. But if we recognize that such basic subjective feelings and thoughts may occur in animals, we can often make much better, and more parsimonious, sense of their behavior, as emphasized by Dawkins (1993). In a thoughtful discussion of the basic challenges of investigating animal mentality, Russell (1935) concluded that "perception or imagery which does not issue in action must remain unknown to us, unless of course the subject can in some way communicate such perceptions and images to us" (97). This is just what communicative behavior sometimes does. Yet psychologists have paid little attention to the communicative behavior of animals, for reasons that are not entirely clear. Could this lack of interest stem from the pervasive inhibitions of behaviorism for the very reason that communication does suggest conscious thinking?

One reason that ethologists have been discouraged from using the communicative behavior of animals as a source of evidence about their feelings and thoughts is that they are convinced that all animal communication is a direct result of internal physiological states that are not under any sort of conscious control. Animal communication is thus held to be comparable to human eye blinks, blushing, gasps of surprise, or groans of pain. These do of course serve to communicate to others the state of irritation of the eye, embarrassment, surprise, or pain. But they are not intentional signaling employed for some perceived purpose. I have called this general view of animal communication the "groans of pain," or GOP, interpretation (Griffin 1985).

A related view is that threats are not signals that an animal wants or intends to attack but predictive information that leads to an appropriate response on the part of the animal that is threatened. This viewpoint considers animals to be simple-minded "behaviorists" that care only about what other animals do. On the other hand, insofar as conscious thoughts and subjective feelings affect subsequent behavior, it must be more efficient for both sender and receiver to recognize them by means of the communicative signals that report them.

How can we hope to tell whether a given sort of communicative behavior does or does not fall into the GOP category? One important indication is the effect of an audience. Since GOPs are assumed to depend directly on some internal physiological state, it should not matter whether any other animal is present. Eyelids blink when the cornea is irritated regardless of audience. But as discussed below, the communicative signals of many social animals are often dependent on the presence of other animals, and they are often modified in response to communicative signals received from others. The important basic point is that reasonable and appropriate interpretation of communicative signals exchanged by animals may provide significant, though not conclusive, evidence about their thoughts and feelings. Analysis of this evidence can, at the very least, provide an entering wedge into what has previously been held to be territory beyond the reach of scientific investigation.

Semantic Alarm Calls

Ground squirrels give different alarm calls on seeing aerial and terrestrial predators, but the information conveyed appears to be the urgency of the threat rather than the nature of the predator, as reviewed by Macedonia and Evans (1993). Slobodchikoff et al. (1991) and Ackers and Slobodchikoff (1999) have found that the alarm calls of prairie dogs differ in fine details according to differences in the shape and appearance of human intruders and other stimuli that elicit alarm calling. But there is as yet no evidence to show whether these rodents react differently to such differences in alarm calls.

One of the clearest examples of natural animal communication that suggests conscious thinking stems from studies of the alarm calls and other vocalizations of vervet monkeys *(Cercopithecus aethiops)*. These African monkeys, about the size of a small dog, live both in forests and in open areas where they can be observed more easily. They spend most of their lives in stable groups consisting mostly of close relatives who recognize each other as individuals. When they see dangerous predators they emit at least three types of alarm call, originally described by Struhsaker (1967). One type is elicited by the sight of a leopard or other large carnivore. On seeing a martial eagle, one of the few flying predators that preys on vervets, they give an acoustically quite different alarm call. And when the monkeys see a python they give a third call that is clearly different from the other two, as reviewed by Cheney and Seyfarth (1990).

This differentiation of alarm calls according to the type of danger leads to clearly distinct responses. The immediate response to the leopard alarm call is to climb into a tree, and since leopards are good climbers, monkeys can best escape from them by climbing out onto the smallest branches. But this would make them vulnerable to a martial eagle, and the response to eagle alarm calls is to move into thick vegetation close to a tree trunk or at ground level, where they would not be at all safe from a leopard. In response to the snake alarm call, the vervets simply stand on their hind legs and look around at the ground. Once they see a snake they can easily run away from it, although pythons do take vervets by surprise. Thus the best ways to escape from the three principal predators of these monkeys are mutually exclusive, and it is very important that the alarm calls inform other members of the group *which* danger threatens. A generalized escape response would be inefficient; there is no need wasting time climbing into a tree if the danger is from a python, and mistaking a martial eagle for a leopard or vice versa could easily cause the monkeys to do just the wrong thing.

Although it is clearly advantageous for vervet alarm calls to convey the information that one of the three types of predator has been sighted, many scientists did not accept the differences in the alarm calls as proof that an animal can convey semantic information about the nature of the danger rather than merely its state of fear or arousal. For example, Montagna (1976) claimed that vervet monkeys do not know what predator they are escaping from.

Vervet monkeys spend most of their lives in close-knit social groups, and their first response to an alarm call is to look at the caller. Like many other animals, they can tell in what direction a companion is looking, so they can usually see for themselves what has caused the alarm and respond appropriately. Also, the caller is likely to flee from the danger quickly, so the other monkeys might simply do what he is doing. In view of the deep-seated conviction that animal communication could not convey semantic information, it had seemed more parsimonious to interpret Struhsaker's observations as evidence that the three alarm calls conveyed only the degree of fear, or that they were points on a scale of intensity rather than having specific meanings about the nature of the threat. Yet the three calls vary in intensity with the degree of the caller's arousal, so this interpretation seemed somewhat strained.

Carefully controlled playback experiments by Robert Seyfarth, Dorothy Cheney Seyfarth, and Peter Marler (1980) resolved this uncertainty. The first step was to become so thoroughly familiar with groups of vervets living under natural conditions in eastern Africa that all individuals could be recognized. The next was to habituate the monkeys to the presence of human observers and their recording equipment. Then they played back alarm calls that had been tape-recorded when a member of the group had first seen a predator.

Many precautions were necessary to obtain convincing data. The monkeys might well respond abnormally, if at all, to playbacks of a known companion's alarm calls when he was in plain view and obviously not frightened. Therefore the loudspeaker had to be concealed in vegetation, and since monkeys recognize the calls of individual group members, playbacks were attempted only when the monkey whose calls were to be reproduced had just moved out of sight in the general vicinity of the concealed speaker, when the monkeys were not actively engaged in other behavior, and when they were not reacting to real dangers. Their behavior before, during, and after the playbacks was recorded by means of motion pictures, and the evaluation of responses was made by observers who viewed the films without knowing what call had been played.

These playbacks elicited the appropriate responses. In most cases the vervets climbed into trees on hearing playbacks of leopard alarm calls and dove into thick bushes in response to the eagle alarm calls. Playbacks of the snake alarm call caused them to stand on their hind legs and look all around for a nonexistent snake. Somewhat similar results have been reported for lemurs by Macedonia (1990) and Pereira and Macedonia (1991). Yet many inclusive behaviorists remain reluctant to accept the straightforward interpretation that these three types of alarm call convey information about the type of predator the caller has seen.

One alternative interpretation is that the alarm calls are injunctions rather than statements about the kind of danger. The leopard alarm calls might mean something like "Go climb a tree," or the snake alarm call "Stand up and look around." Even this somewhat strained interpretation recognizes that the calls are more than expressions of arousal. Their meaning might be what to do, rather than what danger threatens, but such injunctions are also semantic messages.

Vervet monkeys emit many other types of sounds during their social interactions. Cheney and Seyfarth (1982, 740) have analyzed by experimental playbacks, comparable to their studies of alarm calls, the "low-pitched, pulsatile grunt, originally described by Struhsaker (1967) . . . given in a variety of social contexts." To human listeners these grunts seem rather nondescript sounds, and like many other animal sounds they were commonly interpreted as a graded series conveying only some state of emotional arousal. But close analysis of their acoustic properties showed very slight differences that could be distinguished by human listeners only after considerable practice. When different responses to grunts were observed, it had been customary to assume that this was due to differences in the situation or context in which an essentially unitary type of sound was emitted. Cheney and Seyfarth suspected, however, that subtle differences among the grunts might be recognized by the monkeys as conveying different meanings. Unlike predator alarm calls, grunts did not elicit any vigorous responses from other vervets, except that they often looked at the companion who grunted.

In planning their experiments, Cheney and Seyfarth reasoned that "if the grunts were really one vocalization whose meaning was largely determined by context, subjects should show no consistent differences in response to the calls. Instead, responses to playback should be a function of the variable contexts in which they were presented. On the other hand, if each of these grunts was different, and if each carried a specific meaning, we should expect consistent differences in responses to each grunt type, regardless of the varying circumstances in which they were played" (740). They therefore selected recordings of grunts emitted in five different social contexts, using only cases when they had been able to record all the interactions that preceded and followed the vocalization. Only grunts of each type that were similar in duration and amplitude were used for playbacks, and grunts that were responses to other grunts were excluded. Numerous playbacks were made of grunts directed at (1) a dominant male, (2) a dominant female, (3) a subordinate female, (4) a monkey moving into an open area, or (5) another group of vervet monkeys.

The monkeys showed some revealing differences in their responses to these playbacks of grunts. One of the clearest differences was that grunts directed at dominants caused none of 12 monkeys hearing the playback to move away from the loudspeaker; they apparently conveyed an appeasing rather than a threatening message. But grunts to subordinates produced movements away from the speaker in 5 out of 12 cases. The vervets spent more time looking toward the speaker after playbacks of grunts to dominants than after grunts emitted on seeing a monkey move into the open, and there were similar quantitative differences between responses to some of the other types of grunts used in these experiments. Although these experiments do not indicate just what meaning the grunts conveyed, they do show that they were not interchangeable. As summarized by Seyfarth (1984, 54), "When a monkey hears a grunt, he is immediately informed of many of the fine details of the social behavior going on, even though he may be out of sight of the vocalizer, and even though the vocalizer himself may not be involved."

In other investigations Seyfarth (1987) and Cheney and Seyfarth (1990) used selective habituation of responses to repeated playbacks of vervet calls to learn something of their meaning to the monkeys themselves. They concentrated on three calls given only in the presence of another group of vervets: a grunt, a chutter, and a call designated *wrr*. Playbacks of any of them caused the monkeys to orient themselves toward the signaler and to look in the same direction as the signaler. When such playbacks were repeated in the absence of another group, the duration of these responses gradually diminished. After such habituation had occurred to repeated playbacks of the *wrr*s of a particular monkey, not only his *wrr*s but also his intergroup chutters elicited a much weaker response. Yet when the same experiment was repeated using the *wrr*s or chutters of another monkey, this elicited a normal intensity of response.

In other words, the monkeys tended to ignore both the type of intergroup vocalization that had been repeated without the presence of another group and the same monkey's acoustically different sound, which presumably conveyed the same or a similar meaning. Thus the habituation was specific to the individual caller, regardless of the category of call used by that animal. The vervets seemed to recognize that if one type of intergroup call by a given companion had proved inappropriate, his other intergroup calls also deserved less attention.

On the other hand, when the same experiment was repeated with alarm calls elicited by leopards or martial eagles, the result was significantly different. Although the vervets did become habituated to the groundless leopard alarm calls of a particular companion, played back from a tape recorder, they still responded to playbacks of the eagle alarm calls of that individual. It seems likely that predator alarm calls are such serious matters that the monkeys cannot afford to ignore them even when the caller has previously "cried wolf" about a different type of danger.

Semantic Screams

An important type of semantic information conveyed by animal calls is the social relationship between the caller and another member of the group to which he belongs. This has been most clearly demonstrated in studies of the free-ranging rhesus macaques *(Macaca mulatta)* that have been studied for many years on Cayo Santiago Island off Puerto Rico. When these and other monkeys engage in aggressive encounters, they often call loudly, and this calling sometimes serves to enlist the aid of others against the antagonist. Gouzoules, Gouzoules, and Marler (1984, 1986) selected a well-studied group of these monkeys for detailed studies of the screams given by immature males when exchanging threats or fighting with other members of the group. These monkeys could all be recognized individually, and both their maternal ancestry and social status were known from extensive previous studies. It was thus possible to distinguish whether a young male was interacting with a close relative and whether his opponent was higher or lower in dominance rank.

Screams from one of these young males often brought the screamer's mother to his aid.

Many previous studies of monkeys' calls had indicated that they varied continuously in their acoustic properties. This has been interpreted as a fundamental difference between monkey calls and human language, because the latter consists of discrete words whereas animal calls were considered to be merely emotional signals devoid of any meaning except to convey the state of arousal of the caller. It was therefore a surprise to find that 90 percent of 561 recordings of the agonistic screams of these juvenile males fell into one of five distinct categories.

On the basis of sound spectrograms these five categories were designated as noisy, arched, tonal, pulsed, and undulating screams. The noisy screams had a broad frequency spectrum from about 2 to 5 kHz. Arched screams had narrow frequency bands that rose and fell one or more times. Tonal screams had a wavering but gradually descending frequency from about 5 or 6 to 2 or 3 kHz. Pulsed screams were similar to noisy screams but broken up into pulses, each lasting only about 0.1 sec. Finally, the undulating screams consisted of a series of four or five harmonics that wavered up and down several times for a duration of roughly 0.7 to 1.2 sec. Although these screams varied considerably in detail, these five categories could easily be distinguished by human listeners.

These screams tended strongly to be used selectively toward different categories of opponent. With some exceptions, noisy screams were directed at higher-ranking adversaries when there was physical contact—including biting. Arched screams were given almost exclusively to lower-ranking opponents when no physical contact was taking place. Both tonal and pulsed screams tended to be given more often to relatives of the caller, and undulating screams were directed almost entirely to higher-ranking opponents when no physical contact was involved. Thus the screams contained information about the severity of the encounter and the social status of the opponent.

The mothers of callers also responded differently to these screams, as demonstrated by playbacks of tape recordings. The basic procedures employed in these playback experiments were the same as those used by Seyfarth, Cheney, and Marler in the experiments with vervet monkey alarm calls. The first response of the mothers was to look toward the concealed loudspeaker; they did so for all of the noisy screams given to higher-ranking opponents with physical contact, but less consistently in response to the other types. The duration of their gaze toward the speaker was much longer for noisy and arched screams than for tonal and pulsed screams, and they reacted more quickly to the former. The responses of mothers to screams that were repeated for several seconds included threat displays and charging from a considerable distance uttering loud calls of their own.

In other experiments the same investigators found that mothers respond more strongly to the screams of their own sons than to those of other young males (Gouzoules, Gouzoules, and Marler 1986). Similar studies of a large group of captive pigtail macaques *(Macaca nemestrina)* by Gouzoules and Gouzoules (1989) showed that in this species different types of screams were emitted when fighting or threatening opponents of different social rank. But screams of pigtail macaques did not seem to differ according to matrilineal relatedness, as they did with rhesus monkeys.

The results of these investigations show that the screams of these monkeys, unlike groans of pain, convey considerable information about the caller's situation, information that affected their mothers' behavior. It seems likely that in these highly emotional situations the monkeys think in simple terms about the degree of threat and danger and about the social relationship between the opponents.

Audience Effects

As mentioned above, one way in which ethologists might be able to distinguish communicative signals that do or do not fall into the "groans of pain" category is to study the effect of an audience on the production of communicative signals. Marler and his colleagues have been attempting to do this with the calls emitted by domestic chickens. Chickens emit a wide variety of calls, as described by Collias and Joos (1953) and Collias (1987), and they are of course convenient animals to study. Marler and his colleagues concentrated on two types of call given by adult males of a small strain, the golden Sebright bantam, that seems to be quite similar to the ancestral jungle fowl from which domestic chickens derived. Because some of the most interesting calls are rather faint, they were recorded by temporarily attaching to a cockerel's back a 2 × 2.5 × 5 cm radio microphone weighing 16 g. The birds became accustomed to wearing these instruments, and their behavior did not seem to be appreciably altered by them.

Chickens, like several other species of birds, have two types of alarm calls: a series of short, narrow-band whistles commonly given when they see aerial predators such as hawks and a pulsed broad-band cackle elicited by ground predators such as dogs or foxes. But the type of danger is not the only factor affecting which call is produced, so they are not a highly specific form of signal. The so-called ground predator call is also given to hawks at close quarters, and the aerial alarm call is often given to small and harmless birds or other objects seen against the sky. Yet despite much variability and many calls given when the observers could detect no danger or other appropriate stimulus, there is nevertheless a strong tendency for the aerial alarm calls to be given for hawklike objects seen in the sky and the ground predator alarm call for moving objects at ground level.

Marler and his colleagues arranged conditions in which a single bantam cockerel lived for long periods in a large outdoor cage. Models of hawks were presented by pulling them along overhead wires, much as Lorenz and Tinbergen had done in their classic studies of avian responses to hawklike patterns. Under these conditions the cockerels emitted 509 aerial alarm calls in 400 presentations of moving hawk models, but no ground alarm calls at all (Marler, Dufty, and Pickert 1986a, 1986b; Gyger, Marler, and Pickert 1987; Karakashian, Gyger, and Marler 1988; Marler, Karakashian, and Gyger 1991).

The cockerels gave significantly fewer aerial alarm calls when alone than when their mate or another familiar female was clearly visible and audible in an adjacent cage. But when this experiment was repeated with unfamiliar females the male called no more often than when he was alone. Familiar males elicited almost as many alarm calls as familiar females. Young chicks were almost as effective an audience as familiar females in eliciting alarm calls, but the presence of bobwhite quail did not increase the frequency of alarm calling. Evans and Marler (1991) have refined these experiments by showing that videotapes with soundtracks have the same effects as live birds. This allows improved experiments in which possibly confounding effects, such as variation in activity of the quail and chickens, could be experimentally controlled by careful selection of the sequences played back.

Cockerels also give other calls when food becomes available, and in the experiments they called more when presented with preferred foods such as mealworms or peas than for peanuts or inedible nutshells. Females approached males giving these food calls more than noncalling males, and they were more likely to approach a male calling about a preferred food. Males also called more when hens were present and hardly at all when another adult male was in the adjacent cage. This is doubtless related to the fact that courting cockerels often bring food to females in which they are interested. The amount of food calling by the males also varied with the nature of the conspecific audience. When no other chickens were present, males gave food calls 13 out of 18 times that mealworms were presented and only once in 18 presentations of inedible nutshells. When another male was present, they gave no food calls at all for either mealworms or nutshells, although they ate 15 of the

18 mealworms. When either familiar or unfamiliar females were present the cockerels never ate the mealworms and gave food calls in 35 of 36 presentations. Thus the presence and nature of an audience had a marked effect on food calling.

Marler and Evans (1996) have lucidly reviewed the results of these experiments and clarified the probable reason for the differences in responses to the aerial predator and ground predator alarm calls. The former are emitted at relatively low intensity, and the calling bird also takes cover if possible and acts in a way that makes it less conspicuous. This, together with the fact that aerial predator alarm calls are given much more often when other chickens are present, indicates that they are the intended audience. In contrast, the ground predator alarm calls are louder and are repeated more often, and the calling bird makes itself conspicuous. It seems that these calls are directed as much at the predator as at other chickens, serving to inform the predator that it has been detected and that the calling bird is prepared to flee, as discussed by Zahavi and Zahavi (1997). This helps explain why there is no appreciable audience effect with ground predator alarm calls.

Marler and Evans's general conclusion from these experiments is that these alarm calls are not impulsive and involuntary and that they can be controlled according to the circumstances. Of course, these calls express strong emotions, but they are also controlled by at least some cognition. As cautious scientists, however, Marler and Evans refrain from speculating about any conscious experiences of the birds they have studied. It is significant that these audience effects are different for food calls and alarm calls. Almost as many alarm calls were given in the presence of familiar males as with familiar females, but no food calls at all were given to males. In other words, the cockerels were appropriately selective about their use of these calls. To be sure, other considerations such as the whole context in which communication takes place are also important in interpreting these experiments, as discussed in detail by Smith (1991). But, at the very least, the experiments demonstrate that as more is learned about communicative behavior of animals, it becomes increasingly difficult to fit such behavior into the procrustean bed of GOPs.

Honeyguides

An intriguing type of foraging that suggests intentional planning and communication of simple thoughts is the guiding behavior of the African greater honeyguide *(Indicator indicator)*. These birds feed on insects, including the eggs and larvae of bees, and they are also fond of honey and of the wax from honeycombs, which they are able to digest. They cannot open bee nests, but in an apparent effort to obtain honey and wax they cooperate with men or perhaps animals that tear open bee nests and leave a considerable amount of honeycomb to the honeyguides. The several species of honeyguides and their behavior were studied intensively by the ornithologist Herbert Friedmann and described in his 1955 monograph. His analysis of the cooperative behavior is significant as a classic example of the reductionistic Zeitgeist that was so prominent in studies of animal behavior for many decades and led even the most experienced field naturalists to underestimate the versatility of the animals they studied. Short and Horne (1985) have thoroughly investigated the behavior and ecology of honeyguides with special emphasis on their roles as nest parasites (birds that lay their eggs in the nests of other species). The following review of honeyguide behavior is based on Friedmann's monograph, substantially supplemented (and in important respects modified) by the recent work of Short and Horne, and especially that of Isack and Reyer (1989), discussed in detail below.

Honeyguides are widely believed to lead the ratel *(Mellivora capensis)*, or honey-badger, to bees' nests. They also engage in the same sort of cooperative behavior with people, and in some areas African natives seek out honeyguides to obtain honey from the nests of wild bees. Some, according to Friedmann, imitate the grunting sounds of ratels or chop on trees to simulate the sound of opening a bees' nest. Occasionally a honeyguide has been reported attempting to lead a mongoose, monkey, or baboon, but only baboons have been observed to follow the bird. Most of the detailed observations of the guiding behavior have involved human cooperators, and several recent students of honeyguides suspect that they never cooperated with ratels and that the idea they did so arose because the nocturnal ratels were seen feeding on honeycomb after it had been exposed by human honey gatherers.

Friedmann believed that the guiding habit had decreased markedly from earlier years as European ways of life provided more easily obtained sources of sweeteners than the opening of wild bees' nests. He also suspected that ratels had become more nocturnal, under increased human hunting pressure, so that honeyguides had less opportunity to obtain the contents of bee nests by cooperating with them. The guiding behavior seems always to have been limited to certain areas within the honeyguide's range, so it is far from being a rigidly fixed species-specific behavior pattern. Regardless of this question, there is no doubt that, in many areas of Africa, honeyguides lead human searchers to bees' nests and that after the honey gatherer has opened the nest the birds profit by obtaining honeycomb.

Friedmann (1955, 32–33, 39–41) describes typical guiding behavior:

> When the bird is ready to begin guiding it either comes to a person and starts a repetitive series of churring notes or it stays where it is and begins calling these notes and waits for the human to approach it more closely. . . . If the bird comes to a person to start leading him, it flies about within 15 to 50 feet from him, calling constantly, and fanning its tail, displaying the white outer rectrices. If it waits for the potential follower to approach it for the trip to begin, it usually perches on a fairly conspicuous branch, churring rapidly, fanning its tail, and slightly arching and ruffling its wings so that at times its yellow "shoulder" bands are visible. As the person comes to within 15 to 50 feet from it, the bird flies off with an initial conspicuous downward dip, with its lateral rectrices widely spread, and then goes off to another tree, not necessarily in sight of the follower, in fact more often out of sight than not. Then it waits there, churring loudly until the follower again nears it, when the action is repeated. This goes on until the vicinity of a bees' nest is reached. Here the bird often (usually in my experience) suddenly ceases calling and perches quietly in a tree nearby . . . and there waits . . . until the person has departed with his loot of honeycomb, when it comes down to the plundered bees' nest and begins to feed on the bits of comb left strewn about.

Friedmann continues,

> Guiding may cover a duration of from a few seconds to half an hour, or possibly even an hour, and may involve a distance of from a few feet to over half a mile, and possibly, at times, even a mile. . . . Guiding leads to the vicinity of a bees' nest, not to the exact spot. . . . If a person does not follow a honey-guide that has apparently come to "lead" him, the bird may increase the tempo and excitement of its behavior as if to urge and entice, or it may give up easily and leave. . . . A would-be "guiding" bird may sometimes follow a person for a very long distance (five miles is the maximum known to me) or for a very considerable period of time (half an hour is the maximum I know of) to attempt to get him to follow it.

Recent students of honeyguide behavior have reported somewhat different patterns, and these may well vary from place to place and from time to time. But there is no doubt that the birds exert considerable effort in attempts to attract the attention of people, that they do fly with frequent stops to the vicinity of bees' nests, and that if someone opens the nest they do eat honeycomb.

After describing this cooperative behavior in great detail Friedmann opens a section of his monograph (54–64) titled "Behavioristic level of the habit" as follows:

> Use of the term "guiding," with respect to the behavior pattern that usually results in the follower arriving at a bees' nest, is unfortunate in that it implies a preexisting purpose or plan on the part of the bird, an intelligent activity far beyond the psychological capacity of any bird. . . . The word "guiding" has a

purposive connotation which is applicable to the species but not to any of its members. . . . There are several features of "guiding" which further indicate its stereotyped nature. One is the fact that guiding is ordinarily not direct. The bird frequently leads in a most erratic course, often actually going a considerable distance beyond a bees' nest and then coming back to it. . . . If guiding were purposive in the individual this would be difficult indeed to explain, especially since there were no obstacles or barriers such as hills, ravines, etc. to be bypassed.

Friedmann illustrates several very indirect routes taken by honeyguides between the place where the leading began and the bees' nest where it ended, and he found he could cover the direct distance between these two points in half the time he had taken when following the bird. Yet, as pointed out by Isack and Reyer (1989), these birds did not move at random but tended toward the location of the bees' nest, though their approach was indirect. Friedmann advanced as another reason for concluding that the guiding is not intentional "the fact that on occasions the bird will lead not to a bees' nest but to a dead animal or to a live snake, leopard, rhinoceros, etc." (56). One reason for this may be the numerous flies or other insects that are often present near animals such as those near which the honeyguide ceases its leading behavior. Friedmann also cites the following episode as evidence that the honeyguides were not leading intentionally to a bees' nest whose location they knew beforehand:

Captain Davison . . . had a group of his natives at one of the rest camps in the Reserve when a honeyguide came to them and chattered and went through all the motions of trying to get them to follow it. Davison refused to let any of his boys go, but got them all on a truck and drove off to the next camp some five miles away. The bird followed them all the way and then Davison told one of the natives to get an axe and follow the bird. The honey-guide "led" this native to a bees' nest less than half a mile from the second camp, but which must have been at least 4 miles from where the bird first began calling to them. (59)

Elsewhere in his 1955 monograph, and especially in later reviews of the same material, Friedmann claimed that these observations *prove* that the guiding behavior is carried out without conscious intention on the bird's part. For instance: "It is now known that this guiding behavior, which looks so purposive, is actually a form of excitement reaction on the part of the bird when meeting a potential foraging symbiont, and that the excitement dies down when the bird, with its symbiont near at hand, sees or hears swarming bees" (Friedmann and Kern 1956, 19). In his 1955 monograph he states that "the releasers of the instinctive behavior constituting 'guiding' are the sight or sounds of ratels, baboons, and humans (away from villages). The stimulus which apparently brings these actions to a halt is the sight or sound of bees"

(59). And, later: "There is no occasion whatever to assume anything involving planning or intelligence on the bird's part. The behavior is wholly on an instinctive level, but it is sought for by the bird, not merely something it does automatically when the necessary stimuli are present" (163).

These statements are representative of the widespread efforts on the part of psychologists and biologists to account for all animal behavior without allowing any role for conscious thinking or intentional planning. The observation that honeyguides do not go directly to a bees' nest is important and certainly argues against a detailed and accurate memory of the most direct route to the goal. But it scarcely proves the absence of any conscious intention. One obvious alternative would be an imperfect memory of where a bees' nest was located, even though the bird might recall that there was one in the vicinity. Or the bird might wait until it has enlisted a follower and then begin searching for a bees' nest. The available data are far from sufficient to confirm or disconfirm such hypotheses, but the important point is that scientists have been very quick to seize upon any failure of an animal to perform with perfect efficiency and to then offer such failure as evidence that it is a totally thoughtless robot.

In a truly revolutionary paper Isack and Reyer (1989) describe an intensive three-year field study of the greater honeyguide in northern Kenya, in an area where honey from the nests of wild bees is an important part of the diet of the Boran people. This is dry bush country quite different from the forested areas where Friedmann had conducted most of his observations of the same species, but the guiding behavior was very similar to the displays and calls Friedmann had described. Some of the Boran people are professional specialists in honey gathering, and they rely on honeyguides to a considerable extent. Considering only days when they did find at least one bees' nest in an unfamiliar area, Isack and Reyer (1989, 1344) report that "their search time per bees' nest was, on the average, 8.9 hours when not guided and 3.2 hours when guided." Very few of the bees' nests are located where the unaided birds can reach them, so the guiding-following behavior is an effective form of behavioral symbiosis that has important benefits for both participants. The extensive observations of these birds by Short and Horne (1985) are quite consistent with the behavior described and analyzed by Isack and Reyer.

The Boran honey gatherers interviewed in their own language by Isack, who is himself a Boran, stated that the guiding behavior of the honeyguides "informs them about the direction of, the distance to, and their arrival at the colony [of bees]" (1344). Isack and Reyer tested this surprising statement by mapping several of the routes honeyguides followed while guiding the professional honey gatherers. The results clearly confirmed that the starting direction was indeed almost always correct within 20 or 30 degrees, and the mean direction of all initial flight directions was within less than one degree of the bearing of the bees' nests to which they eventually led their human cooperators.

As the bird travels toward the bees' nest, three properties of its guiding behavior decrease progressively: the distance between perches where it lands and waits for the man to catch up, the height of these perches above the ground, and the duration of periods when the bird flies off toward the bees' nest and then returns. These periods vary between about half a minute and two minutes, and though it is not possible to follow the birds, it seems likely that they fly part or all of the way to their objective and then return to the man they are guiding.

The human honey gatherers communicate with the honeyguides. They use a loud whistle to attract the birds, and as they follow them "they whistle, bang on wood, and talk loudly to the bird to keep it interested in the guiding" (1344). When it has reached the vicinity of a bees' nest the bird emits an "indication call," which is softer in tone than the calls during guiding, with longer intervals between successive notes.

Isack and Reyer observed the vicinity of bee nests from "camouflaged observation positions occupied before dawn," and they saw honeyguides inspecting the nests, remaining only for about a minute before flying away. On cloudy and cool mornings when the bees were not aggressive, "the bird would fly straight into the entrance of the nest and peer into it" (1345). Isack and Reyer could not gather data adequate to test two additional claims of the Boran honey gatherers: "(i) that a bird, flying lower than the treetops, will guide to a colony close to the ground, and (ii) that when nest distances become very long (about 2 km or more), the birds 'deceive' the gatherers about the real distance by stopping at shorter intervals. However, having found all the other Boran observations to be true, we see no reason to doubt the statements of these excellent 'ethologists' " (1345).

These recent discoveries by Isack and Reyer show that Friedmann's interpretations, reached in the heyday of behaviorism, clearly failed to do justice to the versatility of which the honeyguides are capable. The close correlation of the birds' initial directions of flight and the straightness of the routes along which the honey gatherers are led demonstrate clearly that the birds knew the location of the nest to which they were leading their cooperators. Whether the variations in perch height and in the length of flights between perching are intended to inform the follower is more difficult to ascertain. But even if they are not, they provide evidence that the bird is paying attention to its memory of the nest location. Behaviorists can translate these observations into a series of stimulus-response contingencies, but the resulting positivistic account becomes more and more unwieldy as more is learned about the details of the birds' actual behavior. To assume a simple conscious intent to lead the follower to the bees' nest and get food after he has opened it seems a more parsimonious and reasonable interpretation.

Parrots Who Mean What They Say

Parroting has become a term for imitation of speech without understanding what it means. African grey parrots *(Psittacus erithacus)* are especially proficient at mimicking human speech, but other parrots, mynah birds, starlings, and other species can also mimic words well enough that they can easily be recognized by human listeners. The effectiveness with which parrots and other mimetic birds can imitate a wide variety of words, as well as other sounds, has led their owners to teach them whatever comes to mind. The results are often entertaining, but the meanings of the words imitated are mostly so remote from anything a bird could possibly comprehend as to reinforce the widespread conviction that avian mimicry entails no understanding whatever. In 1988 an African grey parrot could undoubtedly have been trained to say "Read my lips!" just as one trained by Stevens (1888) learned to say "Hurrah for Blaine and Logan" during the campaign of 1884.

Yet some words that parrots learn do have a simple and direct relevance to their situation, as when they learn to ask by name for particular foods. Although this is grudgingly recognized, a sort of simplicity filter has tended strongly to rule out of scientific thinking the notion that a bird might understand that a simple meaning is conveyed by a sound it has learned to imitate. This type of simplicity filter was made to seem more plausible by the failure of several efforts to train parrots to emit particular sounds in order to obtain a food reward. The most often cited examples are the work of Mowrer (1950, 1960a, 1960b, 1980) and Grosslight and Zaynor (1967).

It is perhaps no accident that in the 1950s, when behavioristic learning theory was dominant in psychology, it was O. H. Mowrer who became interested in "talking" birds. For he was sufficiently uninhibited by behaviorism to write: "If . . . we sometimes speak of 'consciousness' (a tabued word for the behaviorists), this is not just a friendly gesture to the past or concession to common sense; it represents instead the growing conviction that the objective study of behavior has now reached the point where some such concept is essential . . . if consciousness were not itself experienced, we would have to invent some such equivalent construct to take its place" (Mowrer 1960b, 7).

Mowrer (1980, 51–54) described how he "acquired a collection of parrots, mynah birds, parakeets, magpies and crows, and set about learning how to teach them to 'talk.' " It is not clear from his publications the extent to which he actually tried to elicit vocalization by means of conditioning with food as a reward, but his efforts were apparently unsuccessful, for he continues:

> The only systematic and extensive investigation of this problem thus far reported, in so far as I am aware, is that of Grosslight and Zaynor (1967). The mynah birds which were used as experimental subjects were put into soundproof boxes, and there periodically heard a tape-recorded word or phrase which was then followed by a pellet of food. . . . These subjects should have learned to reproduce the "conditioned stimulus" (word), that is, to imitate. They did not. . . . Perhaps we get a clue from the fact that there were two or three mynahs around Grosslight's laboratory which his assistants had converted into "pets," and they were all fluent talkers. (1980, 51–54)

It was characteristic of the behavioristic Zeitgeist of the mid-twentieth century that Mowrer's work has since been cited primarily as evidence that because birds cannot be trained to imitate speech by means of standard conditioning methods, their imitation cannot entail any understanding of the meanings conveyed by the words they mimic. This was clearly not what Mowrer or Grosslight and Zaynor concluded. Instead they recognized that interactive exchanges of sounds with companions was probably a necessary condition for the development of "talking" in birds.

Ethologists have subsequently studied the use of complex vocalizations by parrots, Indian Hill mynahs, and other birds under natural conditions and found that imitation of companions plays a large role in their social behavior (Thorpe 1972, 1963; Thorpe and North 1965). Young birds pass through a stage when they emit varied and imperfect versions of the sounds they will use as adults, and they seem to enjoy repeating and imitating sounds they hear around them. The social context and the responses of other birds are important factors in song learning, as demonstrated, for example, by West and King (1988) with cowbirds. Females of this species respond to a small subset of male songs with a rapid wing flick, and this is apparently recognized by the males as a sign of readiness for copulation.

An important advance was made by Todt (1975), who developed an effective training procedure called the model-rival approach. His procedure was much closer to the natural social exchanges by which birds learn their vocalizations, although he used human companions rather than other birds. He exposed a mimetic bird such as the African grey parrot to cooperative human trainers who talked to each other in the bird's presence, one acting as a rival, in effect competing with the parrot for the other's attention. Irene Pepperberg (1981, 1999) improved this procedure by having two trainers talk about objects in which their parrot seemed interested, one asking for one of these objects, the other giving it to her or withholding it according to the correctness of her verbal requests. The trainers exchange roles from time to time. Using this method she succeeded in training a male African grey parrot named Alex to use imitations of several English words in an appropriate fashion.

In his first 26 months of training Alex acquired a vocabulary of nine names, three color adjectives, and two phrases indicating simple shapes, and he came to use "no" in situations where he was distressed and seemed unwilling to do what his trainers wanted, or when rejecting something offered to him. Alex subsequently learned numerous uses of his imitated English words. When shown familiar objects and asked "What color?" or "What shape?" he learned to answer correctly more than 80 percent of the time, which is far better than chance because he had to select one of five colors or four shapes (Pepperberg 1987b). In further experiments Alex learned to say the numbers two to six plus the name of the objects when shown sets of two to six familiar things. His overall accuracy in these tests was 78.9 percent correct, but more than half of his errors were responses in which he named the objects correctly but omitted the number. In such cases the further query "How many?" produced the correct response 95 percent of the time (Pepperberg 1987a).

In other experiments Alex was shown two objects (which might be either familiar objects or things he had never seen before) and required to say what was the same or different about them (Pepperberg 1987b, 1988, 1991). They differed in color, shape, or material, and Alex usually gave the correct response—"color," "shape," or "matter" (which he pronounced "mah-mah")—to designate the material of which the object consisted (paper, wood, cork, or rawhide). His responses to "What's same?" or "What's different?" were 82–85 percent correct when there were three options—color, shape, or material—so a chance score would be only 33 percent correct. In further tests of this general type, some pairs were identical, whereas others were totally different. In the former case Alex learned to respond to the question "What's different?" by saying "None." When the test objects were very different and Alex was asked "What's same?" he would also answer "None." His accuracy in these tasks was about 80 percent.

In still more recent experiments Alex is shown any collection of objects and asked any of the following four questions: "What color is X?" "What shape is X?" "What object is Y?" or "What object is shape Z?" where X might be the name of any of several familiar objects, Y might be any of seven colors, and Z any of five shapes. He had learned all these words and used them correctly in previous experiments. His accuracy in responding was about 81 percent. These questions were mixed with other tests, so that each one was presented only at long intervals. To answer correctly Alex had to understand all the words in the question and use this understanding to select the correct reply from his vocabulary (Pepperberg 1990, 1991, 1999). For many years Pepperberg's pioneering work was based on the behavior of a single bird, but she has recently succeeded in training two younger African grey parrots to match the early stages of Alex's acquisition of meaningful communication.

These findings constitute a truly revolutionary advance in our understanding of animal mentality, comparable to von Frisch's discovery that honeybees use symbolic gestures to communicate to their sisters the direction, distance, and desirability of distant objects, as described in the next chapter. Because they are such startling extensions of what we had previously believed possible for any bird, it is very important to consider carefully whether there might be some flaw in these experiments. As with any surprising discovery, it is very important that it be replicated by other scientists. But to date no one else has apparently attempted to replicate Pepperberg's experiments, presumably because they are very time-consuming. But pending such essential replication we can inquire as critically as possible what flaws might have escaped our notice.

The first thought of any student of animal behavior is that some sort of inadvertent cuing may have taken place, that Alex responded not to the actual properties of the objects he was shown but to some unrecognized behavior of the trainer who presented the objects and asked the questions, something analogous to the cuing that explained the apparent ability of Clever Hans to perform feats of mental arithmetic. But any such inadvertent signals would have had to be quite subtle, some gesture that Alex took to mean he should say "same," "blue," "wood," "none," or any of the many other words he had learned.

Alex is quite sensitive to the presence of his familiar trainers and is frightened by strangers, so to assure his attention and cooperation his primary trainer, Pepperberg, has to be present during critical testing of his discriminatory responses. But during all testing of the parrot's comprehension of the questions posed to him, in contrast to training sessions, Pepperberg, who is keenly aware of the dangers of inadvertent cuing, sits in one corner of the room, does not look at Alex, and does not know what is presented to him by a so-called secondary trainer or examiner. The latter is a person who is familiar to Alex and has trained him in other experiments but has not been present when he was trained in the task under investigation. In other words, someone else had trained Alex to say "same," "different," "none" or whatever was the correct response. Thus Alex has no opportunity to learn any inadvertent cues that might have accompanied the presentation of various objects or combinations by the particular examiner. If he was responding to some unrecognized cue specifying which of numerous words in his vocabulary he should utter on a given occasion, it must have been one that any person would exhibit when showing Alex various objects. This seems highly unlikely, but the examiner did know what the correct answer was and conceivably might have conveyed it to the parrot in some unknown manner.

In a further attempt to control for possible cuing by the examiner, Pepperberg and Brezinsky (1991) presented test objects in a box that was opened in such a way that the parrot could see its contents, but the examiner could not. This procedure was extremely difficult, because it was almost impossible to maintain the parrot's interest in objects in the box, although when the same objects were held close to him he paid close attention to them. To sustain his cooperation the box had two chambers, A and B, and a hinged cover that could be closed to conceal the contents of either chamber.

When testing Alex's understanding of relative as opposed to absolute size of a test object, Pepperberg took the box to a separate room, hid the objects to be tested in chamber A, closed the cover over A, carried the box back to the laboratory, and handed it to the examiner, telling her not what object was in A but what question to ask, for example "How many?' or "What color bigger?" Pepperberg sat with her back to Alex and the examiner, and Alex was asked what he wanted, such as a cork, a nut, or a grape. He watched as the examiner placed the desired object in chamber B. Then the box was turned so that side A faced Alex and the cover was opened so that he could see the contents of chamber A although the examiner could not. Only after these preliminary steps did the examiner ask Alex the test question. Only if he answered correctly was he allowed to take the desired object from chamber B.

Because of these practical difficulties the box was used only in tests of Alex's ability to judge relative sizes of objects, where there were only three correct answers: larger, smaller, or none (meaning no difference in size). Thus inadvertent cuing by the human examiner was more likely than when Alex was required to select one out of several dozen words that he had learned to use in answer to a variety of questions. As in most of Pepperberg's other experiments, the crucial test questions were inserted rarely and in random fashion among a long series of questions about other topics. In responding to the question "Which bigger?" Alex had been trained to say the name of the larger of the two objects presented in chamber A. If they were of the same size, he said "None."

In transposition tests Alex was shown two novel objects that had not been used in preliminary training, both of which were larger or smaller than any of those shown to him previously. Only the first presentation of each pair of objects was counted in scoring the parrot's accuracy, and Alex correctly named the larger or smaller test object in 8 out of 10 trials. When shown objects of equal size he answered "None" in 9 out of 11 trials. When asked "What matter bigger (or smaller)?" he was correct in 8 out of 10 trials, using his names for one of the several materials used for test objects. When such questions were asked about novel objects that Alex had not seen before his answers were correct in 34 out of 44 tests. Thus in these experiments his answers were 77–82 percent correct even when the examiner could not see the test objects. Since she did not know the right answer, it is difficult to imagine how she could provide effective inadvertent cues to Alex.

In a series of recent experiments with Alex and two young parrots Pepperberg has attempted to train them by means of video images. But the two juveniles have learned almost nothing from audio or videotapes showing Alex interacting with his trainers in the model-rival procedure. On the other hand, the model-rival procedure involving two interacting humans has succeeded with all three birds. Pepperberg's general conclusion is that for parrots to learn to use their imitations of human speech meaningfully and with understanding, the training must include abundantly three basic elements: (1) reference, (2) functionality, and (3) social interaction. Reference is what a communicative signal represents. Functionality is what a signal can cause to happen, and social interaction is the complex process of mutual stimulation and guidance that occurs during effective social learning. The attempts to use video and audio recordings to replace the social interactions in the successful model-rival procedure apparently failed because they did not include sufficient social interaction even though the information about reference and functionality was available in a form that the birds should have been able to learn.

The necessity of intense social interaction for the model-rival procedure to succeed raises again the possibility of inadvertent cuing. Could something in these close interactions with a human trainer convey in some way that we cannot grasp just what word the parrot should imitate?

As mentioned above, the large number of possible utterances among which the parrot had to choose, together with the precautions described above, especially the "box" experiments of Pepperberg and Brezinsky (1991), make inadvertent cuing appear very unlikely.

On balance it seems reasonable to conclude that a parrot can learn both to understand and to communicate about several simple properties of familiar objects (color, shape, material), as well as such basic relationships as sameness or difference. Pepperberg makes no claim that Alex has learned anything approaching the versatility or complexity of human language, but he does seem to have demonstrated some of the basic capabilities that underlie it. More important is the strong indication that these parrots can think about colors, shapes, sameness, and so forth. To be sure, those inclined toward a sleepwalker view of nonhuman animals may insist that they are all unconscious robots so contrived that they give the correct verbal responses. But this opinion would have to be based on other considerations than the ability of these parrots to use words meaningfully. In short, they give every evidence of meaning what they say.

CHAPTER TEN

Symbolic Communication

As discussed in the previous chapter, Irene Pepperberg's parrots have obviously learned to use their imitations of human words for a simple type of symbolic communication. And chapter 12 will analyze the extent to which dolphins, chimpanzees, and other apes have learned to use various types of symbolic communication devised by human experimenters. The predator alarm calls of vervet monkeys convey semantic information about the type of danger. And in the normal course of their lives under natural conditions many other animals communicate specific and important information, as well as emotional states such as fear. This chapter will analyze a few specific cases in which communicative behavior is surprisingly versatile and suggests conscious thinking.

Exploiting newly discovered sources of food is a very common problem for animals that live in cooperating social groups, and it can often be facilitated by communication. It is helpful to begin by considering a simple hypothetical example involving only two animals. When insectivorous birds are feeding a nestful of young, they need to locate and capture quite large numbers of insects. They are not alone in this search; other birds and other insectivorous animals are also busy hunting, and the insects themselves are not exactly cooperative in making themselves available. When both parents are feeding nestlings, one parent sometimes finds an aggregation of edible insects at some distance from the nest when the other is less successful. In this situation it would clearly be advantageous for the former to convey to its mate the location of this newly discovered source of food.

However, we do not know whether parent birds actually do inform each other about good sources of food for their young. Such behavior has not been observed, but the types of observation that would be necessary to reveal it are difficult and have not been seriously undertaken, to the best of my knowledge. One would have to follow both foraging parents, note when one of them was having better success than the other, see whether they interacted back at the nest, and, if so, whether the less successful parent then flew to the food source discovered by its mate.

This would be a special case of an "information center" where animals might learn from their companions where to locate food, as suggested by Ward and Zahavi (1973) and recently reviewed by Zahavi and Zahavi (1997). Richner and Heeb (1995), Richner and Marclay (1991), and Shettleworth (1998) have doubted that most animals learn from each other about the location of food. The principal reason for this skepticism is that the information center hypothesis suggests group selection rather than assuming that each animal behaves in a way that increases its inclusive fitness, that is, the likelihood that its genes will be propagated in future generations. Yet, as Shettleworth recognizes, there are some known cases in which animals learn about food sources from companions that are genetically related, so this process could well increase inclusive fitness. Perhaps as ethologists begin to devote more attention to animal cognition and intentional communication, the necessary studies will be carried out to reveal whether or not this sort of communication about newly discovered food sources does occur in birds.

Some of the social insects certainly do share information about newly discovered food sources, as reviewed by Hölldobler and Wilson (1990). One of the best examples involves the weaver ants, whose ingenious cooperative construction of leaf nests was described in chapter 5. Gathering the food needed by any large colony of ants requires a great deal of effort on the part of many of the nonreproductive workers. They often search a relatively wide area and after they have found food return to the nest and recruit many of their sisters to join them in gathering this food. Most species of ants lay odor trails along the route from the food source back to the nest. But this in itself may not be enough, because workers that have not been to the food must be induced to follow the odor trails. Chemical signals are transferred from returning foragers to other members of the colony as they feel each other with their antennae. The forager often regurgitates food from her stomach that others take up in a behavior pattern known as trophallaxis.

In some ants the returning forager moves her body rapidly from side to side while engaging in mutual palpation and trophallaxis with one of her sisters. In *Campanotus serviceus* this behavior induces the recruited ant to grasp the abdomen of the forager, and the latter then moves out along the odor trail she laid down on her return to the colony (Hölldobler 1974, 1977). This results in tandem running out to the food source. Another very important form of recruitment occurs when the colony is threatened by aggressive or competing members of other insect colonies. The weaver ants show a significant difference in their recruiting behavior according to whether recruitment is for food gathering or for fighting off intruders.

Hölldobler and Wilson (1978) have studied in detail the behavior of weaver ant workers after they have returned to the colony either from food sources or from encounters with intruders. The intruders may be members of another colony of the same species that are more numerous and aggressive, or they may be insects of other species. The recruiting weaver ant engages in a series of face-to-face encounters with nestmates at or close to the leaf nests. The recruiter makes lateral movements of her head, which often induce the other ant to follow the odor trail she has just laid down. These recruiting gestures differ according to what it is the recruiter has returned from. The principal difference is that when recruiting to food sources, she moves the head in lateral wagging motions. But when returning from an encounter with intruders, she recruits other workers by jerking her body back and forth toward and away from the nestmate rather than from side to side.

Hölldobler and Wilson also described other communicative gestures used when stimulating nestmates to move to another location, but the distinction between gestures for recruiting to food and for combating intruders is especially clear-cut. The gestures used in recruiting for fighting resemble in some ways the movements employed in actual combat. This may be a ritualized imitation of actual fighting, a sort of pantomime. Whether intentional or not, this differential recruitment is effective, and numerous workers are induced to move rapidly out, either to gather the recently discovered food or to fight the intruders. If we allow ourselves to postulate, tentatively, that these ants might be thinking in simple terms about such important matters as gathering food or fighting intruders, these somewhat specialized gestures may be interpreted as evidence of what they are thinking about.

Why are different gestures used for these two purposes? If the only function of the communication were to recruit nestmates, it would seem unnecessary to use a different gesture in the two situations. Perhaps the recruited ants are better prepared for either food gathering or fighting if informed which to expect as they move out from the nest. Are different types of nestmates recruited by these two sorts of gesture? These recruiting gestures are undoubtedly accompanied by the transmission of odors and chemical signals, and for all we know these may also provide information about what is located at the end of the odor trail.

Some of the ants receiving the recruiting gestures do not follow the odor trail but turn to other workers and repeat the recruiting gestures even though they have not been directly stimulated by intruders or newly discovered food. This sort of chain communication whereby one original recruiter indirectly stimulates large numbers of nestmates is especially significant, because the ant that has been stimulated by recruiting gestures and then repeats these gestures may be expressing a simple thought that has been conveyed to her through the communicative process itself rather than by some external stimulation.

These two kinds of recruiting gestures are an extremely limited form of somewhat symbolic communication. The similarity remarked upon by Hölldobler and Wilson between the gestures used to recruit for fighting and the actual motions of fighting suggests that the communication is iconic rather than symbolic. That is, the signal resembles the action or thing that it represents. Thus, making a sound characteristic of a particular sort of animal as a symbol to represent that animal, such as "bow-wow" for dogs, is iconic, whereas the word *dog,* which has no particular resemblance to the animal in question, is not. On the other hand, the recruiting gestures for food sources do not seem to be iconic. It will be of great interest to learn more about these processes of communication when they can be studied with attention to their possible significance as indications of the thinking of the communicating animals.

Humphrey (1980) extended an earlier suggestion by Jolly (1966) that consciousness arose in primate evolution when societies developed to the stage at which it became crucially important for each member of the group to understand the feelings, intentions, and thoughts of others. When animals live in complex social groupings in which each one is critically dependent on cooperative interactions with others, they need to be "natural psychologists," as Humphrey put it. They need to have internal models of the behavior of their companions, and perhaps it is still more useful to feel with them and thus to think consciously about what the other one must be thinking or feeling. Although Humphrey restricted his criterion for consciousness to our own ancestors within the past few million years, it could apply with equal or even greater force to other animals, including the social insects, that live in mutually interdependent social groups.

The Symbolic Dances of Honeybees

The most significant example of versatile communication known in any animals other than our own species is the so-called dance language of honeybees. This type of communicative behavior is so strikingly different from all other known kinds of animal communication that it has been difficult for inclusive behaviorists to integrate it into their general understanding of animal behavior. For instance, the behavioral ecologist Krebs (1977) called these dances an "evolutionary freak" (792).

Honeybees are quite versatile, and perhaps we have underestimated their cognitive capabilities. Their learning ability rivals that of mammals, as reviewed by Gould and Towne (1988). Many of the complex aspects of learning studied by psychologists in rats and pigeons have also been found in honeybees, as discussed by Bitterman (1988), Shinoda and Bitterman (1987), Couvillon, Leiato, and Bitterman (1991), Lee and Bitterman (1991), Bitterman (1996), Couvillon, Arakaki and Bitterman (1997), and Srinivasan and Zhang (1998).

Beekeepers and students of bee behavior had noticed for centuries that worker honeybees sometimes move about over the surface of the honeycomb in agitated patterns called dances. It was also well known that once a single foraging worker has discovered a rich source of food, such as flowers that have just come into bloom, many other bees from the same colony often arrive a few minutes later, so rapidly that they could not all have found the food by individual searching. This suggested that some sort of recruiting communication occurred, but how this was achieved remained almost totally unknown until the work of Karl von Frisch, who has been mentioned in earlier chapters.

Von Frisch was a brilliant Austrian zoologist who carried out most of his research at the University of Munich, beginning about 1910. Quite early in his career he proved by elegantly simple experiments that bees were capable of discriminating colors. He was led to this discovery by the simple naturalist's belief that the striking colors of flowers must be perceptible to the insects that visit them to obtain nectar and pollen. Early in the twentieth century, prevailing scientific opinion was strongly negative about color vision in invertebrates. But von Frisch developed simple and ingenious experiments demonstrating conclusively that honeybees have excellent color vision.

In the 1920s, when studying the sensory capabilities of honeybees, von Frisch noticed that the agitated dances were carried out by workers that had visited rich sources of food at times when the colony was in severe need of food. In order to see what bees did on returning to the hive, he constructed specialized beehives with glass windows that allowed a clear view of the bees as they crawled over the honeycomb. To study foraging behavior he set out dishes containing concentrated sugar solutions, which bees visit and take up eagerly just as they take nectar from flowers. To identify individual bees he marked them with small daubs of paint on the dorsal surface while they were sucking up sugar solution. These bees performed what von Frisch called round dances, moving in circles with alternating clockwise and counterclockwise directions. At the same time he noticed that others returning with loads of pollen were carrying out a very different type of dance. He could tell that they had gathered pollen because honeybees carry substantial amounts of pollen grains packed between hairs on their legs.

These pollen gatherers performed what are called *Schwanzeltanzen* in German, customarily translated as waggle dances. In a waggle dance the bee walks rapidly in a straight line while moving her abdomen back and forth laterally at about thirteen or fourteen times per second. Then at the end of this straight wagging run she circles back and repeats the straight part of the dance, followed by alternating clockwise and counterclockwise returns to the starting point of a series of straight wagging runs. Although the bee may move a short distance between successive cycles of the waggle dance, the basic pattern is relatively constant under given conditions. Von Frisch thus concluded that the two types of dance were somehow related to the sort of food being brought back to the colony, a reasonable interpretation of the observations he was able to make at the time.

Only much later, during World War II, when his laboratory in Munich had been seriously damaged and he was studying bees at his country estate at Brunnwinkl in the Austrian Tyrol, did he have occasion to move the artificial feeding dishes a considerable distance from the observation hive. When he did this he discovered that the bees gathering sugar solution from more than about 100 meters performed waggle dances (von Frisch 1950, 67–72; 1967, 57–235). This had escaped his notice previously, because for reasons of simple convenience he had set out his dishes of sugar solution relatively close to the observation hive so that he or his assistants could remain in touch while marking bees at the feeder or observing them in the hive. It then became clear that bees returning from a considerable distance perform waggle dances whether they are bringing pollen or nectar.

It is very important to appreciate that these dances occur only under rather special conditions. They are part of an elaborate nexus of social communication that goes on almost continuously in a beehive, as described by Lindauer (1955, 1971) and Seeley (1985, 1986, 1989a, 1989b, 1995). The workers move about a great deal and interact with their sisters by feeling them with their antennae and being felt in return. There is a sort of mutual palpation in which the two bees face each other and feel each other's antennae and head region. At this time one of the bees often regurgitates a small portion of her stomach contents, which is taken up by the other.

This is similar to the trophallaxis of weaver ants and many other social insects; it serves to convey not only food material but also the odors that accompany it. In colonies of specialized social insects such as honeybees, the queen, the larvae, and the younger workers obtain their food in this way. Trophallaxis is so widespread that a given molecule of sugar ordinarily passes through several stomachs before it is finally regurgitated into one of the cells in the honeycomb. By this time the original nectar gathered from the flowers has been modified into honey. Pollen grains transported in specialized pollen baskets formed from stiff hairs on the legs are also transferred to other workers before being stored. During round and waggle dances other bees cluster around the dancer and follow her movements.

In spite of all this activity, worker honeybees seem to be doing nothing at all much of the time, but it is difficult to determine by simply watching them whether their leisurely moving about the hive is idle loafing or whether they are sampling odors and other conditions in ways that will later affect their behavior. The workers often take food from partially filled cells, into which other workers are still adding honey or pollen. Thus the food stores of a beehive are in a constant state of flux, with new material being added after foragers have brought it back, but also constantly being drained by workers, which obtain much of their food in this way.

When a forager returns with a stomach full of nectar to the hive that is in need of food, she ordinarily finds other workers ready, after a brief period of mutual palpation, to take her stomach load by trophallaxis. These other workers then store the somewhat modified nectar in partly filled cells, or they may transfer it to other workers before it is finally stored. This widespread process of mutual palpation and trophallaxis serves also as a sort of communication, because the ease or difficulty with which a returning forager can transfer her load provides information about the general situation in the colony.

This is particularly important when conditions are not optimal and when something is in short supply. The most common shortage is of carbohydrate food. The workers are informed of this by the relative emptiness of storage cells, and when conditions are truly severe, capped cells may be opened and the honey consumed. This results in an eagerness to receive regurgitated nectar during trophallaxis with returning foragers. When, on the other hand, nectar is abundant but pollen is in short supply, foragers returning with stomachs full of nectar have more difficulty finding a sister to whom they can transfer their load. Whether there is some chemical or other signal that conveys more than a reluctance to receive one type of material is not clear. But somehow foragers are induced to change what they seek outside the hive.

This distinction becomes all the more clear under special conditions of overheating. When the hive temperature rises above approximately 35°C, workers returning with either sugar or pollen have difficulty unloading. But workers that have gathered water regurgitate it in small droplets, and other workers fan vigorously with their wings, producing a circulation of air that cools the hive by evaporation. Under these conditions foragers shift from gathering nectar or pollen and visit places where they can take up water. It is not clear whether the high temperature, along with her difficulty in unloading whatever she was bringing in previously, stimulates the returning forager, or whether some other sort of information is transferred from other workers. The important point is that the older workers that fly outside the hive searching for things needed by the colony shift their searching behavior between different commodities according to the needs of the colony.

This network of social communication conveys to the older workers not only what is required by the colony but how badly it is needed. Thus when a forager leaves the hive, she has been induced to search for some particular thing, and this motivation clearly varies in intensity. The most common need is for carbohydrate food, so the nectar of flowers is the usual target of this searching. But sometimes it may be pollen grains or water. Under other special conditions the need may be for waxy materials used in building honeycomb. This need is relatively rare in agricultural beekeeping practice because beehives are equipped with frames built to provide an ideal foundation on which bees can build their cells. But under natural conditions honeybees and other bees must build their own wax foundation or plug holes in a natural cavity.

Dances do not occur at all when everything is going nicely and nothing is in short supply. Foragers return with nectar or pollen, these substances are transferred to other workers, and the net stores of both carbohydrate and protein food are either constant or slowly increasing. Under these favorable conditions it seems that workers can find adequate supplies of nectar and pollen by individual searching efforts. This point is often overlooked in elementary discussions of the bee dances, which occur only in the presence of an audience of other bees.

When something important is in short supply, older workers fly outside the hive and search for what is needed. When they have filled their stomachs from a rich source of nectar they return to the hive and engage in communicative dances. After walking a short distance in from the entrance of the hive, and usually after antennal contact with other workers, the returning forager begins to perform round dances or waggle dances.

The honeybee communication system is partly chemical, for the odors of flowers are conveyed along with the nectar or pollen. And when bees visit desirable things, they often mark their location with secretions from the Nasanov gland, which produces a long-lasting odor that attracts searching foragers. But in addition to the specific odors that are transmitted, the dances convey the direction and distance to the food by a sort of geometrical symbolism.

Round dances are performed when food has been discovered relatively close to the hive, and waggle dances are used for desirable things located at a greater distance. The transition from round to waggle dances is gradual, and it occurs at distances ranging from 2–3 m in the Indian species *Apis florea* to 50–100 m in the widely used Carniolan strain of European honeybees. In a typical round dance a bee circles alternately clockwise and counterclockwise, although occasionally more than one cycle may be executed in the same direction. When von Frisch originally discovered that round dances were used for food sources at relatively short distances and waggle dances for those farther from the hive, he concluded that the round dances contained no directional information and simply informed recruited workers to search in all directions close to the hive. He also found that bees stimulated by round dances arrived in approximately equal numbers at experimental feeders located in different directions within a few meters of the hive.

Many interested scientists have observed and made motion pictures and video recordings of round and waggle dances, but it was nearly forty years before anyone noticed that even at very short distances the point at which the circling reverses does contain directional information of the same kind conveyed by the waggle dances discussed below (Kirchner, Lindauer, and Michelsen 1988). Even in round dances there is a very brief lateral vibration of the bee's body just at the moment when she has completed one circle and is about to begin circling in the opposite direction. But there is no evidence that bees use this information in locating food sources that are close to the hive, where odors probably suffice. The fact that it took so long to appreciate this simple fact indicates how easily we overlook matters that do not readily fall into our preconceived patterns of expectation.

As the distance to the food increases, this lateral wagging motion lasts longer, and the bee walks in a straight line for a gradually increasing distance. This can be observed with the same bees by inducing them to gather concentrated sugar solution from an artificial feeder that is gradually moved away from the hive. When the distance is increased, beginning at a very few meters from the hive, there is a gradual transition from an almost instantaneous lateral vibration at the moment when the round dance is reversed in direction to an increasingly long straight wagging run. At distances of a kilometer or more the wagging run is ten or eleven millimeters in length.

The most significant of von Frisch's discoveries was that the direction of the straight wagging run communicates to other bees the direction that the dancer has flown from the hive to the source of food. When the food is in the direction of the sun, the waggle dances are oriented approximately straight upward on the vertical surface of the honeycomb. If the food is located in the opposite direction from the sun, the dances point straight down, and if it is 90° to the right of the sun, the dances point 90° to the right of vertical. In other words, the orientation of the wagging run relative to straight up approximates the angle between the direction toward the food and toward the sun. When the sun is hidden by clouds that cover only part of the sky the honeybee visual system is able to achieve the same orientation by utilizing patterns of polarization in the blue sky that are correlated with the sun's position.

These discoveries were possible only when von Frisch began to study waggle dances performed by bees that had returned from known food sources located at considerable distances in various directions. In his early experiments, there was no reason to move his artificial food sources more than a few meters from the hive. In the 1920s it would have required a truly superhuman level of enterprising imagination to suggest that an insect might indicate the direction to a food source by some form of communicative behavior. Even a brilliant scientist who had already challenged some of the established "nothing but" dogmas of his time did not make the leap of inference necessary to imagine that the waggle dances that he and many others had observed could conceivably communicate the direction of a source of food. This leads one to wonder what other versatile ingenuities of animals might be staring us in the face and waiting to be correctly interpreted.

This relation between the direction relative to the sun taken by a flying bee outside the hive and the direction of its communicative wagging run inside the pitch dark hive is more truly symbolic than any other known communication by nonhuman animals. The direction of the dance stands for the direction of flight out in the open air.

But this symbolic correlation does not mean that the directional communication is perfectly precise. The accuracy of this distance and direction information has been studied by von Frisch (1967) and by Towne and Gould (1988). They set out test feeders with the same odor as that associated with a rich source of concentrated sucrose and, after removing the original source of food about which bees had been dancing, measured how many recruits arrived at the test feeders. In both von Frisch's original experiments and these more carefully controlled tests, the majority of the recruits went to feeders in approximately the same direction and at approximately the same distance as the location indicated by the dances. There is, however, considerable variation, and the situation is complicated by the possibility that the odors marking the location of the original feeder or the test feeders may diffuse widely enough that bees are attracted to them even though they may not have flown very accurately to the distance and in the direction indicated by the dances they have followed. After carefully controlling for the complicating effects of odors, Towne and Gould concluded that the majority of the recruited bees came to feeders within ±15–20° and ±10–15 percent of the distance indicated by the dances.

One type of imprecision in the direction indicated by waggle dances may be advantageous to the bees. At gradually increasing distances, after the bees have changed from round dances to waggle dances, the straight runs do not at first point directly in the appropriate direction. Instead they alternate between being several degrees to the right and being several degrees to the left of the correct distance. As the distance increases this deviation diminishes, so that by several hundred meters each waggle run points in almost exactly the same direction. Towne and Gould suggested that this decreasing angle of difference between alternate waggle runs results in an indication of a roughly constant size of the goal. This in turn might correspond to the typical size of a patch of flowers.

Recently Weidenmüller and Seeley (1999) have found that dances reporting a source of sugar have a greater degree of variability than those reporting a cavity where a swarm of bees might establish a new colony, as discussed in detail below. Furthermore, the variation in direction of dances to cavities did not decrease as the distance changed from 200 to 400 m. Cavities where colonies are established are of course virtually point sources compared to patches of flowers.

Although the system is far from perfect, the waggle dances do communicate three types of information that are all important to the bees. These are (1) the direction toward the food, expressed relative to the position of the sun, (2) the distance to the food, which is correlated with the duration and perhaps length of the waggling run, and (3) the desirability of whatever the bee is dancing about. The detailed nature of distance communication has been difficult to determine. The number of waggling movements is correlated with the distance a bee must fly. But since the rate of waggling and the rate of forward movement are nearly constant, both the length of the waggling run and its duration are also closely correlated with distance. Although statistical analysis suggests that the duration is a better indication of distance, it is not possible from currently available data to be certain which property of the waggling run is actually perceived by other bees and used to determine the distance they will fly.

The vigor or intensity of waggle dances is easily recognized by experienced observers. Some dances seem clearly more energetic than others, and these ordinarily result when foragers have found a rich source of sugar solution or something else that is important to the bees. Because the concentration of sugar in an artificial feeder can be more easily manipulated by experimenters, it has been studied more thoroughly. Under relatively constant conditions, when carbohydrate food is scarce and dancing is actively under way, the dances look much more lively when the foragers have visited feeders with a high concentration of sugar. And dances by bees returning from rich sources continue for a long time, whereas those by bees returning from less concentrated sugar solutions usually last for only a few cycles. But Seeley and Towne (1992) found no difference between the effectiveness of individual waggle dances recruiting bees to artificial feeders supplying varying concentrations of sugar solution, so it was not clear whether bees are influenced by the vigor of the dancing.

Seeley and Buhrman (1999) recently discovered how dances differ in liveliness, vigor, and enthusiasm. To appreciate their experiments it is important to recall that a bout of dancing consists of repeated dance cycles. Each cycle consists of a waggle run having a particular direction and duration, followed by a return phase in which the bees circle back to roughly the same starting point. Rather than altering any observable property of the individual waggle run, the bees vary the duration of the return phase. When dancing about a more desirable food source or cavity they hurry, so to speak, in the return phase of their waggle dances, and thus execute more cycles in a given time. This, together with an increase in the number of dance cycles in a bout of dancing, is what varies with desirability and gives the subjective impression of greater liveliness.

Another important point about the waggle dances is that they serve to convey information to other bees inside the totally dark beehive when the subject of the communication is something entirely different from the immediate situation, namely, the direction and distance a bee should fly out in the open air. Thus the communication has the property of displacement; the bees communicate about something displaced in both time and space from the immediate situation where the communication takes place. To be sure, one can interpret the dances as communicating not the location of food but the distance and direction of a flight the bee has just completed. For example, Shettleworth (1998, 538) argues that "if the dance is seen as reporting on a just completed journey, it is no more displaced than an alarm call (of a vervet monkey) to a just glimpsed snake." But Lindauer (1954, 1955) observed that under some conditions bees perform waggle dances about locations they have not visited for several hours. Thus the dancing bee reports something that occurred some time in the past and in a totally different situation from the interior of the dark hive where the communication takes place.

Wenner (1959, 1962) and Esch (1961, 1964) discovered that faint sounds accompanied the waggle dances. These are not ordinarily loud enough to hear through the glass window of an observation hive, and studying them requires removing the glass. This is obviously somewhat hazardous, since several hundred bees are completely free to fly out. But if the glass is gently removed most of the bees stay on the honeycomb, and many detailed experiments have been carried out with observation hives that can be opened. In his first experiments Esch detected these sounds by holding a drinking straw with one end close to the dancing bee and the other at the entrance to his ear canal. Small microphones can record these acoustic signals quite accurately, and as discussed below it is now clear that they differ from ordinary sounds in some important ways. They are brief pulses with a fundamental frequency of about 250–280 Hz; each lasts only for a few waves, but the pulses are repeated as a sort of interrupted buzzing. The fundamental frequency is nearly the same as the wingbeat frequency when the bees are flying, but the wings move only a fraction of a millimeter.

Esch (1963) concluded that the desirability of food or other commodities is conveyed, at least in part, by the intensity or temporal pattern of these dance sounds, which seem to vary with the desirability of the food being gathered. But Wenner, Wells, and Rohlf (1967) did not find such a correlation, and it remains unclear whether the dance sounds convey information about food quality.

Interpretation of these observations was initially difficult because honeybees appeared to be deaf. Despite several attempts to do so, no one had been able to demonstrate any responses to airborne sounds. This suggested that the sounds heard by human observers might be an incidental by-product of a mechanical signal that was transmitted either by vibratory motions of the surface on which the bees were standing or in some other manner. This situation has been considerably clarified through acoustical experiments by Michelsen and his colleagues (Michelsen, Kirchner, and Lindauer 1986; Michelsen, Kirchner, Andersen, and Lindauer 1986; Michelsen, Towne, Kirchner, and Kryger 1987). The most important point clarified by these experiments is that the changes in air pressure that we detect as sounds, either by hearing them directly or via microphones, are only one physical aspect of the signals generated by dancing bees and not the aspect that is important to the bees themselves.

When a solid object oscillates against the air, whether it be a loudspeaker diaphragm or the wings of a dancing bee, air in the immediate vicinity moves back and forth at the frequency of the movement. Close to the vibrating object this motion of the air is the primary physical process, but magnitude of air movement falls off with distance very rapidly. The oscillating movements of the air also generate traveling sound waves, which are areas of very slight compression and rarefaction that spread outward at the speed of sound (approximately 344 m/sec in air). It is of course qualitatively true that in order to produce a region of higher pressure some air molecules must move into such a region, but the amount of air that moves back and forth in a traveling sound wave is very small compared to that which moves about close to the vibrating source. This difference leads to the physical distinction between near field and far field sounds. In the near field the air motion is large and in the far field it is very small, because its magnitude falls off rapidly with distance from the source.

The bees that are stimulated by a dancer are ordinarily within less than one body length, and this is definitely in the near field, because the wave length in air of 250 Hz far field sounds is 1.4 m. The sounds with which we are most familiar are far field pressure waves. Bees and most insects lack specialized auditory receptors for sound pressures, although a few specialized insects that respond to the orientation sounds of bats do have sense organs adapted for responding to sound pressure. What bees and many other insects have instead are very sensitive hairlike sense organs that respond well even to feeble air movements, whether these are unidirectional or oscillating at frequencies of a few hundred Hz. Thus it is not surprising that honeybees do not respond to far field sounds, but such insensitivity tells us nothing about their sensitivity to near field acoustic stimulation, which consists primarily of oscillatory air movements.

Michelsen, Towne, Kirchner, and Kryger (1987) measured the near field component of the dance sounds, using measuring devices located within a few millimeters of the dancer. In further experiments Towne (1985), Towne and Kirchner (1989) and Kirchner, Dreller, and Towne (1991) found that honeybees respond to the near field air movement component of sounds similar to the dance sounds. Using sound sources that generated either primarily near or far field sounds of either 265 Hz or 14 Hz, they were able to show clear responses to the near field signals but not to the far field sound pressure stimulation. The latter had been used in previous attempts to learn whether bees could hear.

Honeybees also transmit acoustic signals as vibrations of the substrate. Michelsen, Kirchner, Andersen, and Lindauer (1986) concluded that the dance sounds are transmitted only very feebly into the honeycomb. But Tautz (1996, 1997) and Sandeman, Tautz, and Lindauer (1996) have reported evidence that transmission through the substrate is more important in conveying the information in the waggle dances than was previously thought.

Another acoustical signal, transmitted through the substrate at about 320 Hz, is sometimes emitted by bees following a dancer; this signal causes a dancer or other bees to "freeze" by stopping whatever movements were under way and is therefore termed the "stop signal." It originally appeared to serve as a request for the dancer to stop and regurgitate food samples. But Niah (1993) found that these sounds were produced not only by dance followers but also by dancers and by bees performing "tremble dances"—trembling and shaking movements that may be oriented in any direction and are continued for many seconds. Furthermore, Niah found that stop signals only rarely elicit trophallaxis but do cause dancers to move away from the dance floor, the area on the honeycomb where most dancing takes place. Visscher et al. (1999) have analyzed these sorts of signals when performed on swarms and suggest that they are best termed vibration signals. They occur under a variety of conditions both in the hive and on swarms, and they are performed on swarms "by specialized subsets of bees, most of whom never perform recruitment dances to nest sites" (759). They seem to stimulate other bees, but their function is not clear.

When the honeycomb was set into vibration by artificial devices at about 320 Hz, dancing bees stopped dancing, as they do when this so-called begging signal is generated by other bees. Seeley (1992, 1995) observed that returning foragers that have not found another bee ready to take up the nectar they have gathered often become tremble dancers, and in the colony studied by Niah it was tremble dancers that produced most of the stop signals. Thus he concluded that the principal effect of stop signals is to cause dancers to stop dancing and thus to reduce the recruitment of new foragers. This may well be important when more nectar is being brought into the hive than can be received and processed effectively.

Another quite different type of acoustical signaling is carried out by honeybee queens. When larvae have been fed appropriately by workers they develop into queens, and often several queens are present in separate cells. At this time, it has long been known that the queens emit two kinds of sounds, called tooting and quacking. These are transmitted through the honeycomb substrate and are sensed by other queens.

The tooting is produced by a queen who has emerged from her cell, and the quacking signals come from queens still confined within cells. Usually the original queen, together with a very large number of worker bees, departs from the hive and forms a new colony. One of the queens remaining in the hive then cuts a hole in the side of the other queen cells and kills the occupant. Thus this exchange of acoustical signals is an important part of the social behavior of a honeybee colony around the time of swarming.

Michelsen, Kirchner, and Lindauer (1989) have succeeded in constructing a model honeybee that can transmit signals which direct recruits to search for food in particular directions. This exciting development is still at a very early stage, and all that can be said is that in a general sense the model does work. The reason it has worked better than previous attempts is probably that the near field acoustical signals have been reproduced more accurately. It will be of great interest to follow developments in this new technology for studying the symbolic dance communication of honeybees.

Skepticism about Dance Communication

It has been so astonishing to find insects communicating in such a versatile and symbolic fashion that some skeptics have remained unconvinced that the system really functions as von Frisch described it. As reviewed by Shettleworth (1998), Rosin (1978, 1980, 1984, 1988), Wenner (1989), and Wenner and Wells (1990) insist that von Frisch's discoveries are suspect because they imply that bees are "capable of human-like communication (language)" (Wenner 1989, 119), or because "a hypothesis which claims a human-level 'language' for an insect upsets the very foundation of behavior, and biology in general" (Rosin 1978, 589).

These and other skeptics do not deny that the pattern of the dances is correlated with the location of a food source, but they claim that recruited bees do not use this information and simply search for the odor that they have learned from the dancer is associated with the food. Newly recruited foragers take substantially longer on their first flights after following waggle dances than they need after several visits. This is probably because the dance information is not precise, and therefore new recruits reach only the general vicinity of the goal and then begin to search for the odor that has also been transmitted by the dancer they have followed. But relying on odors to find food at distances of a kilometer or more would seem highly ineffective, even taking into account the fact that odors may be carried downwind so that the bee might need to search only for the odor plume and then follow it upwind.

Recent experiments by Polakoff (1998) have shown that bees can use the directional information obtained by following waggle dances about food sources a kilometer from the hive regardless of wind direction.

The key question is whether newly recruited bees fly in the approximate direction and for the approximate distance conveyed by the waggle dances they have followed, and then search for the appropriate odor, or whether they search randomly. The practical difficulty of following where honeybees actually fly has so far made it impossible to trace the actual flight paths of new recruits. The development of harmonic radar, which requires only a negligible weight attached to an insect being tracked, might make this possible. Bees and even smaller insects have been followed by this new technique, and Capaldi et al. (2000) have followed honeybees for a few hundred meters, but harmonic radar has not yet been used to trace the flight paths of honeybees making their first flights after following waggle dances.

In some ingenious experiments by Gould (1976), bees were induced to point their dances in a direction different from the actual direction from which they had returned. Test feeders were set up in the form of traps that allowed counting of the number of recruits arriving at different places but prevented them from leaving and possibly introducing complications. The results were that most of the recruited workers flew to test feeders in the direction indicated by the dance rather than the direction from which the dancer had returned. The more recent experiments with an effective model bee, which certainly did not convey location specific odors, abundantly confirm that information about distance and direction can be conveyed from the dancer to her sisters.

In general terms, it seems that the directional indication is accurate within ± 20–30° and the distance indication perhaps ± 10–15 percent. Although finding the exact location requires locating the odors of flowers conveyed at the time of the dancing, the symbolism of communicating direction and distance is very significant. As with many scientific controversies, the short answer to the question whether bees find goals by searching for odors or by using information obtained by following dances is *both.* Odors are used to find food sources near the hive or when close to a distant goal, but the symbolic dances are used to reach the general vicinity of distant goals.

Communication about House Hunting

One misunderstanding of the dance communication of honeybees that is very widespread is the belief that it is rigidly linked to food. As mentioned above, the dances are also used to communicate about water needed to cool the colony by evaporation, but water can be viewed simply as a very dilute sugar solution. Waxy materials are sometimes collected and their location indicated by waggle dances. But the most enterprising use of the dance communication system occurs when bees are swarming. These dances were discovered long ago by Lindauer (1955), but for many years ethologists other than Seeley and his colleagues devoted very little attention to them despite their implications concerning cognition. Recently, however, several important analyses of honeybee communication have been reported both by Seeley and his colleagues, as discussed below, and also by de Vries and Biesmeijer (1998), Anderson and Ratnicks (1999), Camazine et al. (1999), Fewell and Bertram (1999), and Visscher and Camazine (1999a, 1999b).

Swarming occurs when a colony of honeybees increases to the point that the hive is crowded. Workers then feed some larvae a different sort of food, which causes them to develop into queens. Under ordinary conditions the bees also prepare to swarm, and part of the behavioral changes that accompany this sort of preparation is a change in the searching images of the older workers that have previously been gathering food. They now begin to investigate cavities. As new queens develop, the older queen stops laying eggs and usually moves out of the hive along with a large portion of the workers. Initially these aggregate in a ball of bees clinging to the surface of the hive or to vegetation. In normal beekeeping practice the beekeeper either enlarges the hive at the first sign of swarming so that the colony can grow further or else he provides a new hive immediately below the swarm. A beehive is an ideal cavity, and the bees usually move directly into it.

Many colonies of bees flourish away from carefully tended apiaries, and when they swarm no beekeeper provides an ideal cavity in the immediate vicinity. Under these conditions many of the older workers, rather than searching for flowers or other sources of food, begin to search for cavities. Often they must search over a very large area, crawling into innumerable crevices in trees, rocks, or buildings. Their central nervous systems must recognize a search image for an appropriate sort of cavity. A cavity is of course something totally different from food, and the workers that now search for cavities have never in their lives done anything of the kind. Swarming ordinarily occurs at intervals of many months, and workers live only a few weeks during the warmer months, when they are active. Although the queen may have participated in swarming many months before, the workers have never experienced anything remotely like the movement out of the old hive and the aggregation of thousands of bees in the open.

It is not easy to find an appropriate cavity. It must be of roughly the right size and have only a small entrance near the bottom. It must be dry and free of ants or other insects. Seeley (1977, 1995) has studied how the scout bees investigate cavities. He established colonies on a small island where no suitable cavities were present and induced swarming by the simple procedure of shaking the queen and numerous workers out of the old hive and leaving them to their own devices in the open air. He then provided experimental cavities of different types at some distance. The workers found these and eventually induced the colony to occupy one of them. In their preliminary visits these workers crawled back and forth through most of the interior of the cavities and spent considerable time investigating them.

When Lindauer studied swarming bees he observed that workers carried out waggle dances on the surface of the swarm. The waggle dances executed on the swarm indicate the distance, direction, and desirability of the cavity that the dancer has visited. This means that the dance communication system, with all its symbolism, is employed in this totally unprecedented situation. The same code indicates the location and quality of something as different from food or water as one can imagine. Worker honeybees that have been gathering nectar from flowers during the past few days, and that may even continue to do so to provide food for the swarm, utilize the totally different search image of a dry, dark cavity of appropriate size to guide both their searches for such cavities and their communication about one that they have visited. If we accept specialized communicative behavior as suggestive evidence of thinking on the part of the communicating animal, we may infer that these worker bees think about either food sources or cavities, according to the needs they have perceived at the time.

In his classic experiments during the 1950s, Lindauer discovered that the waggle dances executed on a swarm lead to a group decision on the part of the colony about where they should move to establish a new colony. Ordinarily dozens of scout bees locate several different cavities and then dance on the swarm about their locations. Furthermore, the intensity of the dances is correlated with the quality of the cavity. Small, damp, or ant-infested cavities produce only a few feeble dances, whereas others that are dark, dry, and of suitable size lead to prolonged and vigorous waggle dances. Since different scouts have visited different cavities, a wide variety of locations are described by the numerous dances that go on over the surface of a swarm.

Lindauer spent many hours in laborious observation of the dances on the surface of numerous swarms, climbing ladders or doing whatever was necessary to reach a suitable observation point. He found that although a wide variety of locations were signaled in the first few hours after a swarm emerged from the original hive, the dances gradually came to represent a progressively smaller number of locations. As time passed, fewer and fewer cavities were described by dances but more and more bees performed these dances. In other words, the cumulative effect of extensive dance communication was a progressive reduction in the number of cavities described. And those that continued to be danced about were the best ones available.

In further experiments Lindauer varied the suitability of particular cavities. If their quality was lowered, the dances became less enthusiastic, and in some situations at least, dances about other cavities became more numerous. This process of repeated dancing about the more desirable cavities continued for a few days, and finally almost all dances were about a single cavity. After this had been going on for several hours, a different sort of behavior occurred, which Lindauer describes as a "buzzing run." The bees making these buzzing runs moved for fairly long distances over the swarm while emitting a buzzing sound. When this had been going on for some time the swarm took wing and flew fairly directly to the cavity that had been described by the concentration of enthusiastic dances over the past day or so.

These dances on the swarm lead to a sort of consensus whereby the colony selects out of many possible cavities the one that has been judged by the scout bees to be the best. There seem to be adaptive advantages to prolonging this process of evaluation, because cavities may change their desirability as conditions change. For example, one that has been dry in good weather may be damp on a rainy day. Thus it seems that the bees do not reach this crucial decision until dancers have been, so to speak, singing the praises of a particular cavity for a considerable period of time. Many factors probably play a role in the evaluation of cavities by individual scout bees. In addition to size, dryness, and a dark interior with a small opening, distance from the original colony is important. Other factors being equal, the bees prefer a cavity a few hundred meters from the old colony. This presumably has the advantage of avoiding competition for food sources with the thousands of bees that remain in the original cavity and continue to search vigorously for food.

Lindauer also observed another feature of the communication on swarms that has significant implications for a cognitive interpretation of the communicative behavior. In the first day or two, after the swarm had emerged and a number of different cavities were being described by various returning scouts, one might suppose that the less desirable cavities dropped out of the communication process because scouts that had visited them simply stopped dancing, while those returning from the better cavities continued. Yet the individual bees return repeatedly to the cavities after a bout of dancing, so that some individuals continue to describe the cavities they have located over many hours or even a few days.

By marking individual dancers, Lindauer found that something even more interesting was apparently going on. Bees that had visited a cavity of mediocre quality sometimes became followers of dances that were more enthusiastic than their own. Then some of them visited the better cavity, returned, and danced appropriately with respect to the superior cavity they had now visited. Lindauer was only able to observe this in a handful of cases, and it is not clear how large a role this change of reference of the dances of individual bees plays in the whole process of reaching a group decision.

The fact that any bees change from dancing about one cavity to another, after switching roles from dancer to follower, means that this type of communication is not rigid or linked tightly to the stimulation received during visits to the first cavity a particular individual has located. It seems reasonable to infer that under these conditions bees are thinking about cavities and are able to change their "allegiance" from one they have discovered themselves to a better one they have learned about as recipients of symbolic information from the dances of one of their sisters.

Seeley and Buhrman (1999) have repeated and confirmed Lindauer's basic observations with improved methods. They marked all the 3,252, 2,357, and 3,649 worker bees in three swarms and made video recordings of all dances performed during the two or three days before the swarm flew off. The scout bees found and danced about cavities in many different directions and at a wide range of distances up to 3 km. During the process of reaching a near-consensus some bees tended to dance less after some time and then to stop dancing, but others changed the site indicated by their dances. Within about an hour, after almost all dances indicated the same cavity, there was "a crescendo of dancing" and the swarm flew off together.

It is convenient to call the site indicated by the nearly unanimous dances before the swarm flew off the chosen site and others nonchosen sites. Of the 113 bees that danced initially about a nonchosen site, 34 (30 percent) switched to dances indicating the chosen site by the end of the dancing. The remainder had stopped dancing by the time the swarm flew off, while a few switched to other nonchosen sites. But none continued dancing about a nonchosen site until the swarm lifted off. In contrast, 94 (42 percent) of the 156 bees that danced initially about the chosen site continued to dance about it until the swarm flew off. The rest stopped dancing, but none switched to a nonchosen site.

Because Seeley and Buhrman were unable to monitor the cavities visited by these bees, they could not tell whether the bees that switched their dancing from a nonchosen to a chosen site visited the latter before changing their allegiance. It was also not feasible to record what dances these bees had followed, so we do not know the extent to which dance following may have influenced these changes in the dances. An important question is whether, as Lindauer reported, bees that switch from nonchosen to chosen sites have followed dances about the chosen site and visited it before changing their dances. Or do they ever change their dances without visiting the new site, in a form of chain communication comparable to that described for weaver ants earlier in this chapter? During the process of reaching a near consensus most bees tended to dance less after some time and then to stop dancing, but some changed the site indicated by their dances. But the fact that in Seeley and Buhrman's experiments none of the 62 bees that ceased dancing about the chosen site switched their dances to a nonchosen site strongly indicates that dances about the chosen site were more influential than those about less desirable cavities.

Camazine et al. (1999) and Visscher and Camazine (1999a, 1999b) have recently analyzed in great detail dances on swarms similar to those studied by Seeley and Buhrman. They have confirmed Lindauer's observation that a few bees first dance about one cavity, then follow dances describing a different site, and later dance about the latter. They recorded dances on swarms placed in desert terrain where there were virtually no natural cavities. Two empty but previously occupied beehives were located in opposite directions, providing ideally attractive nest sites. On one swarm, 41 of 46 bees that danced also followed other dancers during the last 30 minutes before the swarm took off for the chosen site. Nineteen of these 41 followed dances for the site about which they had danced themselves, 6 followed only dances for the other site, and 16 danced about both.

On a second swarm the dances and other behavior of numerous marked bees were recorded both on the swarm and as they entered and left the two beehives. Individual bees visited and danced about both nest sites, and the sequential patterns of dancing and site-visiting varied considerably, with almost every possible sequence followed by at least one bee. Although only a small fraction of the marked bees visited both sites, many followed dances about both. Several interspersed dancing, dance-following, and site-visiting in a manner that provided considerable opportunity for comparison of dance messages, the sites themselves, or both.

It is convenient to summarize the behavior of these bees using the following abbreviations: DN (or DS) means that the bee danced about the north (or south) site; FN (FS) means that the bee followed such dances; and N (S) means that the bee was observed at the north (south) site. Numbers show consecutive minutes during which a given behavior predominated. One bee behaved as follows: FS, 11DS, FN, 22DS, 2FS, 6DS, FS, DS, 14DN, 2DS, 3FN, FS, FN, 8S, 3FN, 3N, 2FS, FN, FS, 6FN, 2DN, FN, 3DN, and the swarm took off to the north site. At intervening times this bee was often away from the swarm, but it was not possible to tell where she went. Although most bees were not observed to vary their communicative behavior in such a complicated pattern, this example shows that bees dancing on a swarm are not necessarily limited to a rigid sequence of behavior. Dancers often become followers and vice versa, with or without visiting the sites whose location and desirability are symbolized. These exchanges of dance communication resemble conversational exchanges, shifting back and forth between messages about two different goals (one colleague likened them to committee meetings). The increase in dancing about the chosen site appears to result partly from a sort of chain communication comparable to that observed by Hölldobler and Wilson (1978) in weaver ants.

Camazine et al. concluded that "the choice among nest sites relies less on direct comparison of nest sites, and more on inherent processes of positive feedback and attrition by dancers dropping out" (348). By "positive feedback" they mean that increasing numbers of dances for a site result in more and more bees dancing about it, often without visiting the site themselves. But many of these bees followed dances about both sites, and therefore they may well have been comparing the dance messages about them even when the did not visit the actual sites. In this experiment there were only two attractive nest sites, but from normal swarms scouts visit and dance about a dozen or more cavities. It is unlikely that each bee visits all of the sites about which she follows dances. Ideally we would like to have similar data about both dance-following and cavity-visiting by marked bees from swarms whose scouts are reporting numerous cavities, but gathering such data would be extremely difficult.

Even though attrition may be quantitatively more important in the process of decision-reaching, individual honeybees can, so to speak, change their minds as a result of exchanging information through the communicative dances. Thus the dance messages can be based on communicative exchanges rather than being entirely dependent on external stimulation.

Another recent discovery about bee communication stems from the experiments of Niah and Roubik (1995) and Niah (1998) with one of the neotropical stingless bees, *Melipona panamica*. The social foraging behavior of stingless bees has been reviewed by Biesmeijer and Ermers

(1999). These bees recruit nestmates to newly discovered food sources but do not perform waggle dances that indicate distance and direction. In some stingless bees, recruits follow successful foragers as they return to a food source, and scent marks may be used to indicate the route, as described by Kerr (1969, 1994). They also transfer odors of the flowers from which they have gathered nectar and pollen and mark good sources with glandular secretions. Niah has recently found that *Melipona panamica* can recruit more nestmates to a food source near the ground or to one about 40 meters up a tower in the rain forest of Barro Colorado Island in the Panama Canal, depending on which location the recruiting bee has herself visited. These bees also emit acoustic signals inside their hives when recruiting nestmates, but it has not yet been possible to determine conclusively whether these are the means by which they convey information about vertical locations.

From all these observations and experiments we can certainly conclude that some ants and bees, but especially the well-studied honeybees, communicate to nestmates, who are sisters or half-sisters, information about not only food sources but other things that are vitally important to the colony. These include intruders to be combated, in the case of weaver ants, and cavities that can provide a new home for swarms of honeybees. Although we can only speculate about what, if anything, the dancing bees and their sisters who follow the dances on swarms are thinking, their vigorous communication suggests that they are thinking about a suitable cavity, perhaps similar to the one from which they have recently emerged.

The principal basis for our inferences about subjective, conscious thoughts and feelings in humans is the communicative behavior of our companions. And here we find that certain insects also communicate simple but symbolic information about matters that are of crucial importance in their lives, and they even reach major group decisions on the basis of such communicative behavior. As I have suggested throughout this book, it seems both logical and reasonable to apply the same procedure that we use with our human companions and infer that the weaver ants and honeybees are consciously thinking and feeling something approximating the information they are communicating. Only by assuming an absolute human-animal dichotomy does it make scientific sense to reject this type of inference.

One significant reaction to von Frisch's discoveries was that of Carl Jung (1973). Late in his life he wrote that, although he had believed insects were merely reflex automata, "this view has recently been challenged by the researches of Karl von Frisch; . . . bees not only tell their comrades, by means of a peculiar sort of dance, that they have found a feeding place, but they also indicate its direction and distance, thus enabling beginners to fly to it directly. This kind of message is no different in principle from information conveyed by a human being. In the latter case we would certainly regard such behavior as a conscious and intentional act and can hardly imagine how anyone could prove in a court of law that it had taken place unconsciously. . . . We are . . . faced with the fact that the ganglionic system apparently achieves exactly the same result as our cerebral cortex. Nor is there any proof that bees are unconscious" (94).

CHAPTER ELEVEN

Deception and Manipulation

In the previous chapters animal communication has been considered as a straightforward transmission of information from one animal to another, and it has been tacitly assumed that this information is reasonably accurate. That is, when one animal, the sender, emits a signal, this signal is taken as some indication of its intentions, or at least of its disposition to behave in a certain way, that can be reliably interpreted by one or more other animals, the receivers. (The terms *actor* and *responder* are sometimes used instead of *sender* and *receiver*, but the latter are preferable because they do not imply anything about behavior of the receiver, which may not react at all.) I have emphasized the possibility that the production of communicative signals may be consciously intentional, that the sender may want to transmit information to the receiver because it wishes to affect the receiver's behavior in some way. Inclusive behaviorists have avoided any such inferences, preferring to concentrate attention on the functional effects of the communication, especially its contribution to evolutionary fitness, rather than on any mental experiences of either sender or receiver.

In an influential paper, R. Dawkins and J. R. Krebs (1978) have argued against what they call the classical ethological analysis of animal communication, which emphasizes cooperation between individual animals facilitated by transmission of accurate information about the sender's dispositions to behave in particular ways. Instead they emphasize the competition between individuals, of both the same and different species. This emphasis stems from a strong preference for the mechanomorphic view of animals as machines designed to preserve and propagate their genes, as argued forcefully by R. Dawkins (1976). Dawkins and Krebs interpret communicative signals as means to manipulate others, rather than to inform them. From this perspective it seems likely that the information transmitted by communicative signals is often inaccurate and serves to misinform the receiving animal. Dawkins and

Krebs recognize that animal communication requires a transfer of information, in contrast to the claims of Owings and Morton discussed in chapter 2, but both viewpoints emphasize manipulation of the recipient.

As pointed out by Hinde (1981), the difference between the two approaches is not as fundamental as Dawkins and Krebs argued. Ethologists studying animal communication have almost always considered that communicative behavior is adaptive, that it has resulted from natural selection, and that it often enables one animal to alter the behavior of another to the former's advantage. The difference is one of emphasis: Dawkins and Krebs conclude that most animal communication has been selected for its effectiveness in manipulating others for the sender's benefit, whereas many other ethologists recognize that mutual benefits to both sender and receiver are widespread, as discussed by Smith (1986). For, in the long run, "cheating" by transmitting inaccurate information works only if most signaling is reasonably accurate. Otherwise, receivers will tend to ignore or even perhaps "see through" the inaccurate, manipulative signaling, as reviewed by Wiley (1983). The evolutionary significance of honest and dishonest signaling has been extensively discussed by ethologists, as reviewed by Guilford and M. S. Dawkins (1991).

Many ethologists and behavioral ecologists have found the approach espoused by Dawkins and Krebs a congenial one. One reason is that it tends to appear consistent with the "selfish gene" approach to animal behavior, but another reason may be that it seems to avoid the need to suppose that communicating animals might consciously "mean what they say." This viewpoint has been advanced as a reason to reject the suggestion that animal communication could serve as an effective source of data about animal thoughts or feelings. The fact that animal signals sometimes convey inaccurate information detrimental to the receiver but beneficial to the sender complicates our interpretation, since we must consider that the communication may be either accurate, inaccurate, or some mixture of the two.

In either case, however, the signals are expressions of something going on within the sender's nervous system, and they may convey the sender's feelings or thoughts whether these are cooperative or competitive, honest or devious. In fact, deceptive communication may be more likely to require conscious thinking than honest expression of what an animal feels, desires, or believes. Krebs and Dawkins (1984) later modified and elaborated their analysis by adding "mind-reading" to their evolutionary analysis of animal communication, as discussed in chapter 1.

The present chapter will review several cases in which animal communication does not appear to convey accurate information and in which the flexible versatility displayed suggests conscious thinking on the part of sender, receiver, or both. This subject has been reviewed in a symposium edited by Mitchell and Thompson (1986), and I will draw on several specific examples of deceptive behavior described in that volume.

Some discussions of deceptive behavior include in this category morphological mimicry, in which the colors or shapes of animals resemble dangerous or inedible objects, as well as cases in which their behavior conveys inaccurate information. Clearly the former type is of little or no interest in relation to the possibility of conscious deception. Most animals have no choice about their body coloring, although in a few cases, such as squids, chameleons, and many groups of fish, chromatophores in the skin are under nervous control and the animal changes its color pattern. Sometimes this serves to make it inconspicuous by matching the background, as discussed by Winkler (1968). Cryptically colored moths tend to land selectively on surfaces where they will be less conspicuous (Kettlewell 1955, Kettlewell and Conn 1977, Sargent 1981, Partridge 1978), but it is difficult to tell from available evidence how much versatility is required in making these selections of landing places.

In many other cases the changes in color patterns are a major part of social communication. For example, Moynihan and Rodanichi (1982) and Hanlon and Messinger (1996) describe how the social interactions of squids entail elaborate changes in their body color patterns. This behavior seems not to have been studied thoroughly enough, however, to provide much indication of the degree to which it might entail conscious thinking.

Some animals make definite efforts to display patterns that look like dangerous objects. Although they have no control over the presence of these patterns, they do expose them on appropriate occasions. For example, many species of butterflies and moths have eye spots on the dorsal surfaces of their hind wings that look remarkably like two eyes. These are not visible when the wing is folded, but are suddenly exposed by spreading the wings when the moth is attacked. Experiments by Blest (1957) showed that these startling displays often scare off attacking birds. When the scales forming the pattern on a moth's wings were removed, it spread its wings normally, but the display had a much-reduced effect. Moths without eyespots elicited only about one-quarter as many escape responses from birds as did normal moths. Blest was also able to show similar effects when artificial eye-like patterns were presented to birds along with mealworms. But we know too little to judge how flexible and versatile these antipredator displays of moths actually are.

Firefly Communication and Deception

The extensive investigations of Lloyd (1986) have revealed that these luminescent beetles engage in fantastically complex social communication mediated by temporal patterns of their self-generated flashes. Numerous species of fireflies are very similar morphologically, so much so that entomologists can often identify them more readily by their flashing patterns than from examination of captured specimens. Of the 130 species Lloyd has studied under natural conditions, the most intriguing patterns have been analyzed in the American genera *Photinus* and *Photuris.* Male *Photinus pyralis* typically begin to search for females by emitting a half-second flash about every six seconds while flying near bushes where females are likely to be found. A responsive female waits about two seconds and then emits her own half-second flash. The male turns toward her, and they exchange similar flashes until he reaches her position, where mating often takes place. Females usually find a mate within a few minutes and then return to their burrows to lay eggs. Males, on the other hand, may search for several nights before finding a responsive female of their own species.

Life is complex and hazardous for male *Photinus pyralis* fireflies. Their flashes are answered only about half the time by females of their own species. At other times very similar answering flashes come from females of the larger genus *Photuris,* which are predatory and may catch and eat males that come too close. Yet the "femmes fatales" actually capture only about 10–15 percent of the males that approach them, suggesting that at close range the male *Photinus pyralis* can detect differences between the flashes of conspecific females and those of the predatory *Photuris.* It is also possible that other information such as species-specific odors might play a role in these close encounters. Sometimes the predatory females also fly aggressively toward potential prey and capture some in this way.

This sounds complicated enough, but the actual situation is much more intricately hazardous for the males. Several other species of fireflies are often present, flashing in somewhat different patterns. And even one well-studied species such as *Photinus pyralis* changes its flashing drastically under some conditions, such as when aggressive mimics are active. Sometimes several males approach a female but land far enough away that escape is possible if she turns out to be a femme fatale. They seem to be probing with their light flashes to determine from her responses whether she is a conspecific female or a dangerous aggressive mimic. In this situation Lloyd found some indication that males may mimic flashing signals of predatory fireflies, as though to deter rivals long enough to reach a responsive female themselves.

Unfortunately, we do not know enough about the histories and experiences of individual fireflies to judge how versatile all this communicative interaction actually is. They are not readily raised in captivity and studied under controlled conditions, so it is difficult to determine whether their behavior varies in response to changing circumstances or whether each firefly has a relatively constant pattern but such patterns vary widely among individuals of the same species. There is certainly great variability and complexity, but not enough is yet known to distinguish random variation from systematic and possibly rational, even Machiavellian, competition. It has been traditional to assume that insects behave in rigid, stereotyped ways that are genetically predetermined. But recent advances in our understanding of insect social behavior have provided abundant evidence that this view is a serious oversimplification.

Mantis Shrimps

Mantis shrimps of the genus *Gonodactylus* are predatory marine crustaceans equipped with appendages that serve, in different species, as clubs or spears. They occupy cavities and burrows in coral reefs, where they are protected from predators. They use the cavities as ambushes from which to seize prey and for mating and egg-rearing. Cavities are essential for survival, and mantis shrimps often fight over them (Caldwell and Dingle 1975). Larger animals can usually evict smaller ones from cavities, but much of this fighting is ritualized posturing and gesturing. Although they can seriously damage each other, they seldom do so (Dingle and Caldwell 1969). Experiments by Caldwell (1979, 1984, 1986) with adult *G. festai* 40 to 47 mm in length demonstrated that they can recognize other individuals of their species by chemical cues. They avoid empty cavities with the individual odor of a mantis shrimp that has defeated them in a previous encounter, but they enter cavities containing the odor of one they have bested. Many encounters involve a sort of bluffing, and as Caldwell puts it, they are able both to recognize other individuals and to remember their "reputations" as fighters.

It used to be taken for granted that individual recognition was impossible for invertebrate animals, but this relatively complex behavior is typical of the surprises that have resulted from detailed ethological investigations. Yet Caldwell is also typical of inclusive behaviorists in closing his review of these ingenious and revealing experiments with the following disclaimer:

The use of reputation and bluff in stomatopods should not be viewed as conscious acts. Rather, they are the product of natural selection operating on probabilities of performance and response. The selective equation has balanced over many generations the costs and benefits of generating, as well as accepting or discounting, signals that correlate with the probable outcome of a contest. The resulting product is further tuned by experience and by the degree to which information so derived is available and accurate. I hope that by demonstrating the occurrence of such supposedly complex mechanisms as reputation and bluff in relatively simple animals such as stomatopods, I can suggest the existence of similar processes producing "deceptive" interactions in more sophisticated animals whose sensory and integrative capacities make objective analysis much more difficult. (1986, 143)

This is a prime example of the strong inhibiting effect of inclusive behaviorism. Adaptiveness is a completely separate matter from the possibility of conscious thinking. Because mantis shrimps are crustaceans a few centimeters in length, it is assumed in advance that they cannot possibly be conscious. Even when they are shown to engage in moderately complex and versatile behavior, these data are forced into the procrustean bed of mindless automatism. Yet mantis shrimps have well-organized central nervous systems, and although their neurophysiology has not been studied in detail, it is unlikely to be very different from that of crayfish, which have all the basic mechanisms of synaptic interactions that are found in the central nervous systems of all complex animals (Bullock and Horridge 1965, 1187–89; Schram 1986).

The central nervous system of a mantis shrimp is larger than that of a bee, and it is well known that honeybees and other social insects display a variety of versatile behavior patterns, including symbolic communication, as discussed in chapter 10. It is quite reasonable to speculate that mantis shrimps may experience very simple conscious feelings or thoughts about the fights by which they gain or lose the cavities that are so important to them and the antagonists whose odors they learn to recognize from previous encounters. Perhaps on sensing the odor of a larger shrimp that has defeated them previously they do no more than feel fearful and react accordingly. But we have no firm basis for dogmatically denying any subjective experience at all when behavioral evidence suggests versatile adaptation of behavior to challenges that are important to the animal.

Deceptive Alarm Calls

Many animals have calls that signal danger from predators, and these sometimes differ according to the type of predator, as in the case of the vervet monkeys and domestic chickens discussed in chapter 9. The responses of animals that hear alarm calls are usually to flee from the immediate area, so the possibility exists that a sender might use alarm calls to frighten others away from a place the sender would like to have to itself. Because it is vitally important to escape predators announced by alarm calls, such escape responses are likely to be resistant to habituation, as Cheney and Seyfarth found to be the case in experiments described in chapter 9.

Other cases of this type have been observed in mammals, birds, and even ants, as reviewed by Munn (1986a, 1986b). Most of these instances have not been studied in sufficient detail to tell us how consistently false alarms are sounded or to provide very helpful hints as to the likelihood that this behavior is carried out with conscious intent. Moller (1988) has demonstrated that great tits *(Parus major)* gave false alarm calls when no danger was threatening, and that this served to disperse sparrows that were monopolizing a concentrated food source. When food was scarce and the tits had to feed at a high rate in winter, they also gave false alarm calls to frighten other birds of the same species away from limited and desirable food sources.

A striking instance of what appears to be complex deceptive behavior by a hummingbird has been reported by Kamil (1988, 257–58). A male Anna's hummingbird named Spot because of a white spot on his face had detected a mist net in his territory because a heavy dew had made the black threads conspicuous.

> Spot saw (the net) immediately. He had flown along it, and even perched on it. Experience had taught us that once a hummingbird has done this, it will never fly into the net. . . . Suddenly an intruding hummingbird flies into the territory . . . and begins to feed. Male Anna's hummingbirds are extraordinarily aggressive animals. Usually they will utter their squeaky territorial song and fly directly at the intruder, chasing it out of the territory. But that is not what Spot does. He silently drops from his perch and flies around the perimeter of the territory, staying close to the ground, until he is behind the other bird. Then he gives his song and chases the intruder—directly into the mist net.

Although Yoerg and Kamil (1991) argue against inferring conscious intention in animals, Spot's behavior is certainly suggestive of intentional planning.

In the course of detailed studies of mixed flocks of neotropical forest birds, Munn (1986a, 1986b) has documented a type of false alarm calling that is highly suggestive of intentional deception. He studied two groups of birds in the Amazon basin of Peru that live in permanent mixed-species flocks. One group is mostly found in the forest canopy, 15–45 m above the ground, the other in the understory below 15 m.

Although several species sometimes join these flocks, mated pairs of 4 to 10 species form the core of the flock and remain with it most of the time. The understory and canopy flocks may even join together at times, and occasionally as many as 60–70 species may be assembled in a single flock.

In the understory flocks one species, the bluish-slate antshrikes *(Thamnomanes schistogynus)* tend to lead the others when the flock moves for distances of more than about 20 m. They give loud calls that seem to help maintain flock cohesion. In the canopy flocks the white-winged shrike-tanagers *(Lanio versicolor)* play the same lead species role. These two species spend more time perched than the others and obtain most of their food by flying out to snatch insects off leaves and from the air; the others spend more time searching for and flushing insects from the forest vegetation. The lead species in both canopy and understory flocks also serve as flock sentinels; they are almost always the first or even the only members of the flock to give loud alarm calls when they see one of the several hawks of the genera *Micraster*, *Accipiter*, and *Leucopternis* that are serious predators on these insectivorous birds.

On hearing these alarm calls, the other members of the flock look up, freeze, dive downward, or move into thick vegetation. This division of labor in mixed-species flocks seems to be mutually advantageous. The sentinel-leaders obtain at least 85 percent of their food from insects flushed by other species, and in return they provide timely warning of danger from hawks so that the other birds can spend more time searching for edible insects. Typically the sentinel-leaders are located near the center or beneath a group of actively foraging members of the flock, and when an insect is flushed by one of the latter, they fly out and downward in hot pursuit of the quarry. Since they are faster and more acrobatic fliers, they often catch the insect first.

In the course of the multibird tumbling and competitive chases of insect food, the sentinel-leaders often give hawk alarm calls very similar to those elicited by real hawks. Playbacks of alarm calls recorded either on sighting a real hawk or in one of these competitive insect chases clearly startle other flock members. The sentinel-leaders give alarm calls significantly less often when they are the only bird chasing an insect than when others are also pursuing it, indicating that the deceptive alarm calling is employed primarily when it is helpful in competition for food, and not whenever insects are being chased.

Munn (1986a, 174) concludes by interpreting his observations in terms of possible thinking by the birds involved:

Certain facts suggest that some amount of thinking is involved both in sending and receiving the alarm call. That the sender thinks about what its call implies is suggested by one occasion in which a *Thamnomanes schistogynus* began to give the false alarm call as it flew out after a falling insect that was being chased by another bird, but once it became clear that the other bird had captured the insect, the calling antshrike immediately graded its call into a wider-frequency nonalarm rattle call, which functions like a rallying call for other birds. The bird apparently realized that the alarm call was no longer appropriate and switched to the nonalarm call in mid-vocalization.

In addition, the fact that both sentinel species use the false alarm calls more frequently when feeding fledglings might suggest that they are "saving" this trick for a situation in which they are genuinely desperate for extra food. The behavior of receivers suggests that they recognize that one potential meaning of the alarm calls is the approach of a predator. These birds are not simply startled by an alarm call—rather, often they look in the direction of the call. This reaction is especially obvious when birds already in thick cover jerk their heads quickly and look in the direction of an alarm. This looking implies that alarm calls are interpreted as meaning something more like "hawk!" than like "jump!"

Inclusive behaviorists can of course interpret the results of Munn's observations solely in terms of natural selection for behavior that increases the birds' evolutionary fitness or learned behavior reinforced by food. But the flexible versatility with which false alarming occurs primarily under appropriate conditions suggests that these birds are intentionally deceiving their competitors. As in other cases of this sort, the two interpretations are not mutually exclusive. Behavior reinforced by either learning or natural selection may also entail conscious intent. Indeed the capability of such conscious intention may be an important part of what is selected or reinforced.

Predator Distraction Displays

When animals are frightened by the close approach of a larger creature they sometimes exhibit striking forms of behavior that involve ceasing most bodily activity or acting as though badly injured or even dead. The adaptive function of most instances of such behavior is unclear, for it is difficult to understand why a hungry predator about to eat an animal would be deterred by its suddenly becoming immobile and ceasing all efforts to escape. But some of these displays are remarkable in the degree to which they simulate death. Burghardt (1991) discusses the death simulation of the hognosed snake *(Heterodon platirhinos),* which writhes erratically, defecates, turns over, and remains quiet with an open mouth, tongue hanging out, even bleeding at the mouth, and without appreciable breathing. This bizarre behavior has typically been viewed as an uncoordinated, instinctive response. But Burghardt reviews old and new evidence that even in this state of apparent death, the hognosed snake watches the animal that stimulated it. If this animal goes away, the snake recovers, but it remains in the state of death simulation longer if a human intruder looks at it. Thus even in an extreme state of death-simulating display, the snake monitors the animal that stimulated it to perform this display.

Various birds employ a wide range of antipredator behavior, ranging from direct attacks to maneuvers that tend to deflect an intruder's attention to the bird itself instead of its vulnerable eggs or young, as reviewed by Armstrong (1942) and by McLean and Rhodes (1991). When small birds attack larger predators they often call loudly, dive at the intruder, empty on it the contents of the stomach or cloaca, and sometimes even strike it with the legs or bill. Such attacks are usually quite effective, even against hawks or owls that occasionally turn on their tormentors and kill them.

An especially significant and suggestive form of deceptive communication takes place when certain ground-nesting birds perform elaborate displays that serve to distract predators that might otherwise destroy their eggs or young. Plovers and sandpipers typically lay their eggs in simple nests on the ground, where they are especially vulnerable to predation. Their many and varied types of antipredator behavior, reviewed by Gochfeld (1984), include maneuvers that are more complex and devious than direct attacks. Although their eggs are cryptically colored and resemble the substrate on which they are laid, many plovers and sandpipers respond to intruders with specialized distraction displays.

The most striking type of predator distraction behavior is displayed by several species of small plovers, such as the North American killdeer *(Charadrius vociferus)*, piping plover *(C. melodus)*, and Wilson's plover *(C. wilsonia)*. Killdeer typically nest in open fields, and the piping and Wilson's plovers lay their eggs on sandy beaches. The nest itself is little more than a shallow depression. When a large animal approaches the nest of a plover, one common response is for the incubating bird to stand up and walk slowly away from the nest for a few meters. Then it may begin calling, in the case of the piping plover a plaintive peeping has given the bird its name. Often the parent bird then walks or flies closer to the approaching intruder and begins a conspicuous display. This behavior is very different from the ordinary behavior of flying directly away from a frightening intruder when the bird has no eggs or young. The displays themselves vary widely but usually involve exposing conspicuous patterns and moving in atypical ways.

One type of display is called a crouch run: the bird holds its body close to the ground, lowers its head, and runs in a way that makes it resemble a small rodent such as a vole. Sometimes the plover even emits rodent-like squeaks during these crouch runs. Most predators that would eat plover eggs or chicks are also likely to pursue a vole, so the rodent-like crouch run may divert attention from a search for plover eggs or chicks. Similar displays have also been observed in a small passerine bird by Rowley (1962). At times plovers threatened by an approaching intruder settle down into a small depression in the sand as though incubating eggs. Such false incubation may lead a predator to search at the empty depression rather than the actual nest.

Other common distraction displays involve postures and movements that render the bird conspicuous or cause it to appear injured. The tail feathers may be spread abnormally wide, and one or both wings may be extended or even dragged on the ground as the bird moves over an irregular course. In the extreme form of this display, called the broken wing display, the bird falls over and flops about as though badly injured, perhaps with a broken wing. It looks just like a bird that has been winged by a hunter's shot. Predators are especially alert for signs of weakness or injury in potential prey, and these displays must make the bird appear especially vulnerable.

The customary explanation of these distraction displays used to be that the bird is in severe conflict behavior, being motivated both to flee and to attack, with the result that it is thrown into a state bordering on an uncoordinated convulsion (Skutch 1954; Simmons 1955; Armstrong 1949). The strong influence of inclusive behaviorism has led most ornithologists and ethologists to deny emphatically that a bird might intentionally attempt to deceive a predator and lead it away from its young. For example, Armstrong (1949, 179) asserted that "it is ludicrous to suppose that injury-simulation arose through birds deciding the trick was worth trying." Although Armstrong was speaking of evolutionary origins, he and other ornithologists seem equally certain that the birds do not display with the intention of leading the predator away.

One reason to question whether the displaying birds are consciously trying to lead a predator away is the fact that they sometimes continue displaying even after it has killed their young, as described by Drieslein and Bennett (1979). Such continuation of distraction display after the predator has taken the eggs or young may indicate that the bird is not consciously intending to lead the predator away. Although obviously not completely rational, this continuation of the display may be comparable to other emotional expressions that are carried on after they can no longer achieve a desired effect.

Broken wing displays provide an especially suitable situation in which new observations and experiments have thrown some light on the possibility that the birds are acting intentionally. Not that any evidence yet available can settle such a question conclusively; but certain features of the bird's behavior strengthen the plausibility of inferring conscious intent. Almost all data that are available about the details of predator distraction displays have so far come from encounters with human intruders, because it is only rarely possible to observe natural encounters between plovers and mammalian predators. The few observations that have been reported indicate that predator distraction displays in response to natural predators are very similar to those elicited by human intruders who approach a nest where plovers are incubating eggs or areas where young chicks are present.

Armstrong (1942, 89–91) reviewed several observations of an otter, foxes, weasels, dogs, and cats being led away from young birds by parents acting as though injured. Sullivan (1979) observed a black bear being successfully lured away by a blue grouse hen. Pedersen and Steen (1985) analyzed the effectiveness of predator distraction displays of ptarmigan. And Brunton (1990) observed that distraction displays of killdeer were almost always effective in leading potential predators away from eggs or chicks. Byrkjedal (1991) concluded that conflict of motivations is not an adequate explanation of nest-protection behavior in golden plovers. They and other birds performing predator distraction displays may well be thinking in simple rational terms about balancing the risks of nest predation against the effort and risks involved in leaving the nest and carrying out distraction displays, which may be wasted effort or under adverse weather conditions may lower the chances that the eggs will develop normally.

Sonerud (1988) described direct observations of foxes that apparently "saw through" the injury-simulating displays of grouse and responded by searching the immediate vicinity for grouse chicks rather than following the displaying adult. He developed a theory that when small mammal prey is abundant, young foxes do not learn that distraction displays mean young birds are nearby, but that in years when small mammals are scarce, foxes are more likely to learn that such displays are a sign of available prey in the vicinity. Thus there may well be a sort of cognitive interaction between parent birds and predators by which distraction displays are reduced or eliminated in situations where the predators are clever enough to interpret them as signs of edible eggs or chicks nearby rather than an injured bird worth pursuing.

Ristau (1983a, 1983b, 1983c, 1991) carried out systematic experiments with piping plovers and Wilson's plovers nesting on beaches to analyze this selectivity of response to different kinds of intruder. She had human intruders behave in two clearly different ways toward an incubating plover. One category, termed "dangerous" intruders, walked to within two meters or less of the nest, looked about, and acted as though searching. They must have appeared to the parents as more likely to destroy the eggs than others, designated "safe" intruders, who walked past the nest tangentially but did not approach closer than 12–32 m. The two categories of human intruders dressed as distinctively as possible. Both before and after being exposed to one to four such dangerous and safe approaches the plovers were tested by having the previously safe and dangerous intruders walk past at a constant, moderate distance.

In 25 out of 31 tests the plovers reacted more strongly to the intruder who had previously acted dangerously than to the one who had played the "safe" role. Stronger responses entailed looking up or moving away from the nest at a greater distance, staying off the nest longer, and performing more active displays. In three of the remaining six experiments there was no appreciable difference. In the other three cases the response was the opposite of what would be expected if the birds had learned which person was the more likely to threaten the nest, that is, the response to the safe intruder was stronger than the response to the one who had behaved dangerously. Thus in the great majority of these tests the plovers learned in the course of a few exposures to safe and dangerous intruders which one posed the greater threat and called for the more vigorous response.

In other experiments Ristau had human intruders either look at the nest or turn their heads and gaze in the opposite direction as they walked past at a moderate distance. Again, the plovers reacted more strongly to the distinctively dressed persons who were looking directly at their nests than to those who looked the other way. All these observations and experiments indicate rather strongly that plovers learn quickly which types of intruder are dangerous and respond more strongly to them than to others that have previously behaved in a less threatening manner.

One indication that these displays are something other than chaotic convulsions is that the bird frequently looks at the intruder. Even when moving away it repeatedly turns its head and looks back. During broken-wing or other distraction displays the plover does not ordinarily stay in one spot but moves slowly and somewhat erratically over the ground. If the intruder does not follow, the bird typically stops displaying and moves to a different position where it repeats the display, often with increased vigor.

Significant indications of intentional, rather than chaotic or unplanned, displaying also resulted from Ristau's analysis of the spatial relations between the intruder, the nest or young, and the displaying bird. If the displays result from a simple mixture of motivation to attack and to flee, one would expect the parent bird to approach the intruder, then withdraw, and perhaps approach and withdraw alternately as one or the other motivation came to predominate. But there would be no reason to expect that the location of the nest would play any significant role in such movements, except that, having come from somewhere near the nest, a plover that simply approached the intruder and then withdrew would often move back toward the nest.

Intruders are ordinarily detected at a considerable distance, and the parent bird usually walks slowly and inconspicuously away from its nest before flying fairly close to the intruder while it is still many meters from the eggs or young. As the intruder comes close, the displaying plover moves a bit faster, and although appearing crippled and easy to catch, it always manages to stay just out of reach. On one occasion a displaying Wilson's plover led me along a beach for about 300 m with almost constant displays.

When Ristau mapped where the intruder would have been led if it had followed the plover throughout its broken-wing displays, she found that in 44 out of 45 cases the intruder would be led away from the nest. Since the displays began at positions that varied widely with respect to the line connecting the nest with the initial position of the intruder, some routes might be expected to lead initially a little closer to the nest before continuing off in a quite different direction. But in only 6 of the 45 cases would an intruder have moved closer to the nest at any time while it was following the displaying bird.

Another indication of something other than mechanical and random flopping about in chaotic conflict behavior is the degree to which plovers are selective about which intruders will elicit full-blown displays. They habituate to familiar people, especially on heavily visited beaches. Skutch (1976, 1996) and Armstrong (1949, 1956) both observed that when cattle or other hoofed animals approach their nest, instead of behaving as though injured, plovers sometimes stand up conspicuously, close to the nest with spread wings, and may actually fly directly at an approaching cow. This is sensible behavior, because the danger from cattle is not that they will eat the eggs but that they may step on them.

Deceptive Behavior of Monkeys and Apes

Several other examples of deceptive or manipulative communication are described in the book edited by Mitchell and Thompson (1986). Of particular interest is an illustrated English translation of a paper by Rüppell describing deceptive use of an alarm call by a mother Arctic fox whose young were snatching food and keeping her away by urinating at her. Elephants and dogs show clear evidence of deceptive behavior, although it is very difficult to be sure how much conscious intent is involved (1986, chapters 11 and 12). With apes, however, the circumstantial evidence of conscious deception is considerably stronger, as described in detail by Miles (ibid., chapter 15) with respect to an orangutan who had been trained to use a simple form of sign language, and especially by de Waal (ibid., chapter 14) and de Waal (1982).

De Waal's long-term studies of captive chimpanzees documented numerous cases of deceptive behavior. Especially revealing were his observations of subordinate males that often sought, and sometimes obtained, sexual intercourse with adult females, but only when the more dominant adult males did not see them copulating. Both partners in these surreptitious liaisons acted furtively and seemed clearly to be keeping out of sight of the dominant males. If discovered in the act of close sexual advances, such a subordinate male hastily covered his erect penis with his hands. This is in effect deceptive noncommunication. More recently de Waal (1986, 1989, 1996) has described other examples of somewhat deceptive behavior of primates, along with intriguing patterns of friendship and reconciliation.

Byrne and Whiten (1988), Whiten and Byrne (1988, 1997), and Whiten (1991) have reviewed published descriptions of apparently deceptive behavior in many species of primates, and they have suggested how this subject can be more adequately studied in the future. They emphasize the degree to which even experienced primatologists have been inhibited, or even embarrassed, about reporting behavior that suggests conscious deception. The review by Whiten and Byrne (1988) is followed by numerous commentaries from other scientists and scholars, some of whose criticisms demonstrate how strongly the residual influence of behaviorism still limits acceptance of overwhelming circumstantial evidence that primates are quite capable of what Whiten and Byrne call "tactical deception." Yet in their reply to comments on their paper they denied that they are concerned with the "phenomenal worlds" of the animals they study. The behavioristic taboo still lingers on, but it is somewhat relaxed in a more recent sequel (Whiten and Byrne 1997).

On balance, it seems clear that on many occasions animals communicate inaccurate information or intentionally avoid conveying certain types of information to others (Cheney and Seyfarth 1988, 1990a, 1990b, 1991). In many cases the versatility of deceptive or misleading behavior provides even more suggestive evidence of conscious intent than the transmission of reasonably accurate information. Although it is difficult to liken such deceptive communication to involuntary groans of pain, the deceptive tactics of monkeys are often inconsistent. For example, Cheney and Seyfarth (1990a) describe how a vervet monkey emitted a leopard alarm call during a territorial confrontation with a neighboring troop. This appeared to cause the opponents to flee, but the caller then moved into the open himself, contrary to normal, prudent behavior when a real predator is approaching. Intentionally deceptive behavior may thus be executed ineptly, which complicates the task of ascertaining whether it results from conscious planning.

CHAPTER TWELVE

Dolphins and Apes

The whales and dolphins and the monkeys and apes engage in so much versatile behavior that whole books would be required to document the evidence that suggests conscious thinking in each group. I will therefore select only a few of the most striking examples, with emphasis on communicative behavior that provides compelling evidence about what these animals are thinking and feeling.

Cetacean Versatility

The world's largest brains are found in cetacean and not human skulls, and the toothed whales or dolphins have brains that are roughly comparable in size and complexity to our own. Morgane, Jacobs, and Galaburda (1986) pointed out that the large cerebral cortex of the cetaceans lacks some of the organizational complexities of the primate cortex. Yet these marine mammals engage in complex and versatile behavior and communicate with a great variety of sounds both when trained in captivity and spontaneously under natural conditions. It is as counterintuitive to deny that they think consciously about some of their activities as it would be to advance such an absurd claim about the Great Apes. One of the best examples of mental versatility displayed by captive dolphins is imitation of sounds and actions. The bottle-nosed dolphin *(Tursiops truncatus)* is the most commonly trained performer in oceanaria, and it has also been studied most thoroughly by scientists. Experiments with this species have revealed many surprises. For example, Herman (1980, 402) summarizes observations by Tayler and Saayman (1973) of a bottle-nosed dolphin that imitated the behavior of a seal, turtles, skates, penguins, and human divers in its aquarium, employing behavior patterns not seen in this or other dolphins under other circumstances:

> The behaviors imitated included the seal's swimming movements, sleeping posture, and comfort movements (self-grooming). The dolphin, like the seal, at times swam by sculling with its flippers while holding the tail stationary.

> The sleeping posture imitated was lying on one side at the surface of the water, extending the flippers, and trying to lift the flukes clear of the water. The comfort movement mimicked was vigorous rubbing of the belly with the under surface of one or both flippers. . . . Additional imitative behaviors by this dolphin included the swimming characteristics and postures of turtles, skates, and penguins. The dolphin also attempted to remove algae from an underwater window with a seagull feather, in imitation of the activity of a human diver who regularly cleaned the window. The dolphin, while "cleaning" the window, reportedly produced sounds resembling those from the demand valve of the diver's regulator and emitted a stream of bubbles in apparent imitation of the air expelled by the diver. Tayler and Saayman also observed a dolphin using a piece of broken tile to scrape seaweed from the tank bottom, a behavior apparently derived from observing a diver scraping the bottom while cleaning the tank with a vacuum hose. The scraping behavior of this dolphin was then copied by a second dolphin.

Herman (1980, 406) also described spontaneous observational learning by dolphins in the large tanks where they are trained to perform a variety of complex gymnastics to entertain spectators:

> In some cases, the animals may even train themselves. In one interesting episode illustrating self-training, one animal of the group had been taught to leap to a suspended ball, grasp it with its teeth, and pull it some distance through the water in order to raise an attached flag. This animal was subsequently removed from the show and a second animal in the group was trained in the same task, but learned to raise the flag by striking repeatedly at the ball with its snout rather than by pulling it. This second animal, a female, subsequently died and another female of the group immediately took over the performance, without training, and continued to strike at the ball with the snout. Later when this new female refused to participate in the show during a two-day period, a young male in the group immediately performed the behavior, but grasped and pulled the ball with his teeth, in the manner of the originally trained animal.

Pryor, Haag, and O'Reilly (1969) and Pryor (1975) described how a captive rough-toothed dolphin *(Steno bredensis)* learned to perform a new trick every day in order to obtain food. Macphail and Reilly (1989) demonstrated that pigeons can learn to discriminate between novel and familiar stimuli, but creating novel and moderately complex actions is a more demanding task. Pending further experiments comparable to those of Pryor et al., it is difficult to judge whether other animals could learn to do this. But recent experiments by Mercado et al. (1998, 1999) demonstrate that bottle-nosed dolphins understand and comply with the command to repeat one of their previous actions when that action is described by human gestures that designate the action to be repeated in terms of objects or locations, including the command "Repeat what you just did" or "Do something new and different."

Dolphins emit a variety of sounds, including click trains used for echolocation and sounds of longer duration, usually of lower frequency, by which they communicate; and they can be trained to imitate not only actions but also a wide variety of whistle-like sounds (Richards, Wolz, and Herman 1984). They often produce a medley of sounds when engaged in social interactions, but the physical properties of underwater sounds make it extremely difficult to determine which sounds come from a particular animal when they are close to one another. This is because underwater sounds have much longer wavelengths than the same frequencies in air, and they are attenuated scarcely at all over distances of several meters while being strongly reflected from most surfaces they encounter, including the air-water interface.

The sounds emitted by most terrestrial animals are audible to human ears, and a listener can usually judge the direction from which they arrive with considerable ease and accuracy. But underwater sounds cannot be localized nearly so well, even when converted to airborne sounds by hydrophones connected to earphones, primarily because multiple reflections are often at least as intense as the sound waves arriving directly from the source.

One important type of sound emitted by dolphins is the so-called signature whistle, a frequency- and amplitude-modulated sound that is characteristic of each individual. These have been studied in detail by Caldwell and Caldwell (1965, 1968, 1971, 1973, 1979), Caldwell, Caldwell, and Tyack (1990), and Connor, Mann, Tyack, and Whitehead (1998). The Caldwells were obliged to record them primarily from dolphins that were isolated out of the water, because of the difficulties of separating sounds from various sources underwater. But it was nevertheless clear that more than 90 percent of the whistles were specific to the particular dolphin and could easily be distinguished from the signature whistles of other dolphins.

Tyack (1986) developed a small and harmless device that can be attached to a dolphin's skin by a suction cup, which picks up that animal's sounds much more strongly than those of others nearby. This enabled him to distinguish signature whistles from two animals, Spray and Scotty, that had lived together in the same tank for about seven years. Two types of whistle predominated in their vocal exchanges, and each type was produced primarily, but not exclusively, by one of the two animals. Spray produced 73 percent of one whistle type, while Scotty emitted 74 percent of the other type. Tyack suggested that perhaps the whistles of the other animal's characteristic pattern resulted from imitation. It is even possible that these two dolphins were calling the other by name. But much additional evidence would be needed to demonstrate that such use of whistles as the equivalent of names was actually occurring. More recent work reviewed by Smolker, Mann, and Smuts (1993) strongly indicates that individual dolphins recognize each other by their signature whistles and sometimes make sounds much like the signature whistles of another dolphin. Such vocal exchanges are clearly parts of a complex network of social interactions, as reviewed by Roitblat, Herman, and Nachtigall (1993), Reiss, McCowan, and Marino (1997), and Connor et al. (1998).

Pryor and Schallenberger (1991) described how pelagic spotted dolphins *(Stenella attenuata)* have learned about the purse seines used to surround and capture tuna in the tropical Pacific. The tuna tend to stay below schools of dolphins, which therefore act as markers for the presence of the tuna. The tuna fishermen employ speedboats to chase and herd the dolphins into a position where they can be surrounded by the net, which may be as much as a mile in length. These dolphins of the open ocean do not ordinarily encounter any large solid objects, and when this type of fishing was initiated, hundreds panicked and became entangled in the nets. But in recent years they have learned to remain fairly quiet and wait until the crew maneuvers to lower a small portion of the net and allow them to escape. Shortly after escaping from the net the dolphins often leap repeatedly into the air, as though in joyful celebration.

The behavior of these dolphins strongly indicates that they have learned a great deal about the fishing vessels and about what to do when chased by speedboats and surrounded by the net. Experienced fishermen believe that dolphins recognize the fishing vessels from a considerable distance and that on seeing one they often rest quietly at the surface without conspicuous blowing. If the ship approaches, they may swim rapidly in an apparent effort to keep on its starboard side, presumably because they have learned that the cranes and other machinery used to handle the nets are usually on the port side.

Another sort of versatile behavior of dolphins is aiding of other dolphins that are sick or injured. This is relatively uncommon, but it has been observed in detail in several captive groups, and something of the kind occurs occasionally under natural conditions. Aiding behavior is not limited to dolphins, and clear examples in the dwarf mongoose have been described by Rasa (1976, 1983). The instances of dolphins aiding human swimmers presumably result from similar behavior. If a dolphin is visibly weak and sinks below the surface, its companions may swim down and push it from below, lifting it to the surface so that it can breathe air. Occasionally a dolphin carries this aiding behavior to extremes, as with a mother who carried her stillborn baby for days until it had begun to decompose, or the bottle-nosed dolphin that carried a dead shark for eight days without stopping to eat. Dolphins are somewhat selective in this aiding behavior; females and young are much more likely to be assisted in this way, and adult males are sometimes left at the bottom of the tank.

Herman (1986, 1987) and his colleagues have approached the question of dolphin cognition by means of an intensive training program designed to assess the degree to which these animals can understand not only individual signals but simple combinations of signals that are related to one another in a manner resembling the grammatical rules of English. This emphasis stems from the widespread conviction that combinatorial productivity based on the use of rule-governed combinations of words constitutes an essential and unique feature of human language.

Herman's experiments have concentrated on comprehension or receptive competence of dolphins rather than on their ability to produce communicative signals. He trained one female bottle-nosed dolphin, Akeakamai (or Ake), to respond to gestures from a trainer at the edge of the tank, and another, named Phoenix, was trained to respond to underwater whistle-like sounds. Both dolphins learned to respond appropriately to about thirty-five signals, which included the names of objects in the tank and actions such as toss, swim under, jump over, fetch, and put something in or on top of something else. They also learned a few modifiers such as right and left, surface and bottom. A typical combination of signals was RIGHT PIPE TAIL-TOUCH, meaning that the dolphin should touch the floating pipe to her right with her tail.

Both Ake and Phoenix learned to respond appropriately to numerous combinations of these commands, which could be as complex as five-unit sequences such as PLACE BOTTOM PIPE IN SURFACE HOOP. In the system used in these experiments PLACE IN is a single command, and the order of the signals differs from customary English but nevertheless constitutes a set of rules in which the sequential arrangement adds important meaning to the individual signals themselves. This much was not especially novel, because many dolphins have been trained to carry out sequences of behavior of greater complexity. But in their training, these two dolphins were never presented with more than a small proportion of the meaningful combinations of commands; the others were reserved for tests of their comprehension of the rules that governed the combinations. These rules were comparable to those of English, in that the order of the words or commands determined which would be the direct or indirect object and which modifier applied to a particular command.

After the dolphins were responding quite well to these sequential combinations, they were presented with new combinations that they had never encountered before. With two- to five-element combinations, correct responses were 66 percent for Ake and 68 percent for Phoenix (Herman 1986, 232). Although 66 percent correct may not seem very close to perfection, it should be realized that the chance score was very low indeed, since there were a very large number of possible actions among which the dolphins had to choose. Thus these two dolphins had clearly learned not only the meanings of the individual commands but the sequence rules governing which served as direct or indirect object and which modifier applied to a given object name.

Schusterman and Krieger (1984) trained two sea lions *(Zalophus californianus)* to respond correctly to many gestural signs displayed by human trainers. And Schusterman and Gisiner (1988a, 1988b, 1989) trained sea lions to comprehend combinations of commands—demonstrating that this ability does not require the brain volume of dolphins or Great Apes. But they and also Premack (1986) deny that this ability to learn combinatorial rules is remotely equivalent to the versatility of human language. They point out that understanding the following two rules would appear sufficient to account for the performance of both Herman's dolphins and their sea lions (1988a, 348):

Rule 1. If an object is designated by one, two, or three signs (an object sign and up to two modifiers), then perform the designated action to that object.

Rule 2. If two objects are designated (again, by one to three signs each) and the action is FETCH, then take the second designated object to the first.

Comprehension of these rules would seem to require some significant thinking, as discussed by Herman (1988). Of course, this comprehension of both individual commands and the meaning conveyed by their sequential relationships falls far short of the rich versatility of human language. But these dolphins and sea lions have learned to comprehend rule-governed combinations of wordlike signals. Although human languages obviously involve far more than two rules, this is the first demonstration that animals can comprehend *any* syntactical rules at all. In short, two is significantly greater than zero. Furthermore, if dolphins and sea lions think in terms of these rules, they must be capable of thinking in correspondingly complex terms about the relationships between signals and the actions or objects for which they stand. It would be unwise to allow our preoccupation with the vast superiority of human language to obscure the fact that these experiments reveal at least part of what the dolphins were thinking.

The great emphasis on syntactical rules as a fundamental property of language has been developed primarily by scientists and scholars whose native language is English, one of the few human languages in which word order provides almost all of the syntax. But in most languages inflection of words is used to convey the grammatical relationships between them, and English does retain a few vestiges of inflection, for example, the possessive form of nouns and the different forms of pronouns such as *he, his,* and *him*. The fact that so many human languages rely heavily or primarily on inflection of words to convey syntax suggests that inflection may be a more basic and, perhaps, a more easily utilized way to express grammatical relationships. If so, attempts to teach a combinatorially productive language to animals might be more successful if they employed inflection rather than word order. A practical difficulty facing such attempts is the fact that animal signals are often quite variable, and it will take special precautions and controls to establish whether any of the variability might convey syntactic information

A more fundamental distinction between human language and what dolphins and sea lions have accomplished, at least so far, is the ability to switch back and forth between comprehension and production of communicative signals. This is obviously a key attribute of human language, and it is also an ability that has been attained by some Great Apes, as described below. But we do not know whether dolphins could learn to comprehend and produce signals interchangeably, and therefore this question remains open. Obviously dolphins and sea lions cannot be expected to produce human gestures, but Phoenix, who learned to comprehend the rules relating commands presented to her as underwater sounds, could almost certainly learn to imitate those sounds. Appropriate experiments with other dolphins or sea lions trained to communicate by sounds might reveal whether they are capable of using acoustic signals interchangeably in both the receptive and the productive modes.

Primates' Understanding of Social Relations

Humphrey (1980) extended an earlier suggestion by Jolly (1966) that consciousness arose in primate evolution when societies developed to the stage at which it became crucially important for each member of the group to understand the feelings, intentions, and thoughts of others. When animals live in complex social groupings, where each one is critically dependent on cooperative interactions with others, they need to be "natural psychologists," as Humphrey put it. They need to have internal models of the behavior of their companions, to feel with them, and thus to think consciously about what the other one must be thinking or feeling. Following this line of thought, we might distinguish between animals' interactions with some feature of the physical environment or with plants, on one hand, and interactions with other reacting animals—usually their own species, but also predators and prey, as discussed in chapter 4. Although Humphrey restricted his criterion of consciousness to our own ancestors within the past few million years, it might also apply with equal or even greater force to other animals, including the social insects, that live in mutually interdependent social groups.

De Waal (1990) analyzed an intriguing behavior pattern in which a rhesus macaque mother holds her own infant together with another of roughly the same age. In most cases the other infant is the offspring of a dominant female. It seems likely that this behavior increases the likelihood that her infant will later benefit by association with more dominant companions and that these monkeys are intentionally attempting to "promote future associations between their own offspring and high-ranking youngsters" (597).

Recent investigations of the social behavior of monkeys and apes have demonstrated that most species live a large portion of their lives in stable social groups. All members of these groups clearly know each other as individuals. Furthermore, Cheney and Seyfarth (1990) and others have discovered that vervet monkeys, baboons, and many other primates recognize each other not only by sight but also by vocalization. They know a great deal about their companions. They know which youngster belongs with a particular mother, and they appear to know the relative social rank of other group members, as demonstrated by the recent experiments of Zuberbühler, Noe, and Seyfarth (1997), Zuberbühler, Cheney, and Seyfarth (1999), and Seyfarth and Cheney (in press).

Despite the compelling evidence that monkeys know many socially important facts about their companions and their relationships, Seyfarth and Cheney have found no evidence that they are concerned with the subjective thoughts or feelings of other monkeys, as opposed to their behavior. And from this they conclude that "monkeys seem not to know what they know. Consequently, much of their knowledge of themselves remains tacit, without becoming explicitly accessible to consciousness." As discussed in chapter 1, this conclusion that monkeys don't know that they know involves the distinction between perceptual and reflective consciousness. A blanket denial that even monkeys are capable of understanding their own knowledge seems a rather sweeping conclusion to be based on the lack of a kind of evidence that would be very difficult to obtain. In chapter 14 I will return to this important and challenging question.

Ape Language

As emphasized in previous chapters, one of the most promising sources of objective data about animal thoughts and feelings is their communicative behavior, because at least some of their subjective experiences are probably reported in the messages they communicate to others. From this perspective one of the most exciting advances in cognitive ethology was achieved by B. T. and R. A. Gardner at the University of Nevada (1969, 1971, 1984, 1989a, 1989b) when they succeeded in training a young female chimpanzee named Washoe to use more than a hundred signs derived from, but not identical to, American Sign Language (ASL). These signs are gestures developed for communication by the deaf. Based on English, it is one of several sign languages for the deaf adapted for manual gesturing.

The Gardners based their plans for this ambitious attempt to teach a gestural language to a chimpanzee in large part on three previous investigations: (1) the extensive studies of the Great Apes by R. M. and A. W. Yerkes (1925, 1929); (2) the detailed observations of natural communication between chimpanzees described by Jane Goodall (1968, 1986); and (3) the almost totally unsuccessful attempts by the Kelloggs (1933) and the Hayeses (1951a, 1951b) to train chimpanzees to use spoken words. It has now been established that their larynx and vocal tract are anatomically unsuited for producing the sounds of human speech.

Yerkes (1925, 179–80) had suggested that chimpanzees might be able to learn a gestural communication system. His and many other studies involving both laboratory experiments and home rearing of young chimpanzees by the Kelloggs and the Hayeses had abundantly demonstrated that our closest nonhuman relatives can learn numerous complex discriminations. Under natural conditions some populations of chimpanzees use simple tools, and in laboratory studies they can learn to use many human tools. Similar use of human tools by orangutans has been described in detail by Lethmate (1977). Apes communicate with each other by sounds and simple gestures, and captive apes, like many other birds and mammals, learn to recognize simple spoken commands.

Until the Gardners' work with Washoe, however, chimpanzees had seemed almost totally incapable of anything even suggestive of linguistic communication. The intensive and at times heatedly controversial investigations that have followed in the footsteps of the Gardners (and

Washoe) were reviewed by Savage-Rumbaugh (1986) and in the volumes edited by Heltne and Marquardt (1989) and by Parker and Gibson (1990). Extensive recent work reinforcing the earlier conclusions has been reviewed by Savage-Rumbaugh and Lewin (1994), Fouts (1997), and Savage-Rumbaugh, Shanker, and Taylor (1998).

Washoe was acquired by the Gardners at about one year of age, and she was cared for by friendly and familiar people who communicated with Washoe and with each other exclusively in American Sign Language (ASL) when she was present. They signed to her about things likely to interest her, much as human parents talk to their babies. Washoe was taught simple versions of ASL signs, partly by demonstration—use of the sign by her trainers in the presence of the object or activity for which the sign stood—and partly by molding, a procedure in which a trainer would gently hold her hands and guide them through the appropriate motions. In three years she learned to use and respond appropriately to 85 signs (Gardner and Gardner 1971). Subsequently several other chimpanzees and a few gorillas and orangutans have been taught to use signs derived from ASL, and vocabularies of more than 100 signs have been achieved by several of these animals, as reviewed by Ristau and Robbins (1982) and by Gardner, Gardner, and Van Cantfort (1989) and Fouts (1997). Scarcely any behavioral scientists expected that Yerkes's 1925 prediction would ever be so abundantly fulfilled.

Fouts (1989, 250) has found by analyzing videotapes of signing chimpanzees that they sometimes "affect the meaning of a sign by modulating it." This has entailed gesturing and gazing at objects of interest while also signing about them. But it is possible that apes might modify the sign itself in ways that would indicate whether it is meant to denote the direct or the indirect object, for example. The customary efforts to train apes to standardize their signs would probably discourage such inflection, unless it were specifically encouraged. The Gardners (1989b) have indeed observed that their chimpanzees do modulate the signs they have learned and convey additional meaning by doing so. We may perhaps anticipate that in future investigations the possibility of inflectional syntax will be analyzed explicitly, for reasons discussed above.

The implications of the discovery that chimpanzees and other Great Apes could communicate in even a rudimentary form of language were truly shattering to the deep-seated faith in language as a unique human attribute separating humanity from the beasts. As a result, a heated series of controversies continues to rage over the degree to which the communicative behavior learned by apes is actually a simple form of language. Wallman (1992) and Pinker (1994), among others, have argued vigorously that it is not. Many criticisms of the earlier experiments with apes have been significantly constructive and have led to a sharpening of scientific understanding of human language and its acquisition by young children, as discussed in detail by Savage-Rumbaugh (1986). They have also led to improved and better-controlled investigations of the communicative behavior of the apes, and these in turn have served to overcome many of the criticisms of the earlier studies. On balance, it now seems clear that apes have learned to communicate simple thoughts.

Anticipating the appropriate concern that the signing apes were not really communicating linguistically, the Gardners required that quite stringent criteria be satisfied before they accepted a sign as reliably used by Washoe. It had to have been "reported by three different human observers as having occurred in an appropriate context and spontaneously (that is, with no prompting other than a [signed] question such as 'What is it?' or 'What do you want?') . . . each day over a period of 15 consecutive days" (1969, 670). In later vocabulary tests, pictures of objects for which the chimpanzee had learned to sign were projected on a screen visible to her but not to either of two human observers who did not know what picture was shown. One observer was in the same room, and if the chimpanzee did not sign spontaneously when the picture was projected on the screen, she signed the question "What is it?" or some equivalent. The other observer watched from a separate room through a one-way glass window. Both observers noted what sign they judged the ape to have produced, but they could not communicate or influence each other's reading of the signs.

The two observers agreed in their interpretations of 86–96 percent of both correct and incorrect signs produced by three of four chimpanzees. The fourth animal yielded only 70 percent interobserver agreement, probably because she had not mastered signing as thoroughly as the other three. These chimpanzees' signs were correct in 71–88 percent of the trials, and this was far above chance levels because many signs were in use by these animals, so that if they were merely guessing at random, the expected percentage of correct signs would have ranged from 4 to 15 percent (Gardner and Gardner 1984). These "blind" tests seem to rule out the possibility that the data could have been seriously distorted by inadvertent cuing or Clever Hans errors, because neither observer could see the projection screen, and also because any such inadvertent cuing would have had to convey which of numerous signs the chimpanzee should produce in response to the projected picture.

Nevertheless, some critics have pointed out that apes are very adept at detecting inadvertent cues from their human companions, and that it is conceivable that even in these blind tests the ape could have received information about which sign to produce from unintentional sounds or movements of the observers. It is also conceivable that the observers could have inadvertently conveyed information to each other about what sign was appropriate, as suggested by Sebeok and Rosenthal (1981) and by Umiker-Sebeok and Sebeok (1981). Although every sort of precaution and repeated checking of all possible sources of error are always appropriate, these criticisms seem quite far-fetched, and later experiments by Savage-Rumbaugh and her colleagues described below render explanations based on inadvertent cuing extremely implausible.

Washoe and other apes who have learned to use communicative gestures based on ASL seem to use these signals more or less as very young children use single words. They sign spontaneously to request simple things and activities, and they sometimes sign to themselves when alone. Washoe and other signing chimpanzees sign to each other to at least a limited extent (Fouts 1989, 1997; Fouts and Fouts 1989; Fouts, Fouts, and Schoenfeld 1984).

A three-year-old chimpanzee named Ally was trained using the Gardners' methods to use about 70 signs, and he also learned to understand several spoken words and phrases. Next he was taught to use new signs corresponding to 10 of these words that referred to familiar objects, but only the signs and not the objects themselves were presented along with these spoken words during this phase of the training. After this was accomplished, Ally was shown the objects, and he identified them by means of the signs he had learned from their spoken names (Fouts, Chown, and Goodman 1976).

Signing chimpanzees sometimes transfer signs to new situations that are different from those in which they had been trained to use them. For instance, Washoe learned the sign for open to ask that doors be opened, but she then used it to request her human companions to open boxes, drawers, briefcases, or picture books, and to turn on a water faucet. After she had learned the sign for flower, she used it not only for different kinds of flowers but also for pipe tobacco and kitchen fumes. To her, it evidently meant smells.

Stimulated by the Gardners' success with Washoe, other investigators have achieved similar levels of communication not only by several other chimpanzees but also by gorillas (Patterson and Linden 1981) and an orangutan (Miles 1990). Still other investigators have studied the communicative abilities of apes by quite different procedures. Premack and his colleagues concentrated on a type of symbolism based on plastic tokens, which chimpanzees learned to arrange on a sort of bulletin board in order to request desired objects and answer simple questions about them. These plastic symbols were arbitrary rather than iconic in that they did not resemble the object for which they stood.

Premack's star pupil, Sarah, learned not only to select the correct symbol when shown the object for which it stood but to use the symbols to request things she wanted. She could also arrange plastic tokens in strings resembling rudimentary sentences and answer simple questions presented to her through similar arrangements of the tokens. She could answer such questions as "What is the color of ——?" (for instance, an apple) about the plastic representations of objects, even when the colors and other properties of the tokens were quite different from those of the objects for which they stood. For instance, the plastic symbol for apple might be blue (Premack 1976, 1983b; Premack and Premack 1983).

In other ambitious projects Rumbaugh, Savage-Rumbaugh, and their colleagues at the Yerkes Laboratory in Atlanta, Georgia, use a keyboard connected to a computer that records which keys are activated by either a human experimenter or the ape that learns which key it must touch to obtain specific objects or to express simple desires. Each key lights up when pressed, and each has a characteristic pattern to help the apes recognize and select it; but these patterns are not iconic representations of the objects for which the key stands. This system permits two-way communication, and it has the great advantage that an objective record can be kept of every key press. The National Zoo in Washington now includes a public exhibit in which an orangutan is being trained to communicate by means of a similar keyboard.

The initial experiments with plastic tokens and keyboards were interpreted as showing that the chimpanzees were using a simple type of linguistic communication, including a rudimentary sort of grammar in which the tokens or key presses were used in a specific sequence to express relationships between the individual symbols. But, as reviewed by Savage-Rumbaugh (1986), these apes may have learned only to perform specific actions to obtain particular things or activities. They may have been thinking something like "If I do this, he will give me some candy," or "If I do that, she will play with me." They seldom, if ever, tried to use their newly acquired skills to initiate communication with their human companions or with each other, partly because the experimental arrangements in these early investigations allowed rather little opportunity for such spontaneity. On the other hand, Washoe and other apes taught to communicate by manual gestures did often initiate communicative exchanges by spontaneously asking for things they wanted.

All these approaches to teaching language-like communication to chimpanzees have been successful in the general sense that the apes have learned to make requests and to answer simple questions. They have also learned to give the appropriate gestural sign, to press the correct key, or to select the right plastic symbols to label familiar objects. Serious theoretical questions have been raised, however, about the degree to which such communication entails any true understanding on the animal's part of the meanings of the signals and symbols. The alternative interpretation that has been suggested by many critics is that the chimpanzee learns merely to perform certain actions in order to obtain things it wants, including activities or actions on the part of the human companions, such as opening doors or going for a walk. These distinctions, and the evidence that indicates the degree to which the language-trained apes understand what they are communicating, will be discussed in more detail below in relation to the recent studies of Savage-Rumbaugh and her colleagues. And chapter 14 will include a discussion of the relationship between learning and consciousness.

First it is appropriate to consider a fundamentally important aspect of human language and the evidence for its presence or absence in the language-like behavior learned by Washoe and her successors. This is what the psychologist George Miller (1967) has aptly termed "combinatorial productivity." Human speech combines units in various ways to give new meanings not expressed by the units themselves, typically by the use of grammatical rules common to all users of a given language. A simple example is the use of word order in English and few other languages to indicate which word designates the actor and which the object of the action ("John hits ball" versus "Ball hits John"). In most languages many of these relationships are conveyed by modifications or inflections of the words, rather than by word order. But regardless of how it is done, rule-governed combinations of words convey a much wider array of meanings than would be possible if each word were entirely independent and its relationship to other words did not convey any additional meaning. Such grammatical or syntactical rules are so important in making possible the richness and versatility of human languages that linguists and many others often maintain that syntax is a sine qua non of language, and that the use of words or their equivalent is not sufficient to qualify as true language. This view must of course entail denial of true language to young children with vocabularies of only a few words.

The importance assigned to syntax led to great interest in the combinations of signs used by Washoe and other signing apes and to efforts by Premack and his colleagues, as well as the scientists at the Yerkes Laboratory, to determine the extent to which chimpanzees could communicate syntactically. Terrace (1979) replicated the Gardners' training of a young chimpanzee to use signs derived from American Sign Language, and he devoted special attention to series or combinations of signs used by this animal, named Nim Chimpsky. The results were disappointing. A few combinations of two or three signs were used, but only to a very limited extent did Nim attain anything like the combinatorial productivity of human language. When two or more signs were used, there was little consistency in their order, and in only a few cases was sign combination AB used differently from BA. Most series of signs were repetitious, with third or later signs adding almost nothing to the message conveyed. Furthermore, many of Nim's signs were repetitions of the immediately preceding signs of his human companions.

Terrace (1979) and Terrace, Petitto, and Bever (1979) reported that one of the longer series of signs used by Nim was "Give, orange, me, give, eat, orange, give, me, eat, orange, give, me, you." And one of the longest utterances reported by Patterson and Linden (1981) from the gorilla Koko was "Please milk, please, me, like, drink, apple, bottle." Terrace and others concluded that signing apes are not using anything that deserves to be called a language, because of the almost total lack of rule-governed combinations of signs. But however ungrammatical and repetitious these strings of signs may have been, they leave no doubt what Nim or Koko wanted. Terrace's findings have had a widespread and negative impact, causing many to dismiss the whole effort to teach language-like communication to apes as insignificant and misguided. But this dismissal is based on the absence or near-absence of combinatorial productivity, and it does not seriously detract from the significance of signing as evidence of what apes are thinking.

The reasons advanced for denying that the signing of apes serves a function equivalent to human language have included the argument that many of the signs adapted from ASL are similar to naturally occurring communicative gestures (Seidenberg and Petitto 1979). This seems to a biologist quite the opposite of a reason for denying a language-like function. Perhaps chimpanzees have already developed their own types of gestural communication, more versatile than anything ethologists have yet deciphered, and the training achieved by the Gardners and their successors may have simply elaborated on an ability already present. If so, gestural communication is more than an artifact of human training; it may well be a part of the natural behavior of the Great Apes. Seidenberg and Petitto (1979, 199) seemed to believe that natural gestures of chimpanzees are unlearned and therefore different from and inherently inferior to human language. But Jane Goodall's descriptions of chimpanzee society show abundant opportunity for young animals to learn communicative behavior from their older companions. Washoe and other signing apes may not have, literally, learned ASL, but their signing nevertheless conveys some of their thoughts.

In recent years intensive long-term observational studies of chimpanzees living under natural conditions have convinced several leading primatologists that they have a rudimentary form of culture, which Whiten et al. (1999, 682) define as behavior "that is transmitted repeatedly through social or observational learning to become a population-level characteristic." Detailed reviews of chimpanzee culture have been edited by Wrangham, McGrew, de Waal, and Heltne (1994) and by Whiten et al. (1999), and the subject has been summarized concisely by Pennisi (1999) and Vogel (1999). Furthermore, Marshall, Wrangham, and Arcadi (1999) have found evidence that at least some chimpanzee vocalizations are learned. Although there is insufficient evidence to determine how much conscious communication is involved in the establishment and maintenance of cultural behavior, it seems very likely that it is an important component.

Words or their equivalent are obviously basic to any sort of linguistic communication. Without syntax or combinatorial productivity, words are limited and clumsy; but they do suffice to communicate thoughts. Grammar adds greatly to the economy and versatility of human languages, to its refinement and scope, but grammar without words would be empty and useless. Without accepting his claim that animals are unable to think, one can agree with Descartes, as paraphrased by Chomsky (1966, 6), that in a humanlike language "the word is the sole sign and certain mark of the presence of thought." Descartes and Chomsky claimed that nonhuman animals are incapable of using anything equivalent to words, so the key question is whether the signs used by signing apes have the essential properties of words.

Savage-Rumbaugh (1986, 15–32) has clarified these questions by emphasizing that much of the early testing of language-like communication of apes did not suffice to show that their communicative behavior was fully equivalent to human use of words, even those used by young children with limited vocabularies. She agrees with Nelson (1977) and other students of language acquisition by children that "the essence of human language is not found in syntax" but (quoting Nelson) in "the translation of meanings and the expression of these meanings to a social partner for some functional purpose . . . [and] the interpretation of the meaning expressed by others." Thus, to qualify as a true word, a communicative signal must convey meaningful knowledge, and its user must be able to employ it both to transmit and to receive such knowledge. According to this definition, the ability to produce the correct sign when shown the object for which it stands is not enough; to serve as the equivalent of a word the sign must also be used on appropriate occasions both to receive and to convey some sort of knowledge. This leads to the question of naming. Savage-Rumbaugh points out that producing the correct signal when shown an object does not necessarily correspond to human naming of the object. Only if the ape (or other animal) also uses the signal spontaneously to designate an object that is not actually present, and thus cannot be a direct stimulus to the signaling behavior, can one conclude that the animal is naming something.

To qualify as the equivalent of a word, Savage-Rumbaugh believes, a communicative signal must have the following four attributes: (1) it must be an arbitrary symbol that stands for some object, activity, or relationship; (2) it must convey stored knowledge; (3) it must be used intentionally to convey this knowledge; and (4) recipients must be able to decode and respond appropriately to the symbols. Let us leave aside for the moment the stipulation that communication must be intentional, a term that is of course anathema to inclusive behaviorists. But I will discuss this question below in relation to Grice's criteria for linguistic communication.

Many types of naturally occurring animal communication satisfy Savage-Rumbaugh's other criteria, at least in general terms. For example, the honeybee waggle dances include an arbitrary symbolism conveying direction relative to an invisible sun or pattern of polarization of the blue sky; this knowledge has been stored by the forager, and recipients decode the symbols and react appropriately. But Savage-Rumbaugh points out that many of the vocabulary tests reported for Washoe and other signing apes (and also for dolphins and sea lions) were limited to demonstrating what linguists call receptive competence, the ability to understand or at least to react appropriately to a particular communicative signal.

Savage-Rumbaugh argues that only when the ape also uses the symbol to convey to others something they did not already know does it begin to serve as something like a word. The waggle dances of honeybees convey to other bees important information they did not have previously. Signing apes ordinarily begin to use a symbol as some sort of request; they learn to use ASL signs, plastic tokens, or keys on a keyboard to manipulate the behavior of their companions. This is one type of what linguists call productive, as opposed to receptive, communication. Again, most natural animal communication includes both productive and receptive use of communicative signals, but many experimental tests of language-like behavior learned by animals have been limited to one of these two fundamental attributes.

Another sort of objection against equating the performance of the signing apes with human language has been raised by philosophers such as Grice (1957) and Bennett (1964, 1976, 1988, 1991). The former argued that for true language use a speaker must intend to communicate to a listener, must intend to induce a belief, or a change in belief, in an audience, and intend that the communication be recognized by the audience as having such intent. Bennett suggests that apes such as Washoe use signs as injunctions, requests, commands, and the like whose purpose is to produce a desired behavior on the part of the recipient rather than to change the recipient's beliefs, as Bennett believes most human language users seek to do.

It is appropriate to point out that these definitions of language also exclude much human speech that is uttered without any intent to affect the belief of a listener. In the case of nonhuman animals, it is difficult enough to gather even suggestive evidence about their beliefs; it is doubly difficult to obtain any hints as to whether they try to change beliefs as well as behavior of others. Intention to affect the belief of a listener requires some sort of reflective consciousness, that is, thinking about thoughts. As discussed above and in chapters 1 and 14, the available evidence is far too limited to support any definite conclusion about the occurrence of nonhuman reflective consciousness. Thus, here as elsewhere in the area of animal mentality, cautious agnosticism is prudent, and these more philosophical questions about just what animals intend to achieve by their communicative behavior must remain open until better methods are developed to learn more about just what, if anything, they are consciously thinking.

Still another feature of human language that is generally believed to be lacking in all animal communication systems, including those of the signing apes, is creativity, that is, the ability to conceive and convey new messages, different from anything ever thought or said before. It is often claimed that an animal's communication is limited to a few relatively fixed signals that are genetically determined, and that it is incapable of producing novel communicative signals in newly arisen circumstances. But animal signals are not always rigid and invariant; in fact, one contrast with human language that is often emphasized is the graded nature of many animal signals that are believed to be expressions of emotion varying only in intensity but lacking any semantic content. As mentioned above, a few clear exceptions to this generalization are now known, such as the semantic alarm calls of vervet monkeys discussed in chapter 9.

If animals did spontaneously produce novel signals, these would be difficult to identify because they would probably seem to be meaningless variations of known signals. Some signs do seem to have been "invented" by Washoe and other signing apes, but it is very difficult to be sure that no precursors were included in the rich variety of signing and other social interactions with human caretakers and trainers. We have one clear example of complex animal signals that change over time in the songs of male humpback whales, which undergo gradual changes from season to season (Payne and Payne 1985). But it unfortunately is not yet clear what messages these songs actually convey, so there is no way to tell whether they constitute new or altered signals or changes in the acoustic features conveying the same basic message.

Savage-Rumbaugh and her colleagues at the Yerkes Laboratory have improved upon the earlier studies of language-like behavior in apes in several important ways, as described in a monograph dedicated to Sherman and Austin, two of her star subjects (Savage-Rumbaugh 1986). These two male chimpanzees learned to use tools to open containers from which they could obtain desired foods, as other Great Apes have done (for examples see Lethmate 1977). But Sherman and Austin also learned to use the Yerkes keyboard system to ask each other to pass the needed tool through a small window. Previously they had learned to request desired foods, but they required much additional training before moving on to what for chimpanzees is apparently a more difficult task, namely, learning to use the keyboard both to transmit and to receive information. Much ingenious experimentation was required to teach them to use the keyboard for this type of cooperative communication, but at many points in the long and complex series of attempts, both Sherman and Austin seemed suddenly to grasp what they had to do and then proceeded to learn new applications of the same basic type of communicative behavior rapidly and with relatively few errors.

As one reads Savage-Rumbaugh's detailed and meticulous account of the gradual acquisition by Sherman and Austin of the ability to request the appropriate tool needed to obtain food from a particular type of container, it becomes evident that these animals gradually "caught on" to this novel and complex task, and that once they understood what to do, they proceeded with enthusiasm to apply their new skill. Blind tests of their abilities included situations in which Austin watched a teacher place food in a type of container from which it could be removed only by using a particular tool. For example, the food might be placed in the middle of a long, transparent, horizontal tube from which it could be removed by poking with a long, thin stick. The teacher then moved out of the room, and Austin went to a keyboard inside the room and requested the stick by pressing the appropriate key.

The keyboard was some distance from the food container, and by the time the keyboard was turned on, the teacher was out of Austin's sight and hence unable to provide any inadvertent cues as to which key he should press. Six tools were available outside the room: a key used to open various locks, coins that operated a vending machine, straws to obtain liquids through small holes, long sticks needed to push food out from the middle of the horizontal tube, sponges to soak up liquids from vertical tubes, and a socket wrench to open various bolted doors. Both Sherman and Austin had used all these tools extensively to obtain desired foods, but in these tests they had to obtain them by pressing the appropriate key while alone in the room, then walk outside to the tray of tools. Here the teacher was equipped with a projector that displayed which key the ape had pressed while still alone inside the room, and if it was the correct one, he was given the requested tool and could carry it back to the baited container and use it to get the food that was otherwise inaccessible. These tests ruled out inadvertent cuing even more convincingly than earlier experiments by the Gardners and others.

After Sherman and Austin had learned to communicate about tools, the situation was gradually changed so that in order to obtain the food, one of them had to request that the other give him the appropriate tool. The two were located in adjacent rooms, but they could see each other through a window equipped with a small opening through which tools or pieces of food could be passed back and forth. Yerkes Laboratory keyboards were located conveniently beside each window, so that Sherman and Austin could communicate by pressing the keys while watching each other through the window. When the keyboard system was turned on, touching a key caused it to be backlit, so that both chimpanzees and the experimenters could see which key had been activated. In some experiments there was also a large replica of the keyboard on a wall of the room, enabling all concerned to see which key had been pressed.

At the start of each experimental session the window was covered, and food was placed in one of several types of closed containers. This was clearly visible to the chimpanzee in that room, but the tools necessary to open the container were located in the other room. When the cover was removed from the window, the first chimpanzee used his keyboard to ask the other to provide the type of tool needed to open the particular container in which the food had been placed. After the correct tool was passed through the small opening in the window, the first chimpanzee opened the container and passed at least some of the food to the animal in the adjacent room who had provided the necessary tool.

It took a considerable amount of training to teach these procedures to Austin and Sherman, especially the sharing of food, but once they had learned how to engage in this cooperative communication, they repeatedly and efficiently followed these procedures to obtain desired foods. And they both became adept at playing both roles. If the keyboard was turned off so that no keys were illuminated, the appropriate tool was handed over only at about a chance level, indicating strongly that the keyboard communication was necessary and that no other unrecognized form of information transfer would suffice. During the first five days when such cooperative communication about tools was called for, the percentage of correct signs rose from 77 percent correct for Sherman and 91 percent for Austin out of 47 trials on the first day to 91 percent for Sherman and 97 percent correct for Austin out of 90 trials on the fourth and fifth days combined. Somewhat more errors were made in requests than in the tools provided, but the whole process was remarkably effective. On the sixth day the keyboards were turned off so that the second ape could not see which key his companion had pressed, and the proportion of correct scores fell to 10 percent.

Many additional indications that this cooperative communication was fully intentional were provided by the behavior of Sherman and Austin when requests for a tool did not lead to an appropriate response. Then the first chimpanzee often gestured vigorously, tried to direct his companion's attention to the keyboard, and pressed the requesting key repeatedly and emphatically. Savage-Rumbaugh (1986) provides detailed descriptions of the many complex ways in which Sherman and Austin interacted while communicating and sharing food and of the apparent game-playing in which they often indulged. These accounts make it abundantly clear that these two apes were intentionally communicating by means of the Yerkes keyboard as well as by gesturing, and that they not only understood but often enjoyed what they were doing.

Until the 1980s, language-like behavior had been studied primarily in common chimpanzees *(Pan troglodytes)*, with a few studies also showing that similar signing behavior can be learned by gorillas and orangutans. But there is another species of the genus *Pan:* the pygmy chimpanzee, or bonobo *(Pan paniscus)*. One of the two chimpanzees first studied in detail by Yerkes was a bonobo, and although they are less commonly available in captivity, bonobos are clearly more versatile than common chimpanzees. A male bonobo named Kanzi, born at the Yerkes Laboratory, has demonstrated much greater communicative spontaneity than Sherman and Austin or other common chimpanzees, as described by Savage-Rumbaugh (1986), Savage-Rumbaugh and Lewin (1994), Savage-Rumbaugh et al. (1989), and Savage-Rumbaugh, Shanker, and Taylor (1998).

As a dependent youngster, beginning when he was 6 months old, Kanzi accompanied his mother during prolonged efforts to teach her to use the Yerkes keyboard, but no effort was made to teach him which keys had which meanings, or indeed to use the keyboard at all. But the trainers used gestures and spoken English to communicate with him, and he had abundant opportunity to observe how the keyboard was used for communication by his human companions, and to a limited extent by his mother. Beginning at about 18 months, Kanzi spontaneously began to use gestures such as pointing in directions he wished to be carried or making twisting motions toward containers when he needed assistance in opening their lids. When he was two and a half years old his mother was withdrawn from the training, partly because she was making very little progress.

At that time Kanzi began using the keyboard to request desired objects and even to communicate about things that were not present, such as desired foods or locations to which he wished to travel. This took place without any specific training; evidently he had discovered for himself how the keyboard operated by watching the human trainers use it, and also perhaps by watching their relatively unsuccessful efforts to teach its use to his mother. By the time he was 46 months old, Kanzi was using at least 80 keys more or less as words. A symbol was "classified as a member of Kanzi's vocabulary if and only if it occurred spontaneously on 9 of 10 consecutive occasions in the appropriate context *and* was followed by a behavioral demonstration of knowledge of the referent. For example, if Kanzi requested a trip to the 'treehouse' he would be told, 'Yes, we can go to the treehouse.' However, only if he then led the experimenter to this location would a correct behavioral concordance be scored" (Savage-Rumbaugh 1986, 389).

Kanzi's use of the keyboard is much more spontaneous than the signing performed by Nim, which led Terrace and others to conclude that such signing had little in common with human language. The majority of Kanzi's "utterances" via the keyboard are not imitations of key presses by his human companions. He also uses many more, and longer, combinations of key presses than Nim exhibited with his signing, although Savage-Rumbaugh does not feel that these are rule-governed like human grammar. But to a much greater extent than with Nim, the individual key presses add meaning to previous members of a string. Kanzi also uses many gestures, which add meaning to his key presses. In well-controlled blind tests in which the experimenters cannot see pictures shown to Kanzi, he can identify what is shown in the picture by pressing the corresponding key. He can also do this in response to spoken words. Indeed he can answer simple spoken questions by means of the keyboard (Savage-Rumbaugh, Shanker, and Taylor 1998).

Studies described in detail by Greenfield and Savage-Rumbaugh (1990) have demonstrated that, unlike Nim in Terrace's earlier studies, Kanzi learned to use a few very simple grammatical rules in his communication with his human companions. He used these rules productively, that is, in his communicating to others with the keyboard, and he also comprehended their use by others. He even invented a few of his own rules. Kanzi's accomplishments thus call into serious question the generalization that only human language employs meaningful rule-governed combinations of individual communicative elements. Furthermore, his invention of simple but meaningful combinations suggests combinatorial productivity, a property that had previously appeared to be limited to human language, as emphasized by Miller (1967).

In short, Kanzi, and to a lesser extent other bonobos at the Yerkes Laboratory, have learned to use a combination of gestures and the Yerkes keyboard to achieve two-way communication with their human and bonobo companions. The versatility and spontaneity with which Kanzi does this, together with the behavioral concordance between what he asks for and what he does subsequently, make it abundantly clear that he can voluntarily communicate simple desires and intentions. His communication has certainly served as an effective "window" on what he is feeling and thinking. The richness and versatility of both his communication and the thoughts it conveys is greater than that displayed by other animals, even by other Great Apes.

The obvious significance of these discoveries about the acquisition of language-like communicative abilities in the Great Apes should be viewed against a background of many other demonstrations that these animals possess superior mental capabilities. A thorough review of primate intelligence is far beyond the scope of this book, but some recent discoveries are especially significant. Although Caro and Hauser (1992) have reviewed the available data and find very few examples of teaching in nonhuman animals, Boesch (1991, 1993) has observed that occasionally mother chimpanzees actively demonstrate to their young how to open hard nuts by the use of stone tools. This appears to be a clear case of intentional instruction, despite the widespread belief (reviewed by Tomasello and Call 1997, 307) that teaching is a uniquely human ability.

These many discoveries about the mental and communicative skills of both dolphins and apes demonstrate how cognitive ethology can progress by careful and critical observations and experiments, especially by using the "window" of communication. They also confirm that the Gardners were quite correct in their earlier conclusions based on experiments with Washoe and other signing chimpanzees. The original experiments have been greatly improved upon, and more conclusive blind tests and other procedures have answered the many criticisms advanced to avoid the conclusion, which seems so unpalatable to many behavioral scientists and others, that apes can use the equivalent of words to communicate a rich array of feelings and thoughts. Human language and thought may be, in the words of Donald (1991, 136), "light-years removed from Kanzi's accomplishments." But the ape language experiments have clearly demonstrated evolutionary continuity between human and nonhuman capabilities of communication and thinking.

CHAPTER THIRTEEN

The Philosophical and Ethical Significance of Animal Consciousness

Three basic and interrelated reasons for our concern with animal mentality stand out as especially significant. For convenience they may be designated as *philosophical, ethical,* and *scientific.* This chapter will briefly review why the first two are important and suggest how our current ignorance about them can be reduced; chapter 14 will consider the scientific importance of the subject. The philosophical importance of animal consciousness lies in its relevance to the general question of other minds. The ethical importance lies in the widespread belief that causing pain and suffering to a conscious creature is morally wrong in an important sense not applicable to an unfeeling mechanism. And the scientific importance lies in our interest in the animals themselves. We want to understand what the lives of these other creatures are like to them. For it is a profoundly significant question whether only members of our species are conscious.

Previous chapters have discussed several kinds of animal behavior that suggest conscious thinking, especially cases in which animals communicate messages that appear to be expressions of simple thoughts or feelings. These examples have been selected primarily because evidence is available about the versatility of an animal's behavior as it adjusts to challenges that would seem difficult if not impossible to predict. It is unlikely that either evolutionary selection or learning from previous experience could provide specific and detailed prescriptions for coping with the multitude of unpredictable challenges the animal is likely to encounter. If members of a particular species are capable of perceptual consciousness (Natsoulas' Consciousness 3, defined and discussed in chapter 1) under some conditions, it seems likely that they make use of this ability in other situations where it is helpful. And, as pointed out in chapter 1, it is obvious that thinking about the probable outcome of various possible actions must be very useful in a wide variety of situations in which animals must make choices that have important effects on their survival and reproduction. Thus the cases in which suggestive evidence of conscious thinking has become available could well be the tip of a large iceberg.

A conscious organism is clearly different in an important way from one that lacks any subjective mental experiences. The former thinks and feels to a greater or lesser degree, whereas the latter is limited to existing and reacting. One important difference between unconscious and conscious thought is that the latter includes paying attention to internal images or representations, that is, thinking about them to oneself. Such representations may involve any sensory modalities; they may be directly elicited by contemporary external stimulation, they may be based on memories, or they may be anticipations of future events. They can also be literal imagination of objects and events that do not actually exist.

But does it really matter whether any animals are ever conscious, which ones, under what conditions, and what the content of any animal consciousness may be? Strict behaviorists argue that it does not matter, thereby impaling themselves on the horns of a dilemma. Either they must deny the importance of human consciousness, or they must accept its importance but hold that no other species can be conscious to a significant degree. As reviewed in chapter 2, many scientists claim that the distinction between conscious and unconscious mental states is an empty and meaningless one, at least when applied to nonhuman animals, because, they say, anything an animal does might be done without any accompanying consciousness. In one sense this is simply a denial of concern, a confession of limited interests. But it is often combined with an appeal to scientific parsimony and an insistence that consciousness is a needless complication with which scientists should not be concerned. Some contend that it does not matter whether animals, or even people, are conscious of anything at all. This attitude may lead to such sweeping and dogmatically negative pronouncements as "The idea that people are autonomous and possess within them the power and the reasons for making decisions has no place in behavior theory" (Schwartz and Lacey 1982, 16).

Philosophical Issues

Many inclusive behaviorists see no way to determine what thoughts or feelings, if any, are experienced by a member of another species. And philosophical purists of the school known as skepticism make essentially the same argument about us. Perhaps the most appropriate response to both these claims is to point out that total perfection of argument and proof is no more available in science than in general affairs. We can only make stronger or weaker claims with a higher or lower probability of correctness. Paralytic perfectionism discourages investigation of challenging problems. Even when vital practical decisions are at stake, we have no choice but to act on whatever interpretation appears most likely to be correct, based on the most balanced assessment of available evidence of which we are capable. We do not know whether conscious mental experiences are correlated with any specific and identifiable states or activities of central nervous systems, although of course certain parts of the human brain such as the prefrontal cortex are clearly of great importance. But our ignorance of the neural mechanisms of consciousness does not mean that conscious experience is nonexistent.

Midgley (1983), Radner and Radner (1989), and Dawkins (1993) have lucidly exposed the limitations and inconsistencies of behavioristic denigration of animal consciousness. Midgley emphasizes the overwhelming advantages for understanding animal behavior provided by the assumption that animals experience simple feelings, fears, desires, beliefs, and the like. A literal adherence to the behavioristic prohibition against consideration of any subjective mental experiences tends to render much animal behavior unintelligible. Humphrey (1980, 60) has forcefully made a similar point with respect to social animals:

> Academic psychologists have been attempting by the "objective" methods of the physical sciences, to acquire precisely the kind of knowledge of behaviour which every social animal must have in order to survive . . . [but] they have been held up again and again by their failure to develop a sufficiently rich or relevant framework of ideas. . . . Indeed, I venture to suggest that if a rat's knowledge of the behaviour of other rats were to be limited to everything which behaviourists have discovered about rats to date, the rat would show so little understanding of its fellows that it would bungle disastrously every social interaction it engaged in.

As mentioned in chapter 1, we can draw a helpful analogy with the history of genetics. The reality of heredity has always been obvious, but the biological mechanisms by which offspring come to resemble their ancestors to some degree, though of course not completely, was largely explained during the twentieth century by a series of interrelated biological discoveries: gametes, chromosomes, genes, nucleic acids. The gene was originally a theoretical construct, and only much later were genes found to consist of DNA and RNA.

Before the mid-twentieth century, geneticists had no way of knowing what genes actually were, although they had good reason to infer that they must exist within chromosomes. Had the study of heredity been impeded by taboos comparable to those that discourage students of animal behavior from investigating animal mentality, the progress of genetics would have been seriously and needlessly impeded. To be sure, geneticists could gather relevant empirical data from breeding experiments, whereas behaviorists deny that any objectively verifiable data about mental experience can ever be obtained. But expressive communication and versatility of behavior have already provided objectively verifiable evidence, and with further refinement such evidence can lead to even more convincing data on which to base inferences about the mental experiences of animals.

Many philosophers have wrestled with the question of how we can know anything about the minds of others, whether they be other people, animals, extraterrestrial creatures, or artifacts such as computers. Some are convinced that other minds are found only in our species; others consider it possible, or even likely, that they can also be found in animals or computers. Philosophers struggle to devise logical and reasonable criteria for the presence of minds and consciousness, criteria that can be applied to animals or to computer systems as well as to borderline human cases such as newborn infants or persons with severe brain damage. When many of these criteria were first proposed they seemed to be impossible for any nonhuman animal to satisfy. But, as emphasized in previous chapters, and reviewed in two earlier books (Griffin 1976, 1984), increasing understanding of animals and their behavior has often disclosed cases in which the criterion in question is satisfied after all.

James (1890) and Popper (1987) were concerned with the nature of animal minds; and recently Arhem and Lindahl (1997) and Lindahl (1997, in press) have discussed the evolution of consciousness, emphasizing that since it occurs in one species it presumably increases inclusive fitness. The adaptive attributes of consciousness may well be effective for other species as well, as discussed in chapter 12 in relation to Humphrey's suggestion that it enables primates to be effective "natural psychologists." The philosopher Dennett (1983, 1987, 1988) visited Seyfarth and Cheney at their study area in Africa to see how they observed and analyzed the semantic communication of vervet monkeys, as discussed in chapter 9. This has led to a stimulating and helpful exchange of ideas concerning the degree of intentionality displayed by these monkeys in their communication about predators, even though Dennett (1989) has described the inquiry into possible animal consciousness as a wild goose chase.

Sober (1983) lucidly analyzed the limitations of behaviorism and the problems that have discouraged psychologists and others from devoting much attention to investigation of the beliefs, desires, intentions, and the like that appear to influence behavior. He points out that many of the behaviorists' objections to mentalism can be applied with equal force to behaviorism itself. For example, the claim that mentalism is too easy because one can always dream up some hypothetical mental state to explain any behavior can be countered by pointing out that when behaviorists claim that whatever a person or animal does must have resulted from prior conditioning, they are often stating an assumption about such causation rather than being able to identify in any plausible fashion just how the person was conditioned. Thus the argument that statements about mental states cannot be verified or falsified can be applied with equal force to many behavioristic explanations. Sober also points out that even though the beliefs or other mental states of animals cannot have as rich a content as their human counterparts, this is no reason to deny their existence or significance.

Lycan (1987) discussed the philosophical issues surrounding the nature of consciousness from the viewpoint of a "teleological functionalist" who recognizes the likelihood that mental experiences constitute an evolutionary continuity like other biological characters. He considers mental experiences to have the property of *intentionality,* used in the philosopher's sense of "aboutness," that is, they relate to something, real or imagined. He holds that "to be in an intentional state is to host a mental representation, a brain state that bears a natural (causal and teleological) relation to the object represented or, in the case of abstract or nonexistent objects, to linguistic events that go proxy for them" (71). This definition is quite applicable to mental representations in animals, especially if we extend the customary idea of linguistic events to include messages conveyed by animal communication.

Thomas Nagel (1974) stimulated the interest of many philosophers in such questions by inquiring what it is like to be a bat, as discussed in chapter 1. He concluded that because bats are so different from us, and especially because they rely so heavily on echolocation, we can never know precisely what life is like to an insect-eating, sonar-guided flying mammal. But Nagel does not deny that partial understanding and significant, though incomplete, information about the experiences of bats or other animals can be deduced from their behavior. Although he may be quite correct that we cannot hope for perfect descriptions, we can avoid paralytic perfectionism and make substantial progress; and informed inferences about animal thoughts can eventually come to be as well grounded as the conclusions reached in many other areas of biological inquiry.

In the case of bats specialized for echolocation, we can base some preliminary estimates on the similarity between the echolocation of bats and the detection of obstacles by the human blind by means of sounds and echoes, as discussed in chapter 1. Blind people, and blindfolded volunteers who have had considerable practice, can detect and classify objects in their vicinity by emitting audible sounds and hearing subtle differences depending on the presence or absence of the object. But, curiously enough, many of the most proficient do not consciously recognize that they are accomplishing this by the sense of hearing. Instead they report that they simply feel that something is out there. (A common term for this ability is "facial vision" [Griffin 1958; Rice 1967a, 1967b].) Nevertheless, the feeling and the alleged "vision" cease almost totally if they can make no sounds or if their hearing is blocked.

We might guess that when bats detect and identify insects by echolocation they perceive not a pattern of echoes but rather the presence and location of an object with certain properties—an edible insect or a falling leaf. An animal's perceptual consciousness may well relate to objects in the outside world in ways that do not differ greatly whether the sensory information on which perception is based is visual, auditory, or in a modality even more remote from our experience, such as the electric sensing of weakly electric fishes.

Searle (1984, 1990a, 1990b, 1992, 1998) has critically analyzed the cogent reasons for recognizing that conscious thinking is of the greatest significance. Although he discusses primarily human consciousness, several of his arguments are equally applicable to animals. For example, after explaining the philosopher's definition of intentionality as "that feature of certain mental states or events that consists in their . . . being directed at, being about, being of, or representing certain other entities and states of affairs," he writes: "Consider the case of an animal, say a lion, moving in an erratic path through tall grass. The behavior of the lion is explicable by saying that it is stalking a wildebeest, its prey. The stalking behavior is caused by a set of intentional states: it is *hungry,* it *wants* to eat the wildebeest, it *intends* to follow the wildebeest with the *aim* of catching, killing, and eating it" (1984, 14–15).

After emphasizing that the rejection of teleology by seventeenth-century physics, and by the Darwinian account of the origin of species were "liberating steps," Searle comments: "But ironically the liberating move of the past has become constraining and counterproductive in the present. Why? Because it is just a plain fact about human beings that they do have desires, goals, intentions, purposes, aims, and plans, and these play a causal role in the production of their behavior. . . . . Just as it was bad science to treat systems that lack intentionality as if they had it, so it is equally bad science to treat systems that have intrinsic intentionality as if they lack it" (1984, 15).

As discussed in chapter 2, Searle (1990, 585) had thought that

> the major mistake we were making in cognitive science was to think that the mind is a computer program implemented in the hardware of the brain. I now believe the underlying mistake is much deeper. We have neglected the centrality

of consciousness to the study of mind. . . . If you come to cognitive science, psychology, or the philosophy of mind with an innocent eye, the first thing that strikes you is how little serious attention is paid to consciousness. Few people in cognitive science think that the study of the mind is essentially or in large part a matter of studying conscious phenomena: consciousness is rather a "problem," a difficulty that functionalist or computationalist theories must somehow deal with. . . . As recently as a few years ago, if one raised the subject of consciousness in cognitive science discussions, it was generally regarded as a form of bad taste, and graduate students, who are always attuned to the social mores of their disciplines, would roll their eyes at the ceiling and assume expressions of mild disgust.

In a presidential address to the American Philosophical Association titled "Thoughtless Brutes," Malcolm (1973) expressed his dismay at the idea that dogs, and presumably other animals, might experience thoughts that they could not express for lack of language. "The relationship between language and thought," he continued, "must be so close that it is really senseless to conjecture that people may *not* have thoughts, and also really senseless to conjecture that animals *may* have thoughts" (17). But others have dissented from such a necessary linkage between language and thinking. For example, Ferguson (1977) emphasized the important role of pictorial thinking in technology. Thus, as discussed in detail by Alport (1983), many scholars have abandoned the formerly widespread belief that human language is essential for conscious thinking.

Other philosophers have been interested in animal mentality, and some have expressed a general vierwpoint similar to Malcolm's. For example, the neurophysiologist Eccles debated these basic issues with Savage in a book edited by Globus, Maxwell, and Savodnik (1976). Eccles (159) argued that "in the biological world only human beings are endowed with a self-consciousness, and with a cultural creativity, and they are distinguished completely from animals by the ability to think logically, creatively, and imaginatively and to communicate these thoughts in every medium of cultural expression." Savage countered (152) with the opinion that "animals, too, have souls: feelings, desires, purposes, thoughts, consciousness, rights—the same rights to life and to the absence of pain that we accord to humans . . . refusal to make this concession seems, to this author, to be the product of human vanity." In this context Savage seems to mean by the word *soul* something close to conscious mental states rather than anything necessarily immaterial or spiritual.

Johnson (1988, 282, 288) claimed that most animals, except possibly the Great Apes, are incapable of believing something. He agrees with

D. Armstrong (1981) that animals perceive and that their perceptions affect their behavior, but he differs from Armstrong in holding that something more is necessary to qualify as a belief. This additional requirement, Johnson argues, includes a causal role for a true belief in affecting behavior, although it is difficult to see that this distinction is fundamental since he recognizes that perceptions influence an animal's behavior. He supports his position by claiming that animal behavior is determined by "internal processes very different from those humans employ." Although brain mechanisms do not appear to differ greatly between species, Johnson seems to be relying on the widespread assumption that human mental experiences are the only possible kind.

Another interpretation is that Johnson and others are coming to recognize that animals probably do experience simple conscious thoughts, that is, they are capable of perceptual consciousness, although human thoughts are held to be the only ones worthy of serious consideration. In this vein Johnson asserts as self-evident that "flies do not reason on the basis of meanings or goals, but simply react automatically to cues." He further claims that animal thinking "takes no account of meanings" (282), but he gives no reasons for excluding, for example, the possibility that when an animal sees a distant predator this means that appropriate escape behavior is called for. He seems to take it for granted that only our species is capable of internal processes qualifying as beliefs, so his arguments are little more than reiteration of a prior conviction.

Here we see a philosopher basing his arguments about other minds on opinions about animal behavior, in particular the widespread conviction that all insect behavior is rigidly stereotyped. The goals of insects may be simple compared to the theories of philosophers. But this provides no firm basis for dogmatically ruling out conscious perception of simple meanings and desired goals. A vervet monkey's alarm call may well mean, to the monkey as well as to us, that it has seen a particular kind of predator, and a honeybee's waggle dance may mean to her that food is located at a certain distance and in a certain direction.

Johnson goes on to offer as evidence against the presence of beliefs the fact that birds can be fooled when several hunters enter a blind and a smaller number leave it. He denies that they can count but admits that ducks must have some sort of perception of numerousness, so that if the numbers involved are small enough, say, two entering and one leaving, they may not be deceived, whereas a man having counted seventeen entering and sixteen leaving would realize that a dangerous hunter must still be inside (285). But in Johnson's example it seems likely that the duck believes there is or is not a hunter still inside the blind, even though its ability to count may be very limited. There must be some upper bound to human abilities to make such counts; any of us might be confused if 140 hunters entered the blind and only 139 left, so the distinction is a quantitative one. Thus such examples do not constitute valid evidence against the ability of animals to believe something simple and important in their lives.

Johnson continues (288) with the argument that "the crucial distinction between (human) belief and the superficially similar adaptive strategies of animals is that the former but not the latter involves use of explicitly entertained mental representations." The wording "explicitly entertained mental representations" seems to mean thinking consciously about one's beliefs, and Johnson claims that only human beings can do this. But how can anyone be so confident of this dogmatic negative assertion, especially when some animals, such as the dancing bee that seems to believe there is a very desirable cavity at a certain distance in a particular direction, communicate about what appear to be simple beliefs?

Bennett (1964, 10) also considered the question of animal communication from a philosophical viewpoint, with special emphasis on the dances of honeybees. He developed at length a philosophical argument that "honeybees are not rational creatures, and also that their dances do not constitute a language." Most of his argument was based on the fact that honeybee communication is not nearly as flexible and applicable to as wide a range of situations as human language. Bennett also doubted that bees could alter their communication rationally as a result of unusual circumstances, such as the receipt of a message that the follower of a dance knew to be erroneous or implausible (1964, 58). He did not consider the evidence that at least a few bees that have danced about one cavity follow dances that report either the same or a different cavity, as discussed in chapter 10. Some of these then change their previously executed dance communication to follow a more enthusiastic dancer, and some then visit and dance about the cavity she was describing. A change of message on the basis of information received from another bee's dance seems quite rational.

Bennett was especially concerned, however, with the question whether bees could deny a message received by following waggle dances. He assumed that they could not do this, but Gould (1984) and Gould and Towne (1987) reported experiments in which something of the kind may actually have happened, although the data are not sufficient to provide a fully convincing case. Honeybees are reluctant to fly over water, and when hungry bees familiar with the local environment were exposed to waggle dances that signaled a food source in the middle of a small lake, many fewer left the hive than when they were stimulated by similar dances signaling food at equidistant locations over dry land. Other variables may have affected these results, but this type of experiment holds considerable promise for elucidating the degree to which honeybees may combine memories of topographical features with the information received from dances to guide rational responses to the messages they receive in this way.

More recently Bennett (1988), in "Thoughtful Brutes," obviously diverged from the views expressed by Malcolm fifteen years earlier. Recognizing the progress and promise of cognitive ethology, Bennett (1991) nevertheless emphasized the serious philosophical difficulties that are encountered when we attempt to infer what nonhuman animals are thinking.

Premack, whose experiments with chimpanzees have demonstrated many of their mental abilities through carefully controlled experiments (Premack 1976), has argued that the dance communication of honeybees does not qualify as language, primarily because he considers it too rigid and inflexible. One of his comments is especially significant as an indication of the reluctance of inclusive behaviorists to appreciate the versatility of animal communication: "Ordinarily, the contrast between bee and human language is made on the grounds that only one of the two systems is learned, but this is a dubious contrast, since critical aspects of human language, including parts of both syntax and phonology, are probably not learned. More important, even if the bee's unique system *were* learned it probably would not qualify as language. The two systems can be better contrasted by asking if the bee shows any suggestion of representational capacity" (1980, 212). Premack continues,

> Suppose a scout bee were to gather information about the direction and distance of a food source from its hive. The bee encodes this information in its dance, and a second bee decodes the dance; but could the bee, when shown its own dance, judge whether or not this dance accurately represented the direction and distance of the source of food? Could the bee recognize that dance as a representation of its own knowledge? If a bee could judge between the real situation and a representation of that situation, it would be possible to interrogate the bee, just as we can interrogate the ape. A species that can be interrogated, such as the chimpanzee, is well on its way toward being able to make true-false judgments. But I know of no data that even faintly suggests that the bee can recognize the dance as a representation of its knowledge. (1980, 211–12)

When Premack speaks of showing a bee her dance, he implies presenting a visual representation. But bee dances ordinarily occur in the dark, and the information is transferred by tactile, acoustic, and chemical signals. Therefore an appropriate form of "showing" would be to reproduce the nonvisual signals that a bee receives from a dancer. Dancing bees often encounter other bees dancing about the same food source or cavity, and sometimes a bee that has danced follows other dancers. When a model bee such as the one developed by Michelsen, Kirchner, and Lindauer (1989) has been perfected, a bee that has been dancing could easily be stimulated by dances of the model indicating either the same or different direction, distance, and desirability.

In detailed studies of honeybees dancing on a swarm, Camazine et al. (1999, 355) have recently observed that most of these dancers "also followed dances of other scouts, for both their own site and alternate sites, and interspersed this following with their own dances." Thus bees do sometimes follow both "synonymous" dances conveying the same information as their own dances and other dances that convey a different message. This is not exactly "showing a bee a dance" and inquiring whether she judges that the "synonymous" dances really do describe the cavity she has visited. But these bees seem to be doing something equivalent to comparing the two messages.

Camazine et al., as well as Lindauer, observed a few cases where a dancer changed her message and became a follower of more enthusiastic dances of another bee reporting a better cavity than the one she had visited, as discussed in chapter 10. In other words, a bee does sometimes change her behavior as a result of this comparison by visiting and dancing about cavities she would not have known about without receiving information from dances that differed from her own. This difference in behavior with respect to dances that are the same and those that are different from her own indicates recognition that the former are indeed representations of her own knowledge, while the latter are communicating a different message. On balance, it is difficult to understand why Premack is so certain that honeybees lack any representational capacity, when their dances obviously do represent distance, direction, and desirability and convey such representations to others.

The explanation for the dogmatic negativity displayed by Premack and others such as Rosin (1978, 1980, 1984, 1988) or Wenner and Wells (1990) about the symbolic communication of honeybees may lie in a deep-seated reluctance to grant anything remotely comparable to human mentality to "lower" animals, as expressed in the following passage: "Mollusks, spiders, insects—invertebrates generally—differ from humans not in their lack of hard-wired components but in their lack of cognition. Griffin's phrase 'cognitive ethology,' when applied to invertebrates, appears to be a colorful misnomer rather like 'tropical Norway' or the 'nautical jungle.' Vertebrates associate events not only on the basis of contiguity in space or time, but also on the basis of physical resemblance.

Such species categorize or place 'like' items together. . . . Invertebrates, presumably, could not be trained to demonstrate their comprehension of a rule instantiating physical similarity" (Premack 1986, 137–38). Premack seems to be thinking of tasks in which objects are sorted, but equivalent judgments of physical similarity, or of belonging in a given category such as food or danger, must be very widespread in the natural lives of many active invertebrates.

The contemporary philosopher Daniel Dennett (1983, 1987, 1988, 1991) has advocated what he calls "the intentional stance" when analyzing not only human and animal cognition but also many examples of self-regulating inanimate mechanisms. As he defines it, "the intentional stance is the strategy of prediction and explanation that attributes beliefs, desires, and other 'intentional' states to systems—living and nonliving" (1988, 495). His insistence on including such simple devices as thermostats in this extended category of intentional systems leads him to deny any special status to conscious mental experiences. This denial of concern for subjective mental experiences is clearly articulated in responses to commentaries on his two articles in *Behavioral and Brain Sciences* (Dennett 1983, 380–84, and 1988, 538–39).

Dennett appears to be arguing that if a neurophysiological mechanism were shown to organize and guide a particular behavior pattern, this would rule out the possibility that any conscious mental experiences might accompany or influence such behavior. He is clearly uncomfortable with the idea that nonhuman animals might be conscious, and he tends to explain away consciousness rather than trying to explain it. He is closely enough akin to the positivists and the behaviorists to base his inference of intentionality entirely on overt behavior. But, to quote Lloyd (1989, 191), "neurons take care of every need in every brain, so this argument for denying consciousness to the toad opens up to a slippery slope. Surely bats are open to the same counterargument, and . . . ultimately human beings."

Dennett prefers a theoretical framework that encompasses the whole range of systems from thermostats to scientists and philosophers. Yet he applies terms that ordinarily refer to conscious mental states, such as *belief* and *desire*, even to thermostats. This amounts to a sort of semantic piracy in which the meaning of widely used terms is distorted by extension in order to paper over a fundamental problem—namely, the question whether conscious mental experiences occur in other species. Additional limitations of Dennett's position have been discussed by Baker (1987, 149–66).

Other contemporary philosophers have argued that the very notion of such conscious mental states as belief and desire is misguided and obsolete. Stich (1983) and Churchland (1986) have likened all consideration of such mental experiences to the phlogiston theory of combustion and to a belief in witchcraft. They and others apply the derogatory term "folk psychology" to any reliance on conscious mental experience in describing or explaining behavior. They confidently predict that a growing understanding of brain function will lead us to replace such mentalistic terms with specific neurophysiological concepts that refer to the brain functions leading to particular sorts of behavior. A serious weakness with such arguments is that none of the required neurophysiological mechanisms has yet been identified. We are asked to replace valid and useful terms and concepts with what have been called "promissory notes," that is, unspecified physiological mechanisms that, it is said, will be discovered in the uncertain future.

This difficulty has left many philosophers and neuroscientists—for example, Routley (1981), Double (1985), Horgan and Woodward (1985), Jeffrey (1985), Russow (1987), Sanford (1986), Baker (1987), Clark (1987, 1989), Stent (1987), Putnam (1988), and Radner and Radner (1989)—unpersuaded by the arguments of Churchland and Stich. Of course, brains, or central nervous systems in general, are the organs of thought. But to discard useful concepts about mental states because it is anticipated that the neurophysiological mechanisms underlying them will someday be discovered is a misleading device to escape from challenging problems by shifting them to an inaccessible future. Furthermore, the elucidation of the mechanisms by which an important process is carried out does not eliminate the process itself or render it unimportant. For example, the magnificent discoveries of molecular genetics have not eliminated heredity or rendered the term useless or misleading in any way. Likewise, if and when a neural basis for belief or desire is discovered, this will be a comparably significant advance, but it will not render obsolete these important attributes of conscious thinking.

Ethical Significance

Hardly anyone denies that there is a large ethical difference between torturing a dog or a monkey and mutilating even the most elaborate and efficient machine. The latter act may be wasteful or pernicious because it damages something useful or beautiful, but it is not wrong in the same sense as inflicting needless pain or suffering. But we can scarcely escape all responsibility for activities that directly or indirectly cause the injury, death, and suffering of other animals. For instance, even the most confirmed vegetarian would find it difficult to avoid eating vegetables that have been protected by insecticides or other agricultural practices that cause the death of insects that would otherwise have eaten the plants in question.

Thus we are all obliged to make value judgments about what activities are permissible even though they are harmful to some animal, and such decisions are often based on the degree to which we believe that various animals suffer consciously. These are difficult decisions, primarily because we know so little about the feelings and thoughts of other species, but choices must be made about the severity or the importance of animal suffering. For example, we generally exert greater care to avoid hurting mammals and birds than fishes or invertebrates, and most rules about animal welfare apply primarily or exclusively to warm-blooded animals, as discussed by Burghardt and Herzog (1980). But how do we know whether some species suffer more than others? We have so very little firm knowledge on which to base decisions and trade-offs concerning animal welfare that whatever can be learned about the subjective feelings and thoughts of animals has a direct relevance to these sorts of value judgments.

Some of the philosophers who have considered this matter feel obliged to define mental capabilities that are unique to our species and to rely on these as moral justification for treating people very differently from animals. One of the more extreme examples of this position was the argument advanced by Adler (1967, 267–68) that if it should be found that animals differ from men only in degree and not radically in kind, such knowledge would destroy our moral basis for holding that all men have equal basic rights. Like many others, he considered language to be the primary attribute that sets us apart from all other animals. It seemed to Adler that if any animal could communicate intentionally, our moral and ethical standards would be undermined.

Earlier in the same book Adler dismissed the dance communication of honeybees as "a purely instinctive performance on their part [that] does not represent, *even in the slightest degree*, the same kind of variable, acquired or learned, and deliberately or intentionally exercised linguistic performance that is to be found in human speech" (114–15, emphasis in original). He was concerned primarily with early claims that dolphins communicated by means of something resembling human language, but the more recent discoveries about the versatility of communication in both dolphins and the Great Apes pose the same basic question even more emphatically. For there can remain no doubt that these sorts of communicative behavior are variable and acquired or learned, so the general position advocated by Adler and others would now have to rely on the claim that nonhuman animals do not communicate deliberately or intentionally. This distinction has also been blurred by evidence reviewed in chapter 9 that some animals can control their communicative behavior and adjust it to the presence and nature of an audience, as discussed by Marler and Evans (1996). Thus the position advocated by Adler boils down to a denial that any animal communicates with conscious intent.

Garland E. Allen (1987, 158–59) has extended Adler's type of argument by adding a political twist: "To blur the distinction between animal and human, especially by distorting the biological reality (or by claiming for the biological reality more than it can offer), is to play into the hands of a political mood that leads ultimately to fascism; . . . [it] paves the way for relegating some people to the subhuman category on the basis of their biology. Once there, the usual moral restraints and considerations cease to apply, and fascism has arrived." In other words, cognitive ethology should not be investigated, lest the results undermine our moral standards, a view that is reminiscent of the outrage that greeted Darwin's recognition of biological evolution. Fortunately, these imagined threats to morality proved exaggerated in the nineteenth century, and there is no reason to expect a different outcome now. Morals and ethics should surely be based on accurate understanding of the relevant facts, and since they have survived the Copernican and the Darwinian revolutions, strengthened rather than weakened by correction of factual errors, there is no reason to fear a different outcome once evolutionary continuity of mentality is recognized.

More recently Carruthers (1989) has argued that only if creatures are capable of thinking consciously about their own thoughts, and of reporting what they think, do they deserve sympathy and moral concern. If their thoughts and suffering are not accessible to reportable reflective consciousness (Natsoulas' Consciousness 4), Carruthers argues,

> since their experiences, including their pains, are nonconscious ones, their pains are of no immediate moral concern. Indeed, since all the mental states of brutes are nonconscious, their injuries are lacking even in indirect moral concern. . . . Much time and money is presently spent on alleviating the pains of brutes which ought properly to be directed toward human beings, and many are now campaigning to reduce the efficiency of modern farming methods because of the pain caused to the animals involved. If the arguments presented here have been sound, such activities are not only morally unsupportable but morally objectionable. . . . Since their pains are nonconscious ones (as are all their mental states), they ought not be allowed to get in the way of any morally serious objective. (268–69)

Even though Carruthers considers human infants to be nonconscious, that is, to lack reflective consciousness, he grants them moral status because they will later be capable of such consciousness. Logical consistency would oblige him to argue that a newborn baby suffering from an incurable defect that is certain to be fatal before he can achieve reflective consciousness would therefore revert to the nonconscious status to which Carruthers assigns almost all animals, and thus become available for the same justifiable abuses. This is reminiscent of the Port Royal followers of Descartes who are said to have tortured animals with the confident conviction that their cries of agony were comparable to the noises from machinery.

Opposed to the views of Adler, Johnson, and Carruthers are the thoughtful analyses of philosophers such as Routley (1981), Midgley (1978, 1983, 1998), Jeffrey (1985), Rollin (1989, 1990), Radner and Radner (1989), Dupré, (1990), and Colin Allen in Allen and Bekoff (1995), all of whom seriously explore both the indications that many animals are conscious and the philosophical and sociological factors that may help explain the widespread reluctance to give adequate weight to this evidence. These philosophers, and many scientists, see no reason to place such overwhelming emphasis on the distinction between perceptual and reflective consciousness. The latter is a simple form of introspection, and, as discussed in chapter 1, it is ironic that after the long-standing rejection of introspection as a source of reliable evidence about the workings of our minds, this special type of introspection should be elevated to such a crucial status as a litmus test for humanity and moral status.

Some activities have relatively minor adverse effects on animals and pay large dividends in human benefits. To test promising new surgical procedures on deeply anesthetized rats that otherwise have lived reasonably optimal lives seems a justifiable trade-off. But crippling an elephant and leaving it in agony while the hunter enjoys his tea, as described by Midgley (1983, 14–17), is clearly an unwarranted indulgence in minor human satisfaction at the expense of considerable suffering. Most cases fall somewhere between such extremes and can be decided only by weighing the magnitudes of both the human benefit and the animal suffering. This leads to the further and difficult question of the degree to which particular animals suffer when treated in various ways. It is customary to assume that mammals and birds are more deserving of sympathetic treatment than fishes or insects. And even the most extreme advocate of animal rights is unlikely to mourn the extinction of the smallpox virus.

If we grant that some animals are conscious and that we should therefore refrain from causing them pain or suffering, are there other animals for which we need have no such scruples because we can be confident that they are unconscious? Sometimes this question is expressed by asking how far down along some assumed gradient of higher to lower animals such ethical concern is appropriate. Biologists bristle at such a question because biological evolution has been a branching tree rather than a linear progression. There is of course an enormous difference in complexity of animals, and it is reasonable to ask where a line can be drawn between those that do and those that do not suffer. The difficulty is that we simply do not know, and it is not clear how we can find out in the near future.

If we accept communication as evidence of conscious thinking, we must certainly grant consciousness to honeybees. Yet we can scarcely manage to avoid the injury and suffering of all insects, many of which have small but elaborate brains. Recognizing that central nervous systems produce conscious experience, how can we judge how complex a nervous system must be to permit at least simple perceptual consciousness? One response to this challenging dilemma is to assume a continuity of experience, with progressively simpler nervous systems permitting less and simpler conscious content. This approach is sometimes called panpsychism, or panexperientialism, and it has appealed to some philosophers, including A. N. Whitehead. Whatever the approach, there surely comes a point in such considerations at which for practical and ethical purposes it is reasonable to draw a line.

I will not attempt to advocate how best to resolve the conflicting motivations that necessarily confront us with respect to the treatment of animals. These are fundamental moral judgments, which can be helpfully informed by scientific understanding but fall outside the proper scope of purely scientific analysis. No sane person seriously advocates harming animals just for the sake of doing so, although thoughtless cruelty does unfortunately occur. What scientific understanding can provide is evidence against the notion that all animals are incapable of suffering and therefore totally undeserving of sympathy. That idea seems unsupportable on any scientific grounds, and abhorrent as well.

Ethics and morals should be based on positive values, not merely on the exclusion of supposed inferiors. The important and difficult questions do not concern extreme examples but borderline cases and practical trade-offs, as thoughtfully discussed by Bekoff and Jamieson (1991) and Dawkins (1993). How much animal suffering is justified in order to grow crops, eat meat, enjoy hunting or fishing, conduct physiological or behavioral research, test new cosmetics, develop new surgical techniques before they are tried on human patients, or use animals in innumerable other ways that benefit people to varying degrees? I do not feel that scientists have any special right to advocate moral judgments in such difficult matters, but cognitive ethology does hold out the prospect of providing helpful information and understanding that can lead to better-informed decisions.

How can we estimate the degree to which various kinds of animals suffer when treated in particular ways? Only as we learn more about their subjective mental experiences will it be possible to do this on an informed basis. This is a tremendous challenge, and we are at present so extremely ignorant about the conscious mental experiences of animals that it will be a long time before scientific methods can be developed to measure just how much a given animal suffers under particular conditions. But a beginning has been made by the types of experiment reviewed by M. Dawkins (1980, 1990, 1993, 1998) in which animals are allowed to choose between various environmental situations, such as sizes and types of cages. There are many uncertainties in such investigations, but they do point in the right direction. Because we know so little, significant surprises may await future investigators. For example, it might turn out that some treatments that seem at first glance to be detrimental are actually preferred by the animals; and insofar as this may be the case, such information could be appropriately used in weighing the conflicting demands of human benefit against animal deprivation or suffering. But this is almost idle speculation in our current state of ignorance, and I can best conclude this chapter by reiterating that whatever we can learn about the subjective mental experiences of animals has significant potential relevance to the ethics of animal utilization by our species.

CHAPTER FOURTEEN

The Scientific Significance of Animal Consciousness

The whole kingdom of nonhuman animals, comprising millions of species and literally countless numbers of individuals, is clearly an important component of our planet, for the universe would be a very different one if they did not exist. For that reason alone it is important to understand animals as fully as possible; without such understanding we will remain blind to an important aspect of reality. And we cannot understand animals fully without knowing what their subjective lives are like. Until this is possible—and at present it is possible only to a very limited degree—we will remain unable to appreciate adequately either the nature of nonhuman animals or how we differ from them.

The zoological significance of the question of animal consciousness may lack the practical urgency of the ethical questions, and it may not appeal to philosophers as an intellectual challenge comparable in significance to the general problem of other minds. But not only does it bear directly on the philosophical and ethical issues outlined in chapter 13; it is also, in its own right, an important reason to inquire as deeply and critically as we can into the subjects discussed in this book.

Much of twentieth-century science gradually slipped into an attitude that belittles nonhuman animals. Subtle but effective nonverbal signals to this effect emanate from much of the scientific literature. Physical and chemical science is assumed to be more fundamental, more rigorous, and more significant than zoology. Modern biology revels in being largely molecular, and this inevitably diverts attention from the investigation of animals for their own sakes. Part of this mechanomorphic trend may be due to an unrecognized reaction against the deflation of human vanity by the Darwinian revolution. The acceptance of biological evolution and the genetic relationship of our species to others was a shattering blow to the human ego from which we may not have fully recovered. It is not easy to give up a deep-seated faith that our kind is completely different in kind from all other living organisms.

A psychological palliative that may be subconsciously attractive, even to many scientists, is to shift attention away from the embarrassing fact of our animal ancestry by accentuating the aspects of science that are more akin to physics. This may help explain why so many appear to be so certain that consciousness and language are uniquely human capabilities and that the discovery of symbolic communication by honeybees "upsets the very foundation of behavior, and biology in general" (Rosin 1978, 589). Quite the contrary: such discoveries in the field of cognitive ethology extend and improve our understanding of animals; a definition of biology that rules out those discoveries a priori suffers from self-inflicted impoverishment.

Evolutionary biology seeks to explain as many attributes of living organisms as possible in terms of their contribution to survival and reproduction, that is, to their evolutionary fitness. This presents difficult challenges, because so many interacting variables affect the lifetime success and reproduction of most animals. A basic assumption underlying most of evolutionary biology is that various sorts of animals survived and reproduced better than others *in the past.* But there are hardly ever any directly relevant data available to show just how the ancestors of contemporary animals actually outperformed others. For example, it seems clear that placental mammals are generally more efficient than the marsupials. But if a skeptic asks just how we know this, zoologists must fall back on indirect evidence (such as the disappearance of many marsupials from the South American fauna at about the time that a new land bridge at Panama allowed placental mammals to reach that continent) to support this inference. Of course, no zoologist was present to observe just what the placentals did better than the marsupials. But convincing conclusions can be reached even in the absence of ideal data.

Evolutionary biology has been so concerned with identifying how the structure, function, and behavior of animals contribute to their fitness that it has tended to underemphasize those attributes that render animals different in important ways from nonliving systems, plants, or protists. Independent mobility and a heterotrophic metabolism dependent on food materials synthesized by plants are the most obvious distinguishing features of multicellular animals. But animals are also clearly more than mobile metabolisms. They *act*, that is, they do things spontaneously, on their own. What they do is determined in large part by outside influences, yet the complexity and the remoteness of animal actions from whatever external causes may be at work distinguishes them in an important fashion from microorganisms, plants, or physical systems. Most of these spontaneous activities are regulated by central nervous systems, and such systems, together with the adaptable behavior they make possible, are a special feature of living animals not found elsewhere in the known universe. In addition, members of at least one species also experience subjective feelings and conscious thoughts. We cannot be certain how common this additional feature actually is, but suggestive evidence such as that reviewed in this book makes it at least plausible that simple forms of conscious thinking may be quite widespread.

Testability

One criticism of any attempt to study consciousness scientifically is the claim that no testable hypotheses about it have been proposed. This is a serious accusation, for the conventional type of scientific testing of hypotheses is obviously impossible if there are no hypotheses that can be verified or falsified. But there are several ways in which hypotheses about animal consciousness can be tested, with the important qualification that for the time being at least we can anticipate only evidence that increases or decreases the likelihood that a given animal is conscious in a particular situation and what the content of its conscious experience is likely to be. Three considerations show that this apparent lack of testable hypotheses is not the fatal flaw that many scientists assume.

First, when scientists are beginning to attack a truly challenging scientific problem it is premature to require a neat and tidy set of clear-cut hypotheses lined up like ten pins ready to be knocked down or left standing by the bowling ball of hypothesis testing. Perhaps the outstanding example is Darwin's hypothesis that biological evolution had been effected by natural selection. This idea cannot be tested in the literal sense of observing that in the remote past particular animals and plants reproduced successfully while others failed to do so and that this was caused by differences in particular attributes of the two. Although the historical fact of biological evolution is generally accepted, there is still controversy about the relative importance of natural selection and other causal factors. But despite the fact that it cannot be verified or falsified by absolutely definitive tests, Darwin's hypothesis is the bedrock of biology and one of the most significant and influential ideas in all of science.

Second, insofar as animal communication expresses conscious thoughts and feelings, this opens the door to formulating and testing relevant hypotheses, although the specific procedures remain to be worked out. For example, when a vervet monkey gives an eagle alarm call, as discussed in chapter 9, a plausible hypothesis is that it is conscious of a frightening bird approaching from the sky. For present purposes we need not go on to consider further hypotheses about whether it consciously wants to warn its companions of the danger, exhort them to take appropriate evasive actions, or both. The very fact of communication is strong evidence that the monkey is consciously thinking something approximating the message communicated. After all, we assume this when we use human communicative behavior as evidence of human thoughts and feelings. But, as discussed in chapter 9, skeptics can always claim that alarm calling, like blushing or groans of pain, is not done with conscious intent.

Ideally we would like some independent index of conscious as opposed to nonconscious information processing in an animal's brain. But even in the absence of direct neuropsychological data it is appropriate to weigh the kinds of suggestive evidence discussed in chapters 9 and 10, and ask ourselves how likely it is that semantic communication occurs without any awareness on the animal's part of the information it is conveying to its companions. We can put ourselves in Darwin's shoes as he pondered why animals had diverged over the course of geological time. This was surely a worthwhile enterprise on his part, and a comparable approach can be productive for enterprising cognitive ethologists.

Third, can we hope that neuroscientists will discover a clear-cut neural correlate of conscious thinking that can then be looked for in animals? Chapter 8 reviewed several of the promising steps in that direction that have been taken by neuropsychologists in recent years. Especially exciting are the experiments with blindsight in which monkeys with extensive cortical lesions appeared to be blind in part of the visual field, yet could make far better than chance discriminations among visual stimuli presented in these fields. This allowed Cowey and Stoerig (1995) to train them to signal whether or not they saw the stimuli. The result was that the monkeys signaled "don't see anything" when presented with stimuli to which they could nevertheless respond.

Behaviorists such as Macphail (1998, 158–63) can of course argue that these monkeys pressed the appropriate lever although quite unaware of either scene, but this strains credulity, and the experimenters who developed and carried out these ingenious experiments seem to assume that these monkeys are conscious of what they see. Also promising are the experiments on binocular rivalry in which human subjects and monkeys seeing different scenes with the two eyes indicate which of the two they see at a given moment.

At present no tidy litmus test for consciousness is available, but as cognitive ethologists accumulate from any available source evidence bearing on the question of animal awareness they will be in the best of company, namely, that of Charles Darwin. Previous chapters have reviewed suggestive positive evidence, but it is quite appropriate for skeptics to ask what negative evidence would look like. Many kinds of animal behavior *can* be taken as evidence that the animal is acting consciously. What sort of evidence can serve to *lower* the likelihood that this is the case? Failure to adjust behavior when challenged by unexpected situations is one possibility, because it is the opposite of versatile adaptation of behavior. Lack of communication would seem to suggest lack of consciousness only under conditions in which communication to others is clearly called for, because it may often be advantageous not to communicate. None of these possibilities approach the ideal of a concrete all-or-nothing test for consciousness. But we should recognize that we are in the very early stages of exploration into uncharted territory where we will be obliged to make the best of every scrap of evidence that appears to be relevant.

Self-Awareness

Self-awareness is often held to be a capability found only in humans and the Great Apes. And reflective as opposed to perceptual consciousness is often said to be necessary before an animal can be aware of itself. The distinction between perceptual and reflective consciousness is therefore crucially important in this context. If we grant that some animals are capable of perceptual consciousness, we need next to consider what range of objects and events they can consciously perceive. Unless this range is extremely narrow, the animal's own body and its own actions must fall within the scope of its perceptual consciousness.

As discussed in chapter 1, there is no part of the universe that is closer and more important to an animal than its own body. If animals are capable of perceptual awareness, denying them some level of self-awareness would seem to be an arbitrary and unjustified restriction. But those who hold that self-awareness is a unique human attribute often fall back to an insistence that although animals may be perceptually conscious of their own bodies, this is merely "bodily self-consciousness," and no animal can think such thoughts as "It is *I* who am running, or climbing this tree, or chasing that moth."

Yet when an animal consciously perceives the running, climbing, or moth-chasing of another animal, it must also be aware of who is doing these things. And if the animal is perceptually conscious of its own body, it is difficult to rule out similar recognition that it, itself, is doing the running, climbing, or chasing. An example of the difficulties that arise when one denies any sort of self-awareness to animals has been pointed out to me by Lance A. Olsen, who has been impressed by the tactics of grizzly bears in seeking out positions from which they can watch hunters or other human intruders without allowing themselves to be seen, as has been described by Haynes and Haynes (1966), Mills (1919), and Wright (1909). It has also been reported that these bears make efforts to avoid leaving tracks, indicating that they realize that their tracks may be followed by hunters. Concealment by moving behind something opaque would require only simple behavior patterns. A simplistic interpretation might be that the animal moves behind as much vegetation as possible while still being able to see out. But they sometimes seem to do this without exposing any part of their body to view, suggesting some such thought as "I must get *my* whole body behind these bushes."

The question of self-awareness is one of the very few areas of cognitive ethology where we have some concrete experimental evidence. Gallup (1977, 1983) and Suarez and Gallup (1981) demonstrated that some chimpanzees can learn to recognize mirror images as representations of their own bodies. After becoming familiar with mirrors, they were anesthetized, and while they were unconscious a mark was placed on a part of the head that they could not see directly. On awakening they paid no attention to the mark until a mirror was provided, but then they touched it and gave every sign of recognizing that it was on their own bodies. Although orangutans also used mirrors in this way, for many years extensive efforts to elicit such responses from monkeys, gibbons, and even gorillas failed.

Almost all animals react to their mirror image as though it were perceived as another animal, if they pay any attention to it at all. For example, in experiments reported by Povinelli (1989), elephants learned to use mirrors to locate hidden food but showed no signs of self-recognition, according to Gallup's criterion. Gallup (1983, 1994) argues that self-awareness is the criterion of mind but believes that dolphins and elephants can also monitor their own mental states. As discussed in chapter 13, this claim—that recognizing a mirror image as showing oneself is defining evidence of self-awareness—relies on a form of introspection as a crucial criterion. This presents difficulties for behaviorists, inasmuch as Watson and later behaviorists have insisted that introspection cannot provide valid scientific data.

Mirror self-recognition has been extensively studied and discussed in recent years, as reviewed in a book by edited Parker, Mitchell, and Boccia (1994). It is difficult to be certain whether the failure of most animals to recognize mirror images as representations of their own bodies demonstrates that they are incapable of self-awareness, as Gallup claims, or whether they fail for some other reason to correlate the appearance and movements of the mirror image with those of their own bodies. Especially puzzling was the failure of the gorillas studied by Suarez and Gallup to learn mirror self-recognition, since in most other ways they seem just as versatile and intelligent as chimpanzees and orangutans. Patterson and Cohn (1994) later reported that gorillas trained to use communicative gestures similar to those taught to Washoe and other chimpanzees can recognize themselves in mirrors.

Furthermore, only some chimpanzees satisfy Gallup's criteria, and it is difficult to believe that such an important capability as self-awareness is present in only some members of a particular species. Ikatura and Matsuzawa (1993) have trained a chimpanzee to use a modified keyboard similar to the one used by Savage-Rumbaugh to express personal pronouns such as *you, me,* and *him*. This certainly indicates awareness of herself and of others as distinct creatures.

Savage-Rumbaugh, Shanker, and Taylor (1998) describes quite convincing evidence of self-awareness on the part of Austin, one of the chimpanzees who participated in the communication about tools described in chapter 12. For example: "Austin used mirrors to apply makeup to his face and try out fur shawls to make himself look even larger and more intimidating than he already was. . . . When watching super-8 movies of wild chimpanzees, he would interpose his body between the projector and the screen so that his own shadow cast a chimp shadow onto the movie picture: he would then make this shadow chase the chimpanzees in the movies" (35). On balance, it seems most likely that mirror self-recognition as indicated by the Gallup-type experiment does strongly indicate self-awareness, but that failure to satisfy Gallup's criteria is not very strong evidence that the animal is incapable of thinking about itself.

Hauser et al. (1995) have demonstrated that tamarins, small New World monkeys, show some of the simpler signs of mirror self-recognition. These monkeys have white heads, and to make their mirror image especially striking in a Gallup-type experiment, several were anaesthetized, and the fur on their heads was dyed pink, blue, or green. On recovering they did not react to their changed appearance until a mirror was available, but on first seeing their reflection they briefly touched their surprisingly colored heads. The tamarins did not show the full range of mirror-directed behavior displayed by some chimpanzees in the type of experiment pioneered by Gallup, who disagrees with the interpretation of Hauser et al. (Anderson and Gallup 1997 and reply in Hauser and Kralik 1997). But the tamarins did seem surprised at the sudden change in the mirror images of their heads, and this at least suggests that they realized that the mirror image did show part of their own body. If the mirror imaged looked to one of these tamarins like a green-headed companion, it might well be surprised, but touching its own head suggests awareness that the mirror showed its own head.

Both reflective consciousness and self-awareness are often held to be uniquely human attributes. But it remains an open question whether we can ascertain which animals experience even perceptual consciousness, although the weight of evidence reviewed in previous chapters suggests that many of them do. What sorts of evidence might indicate whether or not they think about their own thoughts? One suggestive indication is provided by memories and expectations. When an animal performs complex and demanding learned behavior it may be thinking consciously about the remembered events and what it did to obtain food or avoid something unpleasant. If so, the animal may be reacting to a mental image or representation of its own behavior. Behaviorists will of course object that learned behavior need not be accompanied by any conscious memory at all. But a heron fishing with bait or a pigeon succeeding in the complex experiments described in chapter 7 may well be consciously aware of what it is doing and who is doing these things.

When an animal communicates about something it consciously remembers or anticipates, this communication can appropriately be viewed as a report about some of its thoughts. Does any animal communication include the information that it was the communicator itself that did something or perceived something? Few signs of this have been reported, but it might not have been recognized in our customary ways of studying animal communication. For example, as discussed in chapter 9, Gouzoules, Gouzoules, and Marler (1984) analyzed the distress calls of young rhesus macaques during or after fights with other monkeys, and these calls differed according to the social status of the opponent. Would a finer-grain analysis show any indication of different types of scream when the screamer bit the opponent or vice versa? If so, this might suggest communication of the simple thought "He bit me" or "I bit him back." Such differentiation of communication would indicate that the communicator was thinking about who did the biting, including the thought that it itself did so.

One approach to the question of self-awareness that was initiated by Premack and Woodruff (1978) and discussed in detail by Tomasello and Call (1997) is often called "theory of mind." Insofar as nonhuman animals think about the thoughts or feelings of others, such thinking is termed "having a theory of mind." This wording suggests much more elaborate theorizing than is actually considered a realistic possibility even for apes and monkeys. But much effort has been devoted to inquiring whether even primates ever think about the beliefs or desires of other animals. The alternative view is that animals are "behaviorists" that, if they think at all, think only in terms of the behavior of others—what they do—rather than what they think or feel. One reason for the great interest in this question is the belief that it relates to self-awareness. It is often held, or at least implied, that an animal incapable of thinking about the mental experiences of others could not think about its own thoughts. Most of the experimental evidence from studies of primates supports a negative conclusion: that the animals think only in terms of the behavior of others (Tomasello and Call 1997, ch. 10; Cheney and Seyfarth 1990; Seyfarth and Cheney in press).

Recent observations of small green bee-eaters of India by Smitha, Thakar, and Watve (1999) are relevant to the discussion of theory of mind in Premack and Woodruff's sense. These birds nest in tunnels in mud cliffs and river banks, and they have one or a few favorite perching sites where they land shortly before entering their nesting tunnels. When feeding their young they are reluctant to enter these tunnels when potential predators, including human observers, are present. Smitha et al. arranged conditions that allowed observers to stand in two positions about equally distant from the nest, both clearly visible from the bee-eater's perch but differing in visibility from the nest entrance. In one position the observer could see the nest entrance, but from the other position the observer's view was blocked by intervening bushes, rocks, or a wall.

The bee-eaters hesitated longer before entering the nest tunnel when the human observer could see this entrance than when she could not. This indicates that the birds were aware of the relationships between entrance, barrier, and observer and were therefore much more likely to fly to the nest entrance when the observer could not see its location. This must have required understanding of the fact that the observer's ability to see the entrance was blocked by the intervening barrier. This suggests an awareness of the observer's perceptions, although determined behaviorists can interpret these data as showing only that the birds could understand the fact that an opaque barrier prevented direct vision of the nest entrance from the observer's position.

Instincts and Learning

It is commonly assumed that conscious mental states can only be based on learning and that behavior that has arisen through evolutionary selection cannot entail consciousness. Thus learning and genetic programming are considered to be mutually exclusive alternatives, with only the former allowing for the possibility of consciousness. And when we ponder the likelihood that a particular behavior pattern may imply consciousness, the fact that it is learned tends to tip the scales in favor of consciousness.

On the other hand, our increasing knowledge of animal behavior and cognitive ethology calls into question this basic assumption, which confuses two quite separate issues. An animal may or may not be conscious, and its behavior may be influenced to varying degrees by genetic programming. These are actually quite independent questions, and any combination is theoretically possible. Learned behavior is not always consciously acquired or executed, even in our own species, as discussed in detail by Schacter et al. (1993); and we have no adequate method for estimating how closely it is linked to conscious awareness in animals, primarily because it is so difficult to determine whether a particular animal is conscious.

The same difficulty stands in the way of determining whether genetic influences could lead to a central nervous system that develops conscious thoughts. Conscious thinking about alternative actions, and selecting one believed likely to prove favorable, is an adaptive trait that could well have been selected in the course of an animal's evolutionary history. Once this question is faced squarely it becomes clear that very little, if any, solid evidence supports the customary assumption that if some behavior has been genetically programmed, it cannot be guided by conscious thinking.

Many insects, spiders, and other animals carry out quite elaborately integrated patterns of behavior, and they do so almost perfectly on the first appropriate occasion, without any opportunity to learn what to do. This absence of learning is then taken, almost universally, as proof that the animal has no conscious awareness of its instinctive behavior. The nature-versus-nurture issue is complicated by the great difficulty of teasing out the relative importance of genetic and environmental components influencing a given pattern of behavior. Species specificity, the near-constancy of a given behavior among all members of a species, does not necessarily mean that no learning is involved, because members of the species may be exposed to very similar environmental influences during development. In a few cases in which the evidence is reasonably complete and satisfactory, for example, in Dilger's experiments with genetic influences on nest-building, discussed in chapter 4, the genetic instructions are rather general and the animal learns the specific and essential details.

The assumption that only learned behavior can be accompanied by conscious thinking arises, I suspect, from analogies to our own situation. Human lives clearly require such an enormous amount of learning that many have denied the existence of instinctive, genetically programmed human behavior. It is widely believed that only the simplest human reactions such as eye blinks, knee jerks, sneezing, cries of pain, exclamations one makes when startled, or a newborn's suckling are under predominantly genetic control. Many of these reactions happen automatically, unintentionally, and without any learning, although we may be aware of them as they occur. We do not plan to sneeze, although we certainly know we are sneezing. But we may not even realize that we have blinked in response to a flash of light or the sight of something moving rapidly toward us. From these experiences we tend to infer that when animal behavior requires no learning it cannot be accompanied by conscious thought.

Consciousness of one's bodily activities falls into two general categories: we may consciously anticipate, plan, and intend to perform some action; or the behavior may "just happen" as our bodies do something without any conscious expectation, and perhaps without our being able to affect the action. Yet even in the second case we may be completely conscious of what our body is doing. Simple human reflexes have served as "type specimens" of instinctive behavior that color our view of unlearned behavior. But perhaps it is unwisely anthropocentric to assume that all instinctive behavior is comparable to human knee jerks.

When we say that only learned behavior can entail consciousness, we might mean that an animal behaving in a fashion that has been rigidly programmed genetically is quite unaware of what it is doing, as we are unaware that our eyelids blink in response to a bright flash of light. Or we might mean that the behavior is unplanned and unanticipated, as when we are surprised by a sneeze and yet fully aware that we are sneezing. If animals are perceptually conscious of their genetically programmed behavior, the customary claim that instinctive behavior cannot entail consciousness must mean that the behavior in question is unplanned and unanticipated, although the animal may be aware of what its body is doing.

Does the assumption that learned behavior is more likely to entail consciousness really mean that we believe it is more likely to be planned and anticipated? This is almost certainly not true in all cases, for we carry out many kinds of learned behavior without planning or anticipation, for instance, the split-second braking of our auto on suddenly seeing a child in the road. When animals behave instinctively, they might be fully aware of what they are doing without necessarily having experienced a prior intention, still less understanding of the causes of their behavior or its ultimate consequences. On the other hand, animals might consciously choose *which* of their genetic programs to activate in a given situation, as suggested by Gould and Gould (1994).

The considerations outlined above suggest that genetic influences on a pattern of behavior do not automatically preclude the possibility that the animal in question thinks consciously about what it is doing, even though the action might not be anticipated or planned. Nevertheless, it does seem intuitive that learned behavior is more likely to be accompanied or influenced by conscious thinking. Lloyd Morgan, John Watson, and most early twentieth-century students of animal behavior tended to believe that associative learning was strong evidence of consciousness.

On the other hand, more recent investigations have shown that we can learn some fairly simple matters without any conscious awareness of what we are learning, as reviewed by Velmans (1991), Schacter et al. (1993), and Schacter (1998). Many have argued that since human subjects can learn without knowing what they are learning, animal learning does not demonstrate consciousness. And as learning has been demonstrated in more and more species, often distantly related to us (for example, insects, as reviewed by Papaj and Lewis [1993]), there has been a subtle shift in emphasis so that learning is now often advanced as an *alternative* to consciousness.

Although the word *conscious* is ordinarily avoided in scientific papers about learning and memory, it has become customary to distinguish between two general categories of memory: (1) explicit, or declarative, memory, which in human subjects entails consciousness, so that the subject can say what he has learned; and (2) implicit memory, which entails changes in behavior that have been caused by prior events of which the subject was not conscious. Implicit memory is often termed nondeclarative, because a human subject cannot declare, or state, what is remembered. It has been generally assumed that animals, lacking language, cannot declare anything, so they are incapable of declarative memory by definition. But this view has become less and less tenable as more is learned about animal communication, as reviewed in chapters 9 and 10. Animals may not declare in human words, unless specially trained to do so, but they certainly communicate some of their feelings and thoughts.

Another type of memory that was often considered to involve consciousness and to be limited to humans is episodic memory, that is, remembering not only events but how they were related in time and space. But Clayton and Dickinson (1998) and Griffiths, Dickinson, and Clayton (1999) have recently reported evidence that food-storing jays show evidence of "episodic-like" memory.

Additional evidence indicating that some forms of learning provide significant evidence of consciousness stems, rather surprisingly, from recent analyses of classical, or Pavlovian, conditioning. This used to be considered a primitive form of learning that involves no consciousness. Some types of classical conditioning of human subjects seem to require that they be consciously aware that a stimulus signals that they will shortly receive an electric shock. One is aversive differential conditioning of electrical conductance of the skin. In this procedure two different stimuli are presented, but only one is paired with an unpleasant electric shock. After conditioning this stimulus itself causes the skin's conductance to change.

This surprising finding was reviewed by Dawson and Furedy (1976) and discussed in detail by Shanks and St. John (1994). Later experiments by Ohman, Esteves, and Soares (1995) indicated that conscious awareness is not necessary for successful conditioning if the conditioned stimulus is one that itself induces fear. But arbitrary stimuli that do not arouse appreciable negative emotions *do* seem to require conscious awareness for successful conditioning.

Would comparable conditioning of animals also require conscious awareness, provided the conditioned stimulus was emotionally pleasant or emotionally neutral? If so, such conditioning might provide objective evidence of consciousness. It is not clear that enough differential aversive conditioning experiments with animals have yet been carried out under conditions that are sufficiently similar to the human experiments to allow significant comparisons. But this might be a fruitful avenue for future investigations.

Animal Dreams and Fantasies

When the subject of animal thinking is discussed by scientists they usually think about possible animal thoughts that are objective and realistic, for example, awareness that a certain object is tasty food or that a particular animal is a predator or a known companion. But the content of much human consciousness does not conform to objective reality. Fear of ghosts and monsters is very basic and widespread in our species. Demons, spirits, miracles, and voices of departed ancestors are real and important to many people, as are religious beliefs that entail faith in the overriding significance of entities that lie far outside of the physical universe studied by objective science. Yet when we speculate about animal thoughts, we tend to assume that they are necessarily confined to practical, down-to-earth matters, such as how to get food or escape predators. We also tend to assume that animal thinking must be a simpler version of our own thinking about the animal's situation.

But there is really no reason to assume that all animal thoughts are rigorously realistic. Apes, porpoises, and dogs often seem playful, mischievous, and fickle, and anything but businesslike, practical, and objective. Insofar as animals think and feel, they may fear imaginary predators, imagine unrealistically delicious foods, or think about objects and events that do not actually exist in the real world around them. The young vervet monkey that gives the eagle alarm call for a harmless songbird may really fear that this flying creature will attack. As we try to imagine the content of animal thoughts, we should consider the possibility that their thoughts, like some of ours, may be less than perfect replicas of reality. Animals may experience fantasies as well as realistic representations of their environments.

The recognition that animal thinking may not always be strictly realistic leads to the subject of animal dreams. Darwin and many others have been impressed by the fact that sleeping dogs sometimes move and vocalize in ways that suggest that they are dreaming; their movements resemble those of feeding, running, biting, and even copulation. They sometimes snarl and bark. Some observers of sleeping animals have concluded that these motions and vocalizations accompany dreams related to recent experiences. Damasio (1999, 100) concludes that "in dream sleep, during which consciousness returns in its odd way, emotional expressions are easily detectable in humans and animals."

Human sleepers show two types of sleep when analyzed by an electroencephalogram. The first, a relatively low-frequency pattern, characterizes deep sleep; the second is more irregular and is usually accompanied by rapid eye movements (REMs, hence REM sleep), which can be recorded separately by electrodes near the eyes. When human subjects are awakened from these two types of sleep, they are more likely to report that they were dreaming during REM sleep (Fishbein 1981; Morrison 1983; Hobson and Stickgold 1997). Comparable recordings from sleeping birds and mammals show similar patterns of REM sleep (Hartman 1970; Jouvet 1979, 1999; Cohen 1979), indicating that they also dream. A number of recent theories about the function of dreaming have been reviewed by Winson (1985) without resolving the question definitively. But whatever its functional utility, dreaming seems to be widespread, at least among mammals and birds.

A few rather limited studies of human eye movements during REM sleep suggest that the movements resemble those that would be expected during the activity or experience about which the person is dreaming. For instance, dreaming about a tennis match might produce repeated eye movements from side to side as the dreamer follows the tennis ball. But such experiments are only beginning to provide a reliable procedure for monitoring the content of human dreams, so dreaming dogs or other animals cannot yet be studied in this way. But if it could be perfected, such monitoring might allow us to determine what the animal is dreaming about. We might then be able to study a type of mental experience that exhibits an extreme form of displacement, for nothing in the sleeper's immediate environment ordinarily corresponds to the content of the dream.

Stoyva and Kamiya (1968) proposed that a combined analysis of electrical recording of eye movements and subsequent verbal reports of dream content might eventually lead to objective investigations of human mental experience. But neuroscientists have only begun to develop appropriate methods to accomplish this. Although we can easily imagine experiments along the lines outlined above that might yield verifiable objective evidence, too few have been carried out to permit any firm conclusions. Perhaps a combination of cognitive ethology and cognitive neurophysiology will eventually fill this gap and provide empirical evidence about the reality and content of animal dreams. It would indeed be ironic if evidence that animals think consciously should come to be derived from an understanding of their dreams.

Striving for a balanced judgment based on the whole spectrum of evidence reviewed in this book, I believe it is more likely than not that the emergent property of consciousness confers an enormous advantage by allowing animals to select the actions that are most likely to get them what they want or to ward off what they fear. To paraphrase Karl Popper, animals who think consciously can try out possible actions in their heads without the risk of actually performing them solely on a trial-and-error basis. Considering and then rejecting a possible action because one decides it is less promising than some alternative is far less risky than trying it out in the real world, where, for many animals, a mistake can easily be fatal. We carry out such trial-and-error behavior in our minds, and it is difficult to avoid the conclusion that animals often do something similar at a simple level. This activity, attribute, or capability is a truly marvelous phenomenon. Although almost all discoveries about animal behavior throw some light on animal mentality, scientists have devoted relatively little attention to it, at least in direct and explicit form. Once its significance is appreciated more fully, there is good reason to hope that we can learn much more about it.

This process is already under way, and the shifting climate of opinion has been clearly summarized by Baars (1997, 33):

> All animals engage in purposeful action . . . seeking food, mates, and the company of others. . . . Animals investigate novel and biologically significant stimuli as we do, ignore old and uninteresting events just as we do, and share our limited

capacity for incoming information. . . . Do animals show all the observable aspects of consciousness? The biological evidence points to a clear yes. Are they then likely to have the subjective side as well? Given the long and growing list of similarities, the weight of evidence, it seems to me, is inexorably moving toward yes. . . . Is there still controversy about animal consciousness? My sense is that the scientific community has now swung decisively in its favor. . . . We are not the only conscious beings on earth.

I am confident that with patience and critical investigation we can begin to discern what life is like, subjectively, to particular animals under specific conditions, beginning with the sorts of evidence reviewed in previous chapters. Cognitive ethologists can certainly improve greatly on these preliminary inferences, once the creative ingenuity of scientists is directed constructively toward the important goal of answering Nagel's basic question: What is it like to be a bat, or any other animal? Contrary to the widespread pessimistic opinion that the content of animal thinking is hopelessly inaccessible to scientific inquiry, the communicative signals used by many animals provide empirical data on the basis of which much can reasonably be inferred about their subjective mental experiences.

Because mentality is one of the most important capabilities that distinguishes living animals from the rest of the known universe, seeking to understand animal minds is even more exciting and significant than elaborating our picture of inclusive fitness or molecular mechanisms. Cognitive ethology presents us with one of the supreme scientific challenges of our time, for it constitutes a final chapter of the Darwinian revolution, and it therefore calls for our best efforts of imaginative and critical investigation.

BIBLIOGRAPHY

Able, K. P. 1996. The debate over olfactory navigation by homing pigeons. *J. Exptl. Biol.* 199:124–26.

Ackers, S. H., and C. N. Slobodchikoff. 1999. Communication of stimulus size and shape in alarm calls of Gunnison's prairie dogs *Cynomys gunnisoni*. *Ethology* 105:49–62.

Adler, M. J. 1967. *The difference of man and the difference it makes.* New York: Holt, Rinehart & Winston.

Aeschbacher, A., and G. Pilleri. 1983. Observations on the building behaviour of the Canadian beaver *(Castor canadensis)* in captivity. In *Investigations on beavers,* vol. 1. Berne: Brain Anatomy Institute.

Allen, C., and M. Bekoff. 1997. *Species of mind: The philosophy and biology of cognitive ethology.* Cambridge: MIT Press.

Allen, G. E. 1987. Materialism and reduction in the study of animal consciousness. In *Cognition, language and consciousness: Integrative levels,* ed. G. Greenberg and E. Tobach. Hillsdale, N.J.: Lawrence Erlbaum Associates.

Alport, D. A. 1983. Language and cognition. In *Approaches to language,* ed. R. Harris. New York: Pergamon.

Anderson, C., and F. L. W. Ratnicks. 1999. Worker allocation in insect societies: Coordination of nectar foragers and nectar receivers in honey bee *(Apis mellifera)* colonies. *Behav. Ecol. Sociobiol.* 46:73–81.

Anderson, J. R., and G. G. Gallup, Jr. 1997. Self-recognition in *Saguinus*? A critical essay. *Anim. Behav.* 54:1563–67.

Angell, T. 1978. *Ravens, crows, magpies, and jays.* Seattle: University of Washington Press.

Arhem, P., and B. I. B. Lindahl. 1997. On conscious and spontaneous brain activity. In *Matter matters? On the material basis of the cognitive activity of mind,* ed. P. Arhem, H. Liljenstrom, and U. Svedin. Berlin: Springer.

Armstrong, D. M. 1981. *The nature of mind and other essays.* Ithaca: Cornell University Press.

Armstrong, E. A. 1942. *Bird display: An introduction to the study of bird psychology.* London: Cambridge University Press.

———. 1949. Diversionary display—Part I: Connotation and terminology. *Ibis* 91:88–97; Part II: The nature and origin of distraction display. *Ibis* 91:179–88.

———. 1956. Distraction display and the human predator. *Ibis* 98:641–54.

Armstrong-Buck, S. 1989. Nonhuman experience: A Whiteheadian analysis. *Process Studies* 18 (1): 1–18.

Arnold, A. P., and S. W. Bolger. 1985. Cerebral lateralization in birds. In *Cerebral lateralization in nonhuman species,* ed. S. D. Glick. New York: Academic Press.

Astley, S. L., and E. A. Wasserman. 1998. Novelty and functional equivalence in superordinate categorization by pigeons. *Anim. Learning and Behav.* 26:125–38.

Baars, B. J. 1986. *The cognitive revolution in psychology.* New York: Guilford Press.

———. 1988. *A cognitive theory of consciousness.* New York: Cambridge University Press.

———. 1997. *In the theatre of consciousness: The workspace of the mind.* New York: Oxford University Press.

Baerends, G. P. 1941. Fortpflanzungsverhalten und Orientierung der Grabwespe *Ammophila compestris* Jur. *Tijd. voor Entomol.* 84:72–275.

Baillargeon, R. 1994. A model of physical reasoning in infancy. In *Advances in infancy research,* vol. 9, ed. C. Rovee-Collier and L. Lipsitt, 112–45. Norwood, N.J.: Ablex.

Baker, L. R. 1981. Why computers can't act. *Amer. Philos. Q.* 18:157–63.

———. 1987. *Saving belief.* Princeton: Princeton University Press.

Balda, R. P. 1980. Recovery of cached seeds by a captive *Nucifraga caryocatactes. Z. Tierpsychol.* 53:331–46.

Balda, R. P., and A. C. Kamil. 1998. The ecology and evolution of spatial memory in corvids of the southwestern USA: The perplexing pinyon jay. In *Animal cognition in nature: The convergence of psychology and biology in laboratory and field,* ed. R. P. Balda, I. M. Pepperberg, and A. C. Kamil. San Diego: Academic Press.

Balda, R. P., and R. J. Turek. 1984. The cache-recovery system as an example of memory capabilities in Clark's nutcracker. In *Animal cognition,* ed. H. L. Roitblat, T. G. Bever, and H. S. Terrace. Hillside, N.J.: Lawrence Erlbaum Associates.

Barkow, T. H. 1983. Begged questions in behavior and evolution. In *Animal models of human behavior: Conceptual, evolutionary, and neurobiological perspectives,* ed. G. C. L. Davey. New York: Wiley.

Bartholomew, G. A. 1945. The fishing activities of double-crested cormorants on San Francisco Bay. *Condor* 44:13–21.

Bateson, M., and A. Kacelnik. 1998. Risk-sensitive foraging: Decision making in variable environments. In *Cognitive ecology: The evolutionary ecology of information processing and decision making,* ed. R. Dukas. Chicago: University of Chicago Press.

Baylis, G. C., E. T. Rolls, and C. M. Leonard. 1985. Selectivity between faces in the responses of a population of neurons in the cortex in the superior temporal sulcus of the monkey. *Brain Res.* 342:91–102.

Beck, B. B. 1980. *Animal tool behavior.* New York: Garland STPM Press.

———. 1982. Chimpocentrism: Bias in cognitive ethology. *J. Hum. Evol.* 11:3–17.

Bednarz, J. C. 1988. Cooperative hunting in Harris' hawks *(Parabuteo unicinctus). Science* 239:1525–27.

Beer, C. 1992. Conceptual issues in cognitive ethology. *Advances in the Study of Behav.* 21:69–109.

Bekoff, M., and D. Jamieson. 1991. Reflective ethology, applied philosophy, and animal rights. In *Perspectives in ethology,* vol. 9, ed. P. P. G. Bateson and P. H. Klopfer. New York: Plenum.

———, eds. 1990. *Interpretation and explanation in the study of animal behavior,* vols. 1 and 2. Boulder, Colo.: Westview Press.

Bennett, J. 1964. *Rationality: An essay towards an analysis.* London: Routledge and Kegan Paul.

———. 1976. *Linguistic behaviour.* London: Cambridge University Press.

———. 1988. Thoughtful brutes. *Proceedings and Addresses of the American Philosophical Association* 62:197–210.

———. 1991. How is cognitive ethology possible? In *Cognitive ethology: The minds of other animals,* ed. C. A. Ristau. Hillsdale, N.J.: Lawrence Erlbaum Associates.

Bhatt, R. S., E. A. Wasserman, W F. Reynolds, and K. S. Knauss. 1988. Conceptual behavior in pigeons: Categorization of both familiar and novel examples from four classes of natural and artificial stimuli. *J. Exptl. Psychol: Anim. Behav. Processes* 14:219–34.

Biesmeijer, J. C., and M. C. W. Ermers. 1999. Social foraging in stingless bees: How colonies of *Melipona fasciata* choose among nectar sources. *Behav. Ecol. Sociobiol.* 46:129–40.

Biro, D., and T. Matsuzawa. 1999. Numerical ordering in a chimpanzee *(Pan troglodytes):* Planning, executing, and monitoring. *J. Comp. Psychol.* 113:178–85.

Bitterman, M. E. 1988. Vertebrate-invertebrate comparisons. In *Intelligence and evolutionary biology,* ed. H. J. Jerison and I. Jerison. New York: Springer.

———. 1996. Comparative analysis of learning in honeybees. *Anim. Learning and Behav.* 24:123–41.

Blakemore, C., and S. Greenfield eds. 1987. *Mindwaves: Thoughts on intelligence, identity, and consciousness.* Oxford: Basil Blackwell.

Blest, A. D. 1957. The function of eyespot patterns in the Lepidoptera. *Behaviour* 11:209–56.

Blumberg, M. S., and E. A. Wasserman. 1995. Animal minds and the argument from design. *Amer. Psychologist* 50:133–44.

Boakes, R. 1984. *From Darwin to behaviourism.* London: Cambridge University Press.

———. 1992. Subjective experience. Review of *Animal Minds. Times Higher Education Supplement,* November 22, p. 22.

Boesch, C. 1991. Teaching among wild chimpanzees. *Anim. Behav.* 41:530–32.

———. 1993. Aspects of transmission of tool-use in wild chimpanzees. In *Tools, language, and cognition in human evolution,* ed. K. R. Gibson and T. Ingold. New York: Cambridge University Press.

———. 1994. Hunting strategis of Gombe and Tai chimpanzees. In *Chimpanzee cultures,* ed. R. W. Wrangham, W. C. McGrew, F. de Waal, and P. G. Heltne. Cambridge: Harvard University Press.

Boesch, C., and H. Boesch-Ackermann. 1990. Tool use and tool making in wild chimpanzees. *Folia Primatologica* 54:86–99.

———. 1991. Dim forest, bright chimps. *Nat. Hist.* 100 (9): 50–56.

Bolles, R. C., and M. D. Beecher. 1988. *Evolution and learning.* Hillsdale, N.J.: Lawrence Erlbaum Associates.

Borgia, G. 1985a. Bower quality, number of decorations, and mating success of male satin bowerbirds *(Ptilonorhynchus violaceus):* An experimental analysis. *Anim. Behav.* 33:266–71.

———. 1985b. Bower destruction and sexual competition in the satin bowerbird *(Ptilonorhynchus violaceus). Behav. Ecol. Sociobiol.* 18:91–100.

———. 1986. Sexual selection in bowerbirds. *Scientific American* 254 (6): 92–100.

———. 1995. Why do bowerbirds build bowers? *Amer. Scientist* 83:542–45.

Borgia, G., and M. A. Gore. 1986. Feather stealing in the satin bowerbird *(Ptilonorhynchus violaccus):* Male competition and the quality of display. *Anim. Behav.* 34:727–38.

Borgia, G., I. M. Kaatz, and R. Condit. 1987. Flower choice and bower decoration in the satin bowerbird *Ptilonorhyncus violaceus:* A test of hypotheses for the evolution of male display. *Anim. Behav.* 35:1129–39.

Boswall, J. 1977. Tool-using by birds and related behaviour. *Avicultural Magazine* 84:162–66.

———. 1978. Further notes on tool-using by birds and related behaviour. *Avicultural Magazine* 84:162–66.

Bowman, R. I. 1961. Morphological differentiation and adaptation in the Galapogos finches. *Univ. Calif. Publ. Zool.* 58:1–326.

Bowman, R. S., and N. S. Sutherland. 1970. Shape discrimination by goldfish: Coding of irregularities. *J. Comp. Physiol. Psychol.* 72:90–97.

Boysen, S. T., and G. G. Berntson. 1989. Numerical competence in a chimpanzee *(Pan troglodytes). J. Exptl. Psychol.: Anim. Behav. Processes.* 21:82–86.

———. 1995. Responses to quantity: Perceptual versus cognitive mechanisms in chimpanzees *(Pan troglodytes). J. Exptl. Psychol.: Anim. Behav. Processes* 21:82–85.

Boysen, S. T., C. G. Berntson, T. A. Shreyer, and M. B. Hannan. 1995. Indicating acts during counting by a chimpanzee *(Pan troglodytes). J. Comp. Psychol.* 109:47–51.

Boysen, S. T., and E. J. Capaldi, eds. 1993. *The Development of numerical competence: Animal and human models.* Hillsdale, N.J.: Erlbaum Associates.

Bradbury, J. W., and S. L. Vehrencamp. 1998. *Principles of Animal Communication.* Sunderland, Mass.: Sinauer Associates.

Bradshaw, J. L. and L. Rogers. 1993. *The evolution of lateral asymmetries, language, tool use, and intellect.* San Diego: Academic Press.

Bradshaw, R. H. 1998. Consciousness in non-human animals: Adapting the precautionary principle. *J. Consciousness Studies* 5:108–14.

Bratton, B. O., and B. Kramer. 1989. Patterns of the electric organ discharge during courtship and spawning in the mormyrid fish *Pollimyrus isidori. Behav. Ecol. Sociobiol.* 24:349–68.

Brunton, D. H. 1990. The effects of nesting stage, sex, and type of predator on parental defense by killdeer *(Charadrius vociferous):* Testing models of avian parental defense. *Behav. Ecol. Sociobiol.* 26:181–90.

Buchwald, J. S., and N. S. Squires. 1982. Endogenous auditory potentials in the cat: A P300 model. In *Conditioning representation of involved neural functions,* ed. C. D. Woody. New York: Plenum.

Bullock, T. H., and W. Heiligenberg. 1986. *Electroreception.* New York: Springer.

Bullock, T. H., and G. A. Horridge. 1965. *Structure and function in the nervous system of invertebrates.* San Francisco: Freeman.

Bullock, T. H., S. Karamüsel, and M. H. Hofmann. 1993. Interval-specific event related potentials to omitted stimuli in the electrosensory pathway in elasmobranchs: An elementary form of expectation. *J. Comp. Physiol. A.* 172:501–10.

Bunge, M. 1980. *The mind-body problem: A psychological approach.* New York: Pergamon.

Bunge, M., and R. Ardilla. 1987. *Philosophy of Psychology.* New York: Springer.

Burghardt, G. M. 1973. Instinct and innate behavior: Toward an ethological psychology. In *The study of behavior: Learning, motivation, emotion, and instinct,* ed. J. A. Nevin and G. S. Reynolds. Glenview, Ill.: Scott, Foresman.

———. 1978. Closing the circle: The ethology of mind. *Behav. Brain Sci.* 1:562–63.

———. 1985. Animal awareness, current perceptions, and historical perspective. *Amer. Psychol.* 40:905–19.

———. 1991. The cognitive ethology of reptiles: A snake with two heads and hognosed snakes that play dead. In *Cognitive ethology: The minds of other animals,* ed. C. A. Ristau. Hillsdale, N.J.: Lawrence Erlbaum Associates.

———. 1997. Amending Tinbergen: A fifth aim for ethology. In *Anthropomorphism, anecdotes, and animals,* ed. R. W. Mitchell, N. S. Thompson, and H. L. Miles. Albany: SUNY Press.

Burghardt, G. M., and H. A. Herzog, Jr. 1980. Beyond conspecifics: Is B'rer Rabbit our brother? *BioScience* 30:763–68.

Byrkjedal, I. 1991. The role of drive conflicts as a mechanism of nest-protection behavior in the shorebird *Pluvialis dominica. Ethology* 87:149–59.

Byrne, R. W., and A. Whiten. 1988. *Machiavellian intelligence: Social expertise and the evolution of intellect in monkeys, apes, and humans.* Oxford: Oxford University Press.

Cabanac, M. 1999. Emotion and phylogeny. *J. Consciousness Studies* 6:176–90.

Cade, T. J., and G. L. MacLean. 1967. Transport of water by adult sandgrouse to their young. *Condor* 69:323–43.

Cahalane, V. H. 1942. Caching and recovery of food by the western fox squirrel. *J. Wildlife Management* 6:338–52.

Caldwell, M. C., and D. K. Caldwell. 1965. Individualized whistle contours in bottlenosed dolphins *(Tursiops truncatus). Nature* 207:434–35.

———. 1968. Vocalizations of naive captive dolphins in small groups. *Science* 159:1121–23.

———. 1971. Statistical evidence for individual signature whistles in Pacific whitesided dolphins, *Lagenorhynchus obliquidens. Cetology* 3:1–9.

———. 1973. Vocal mimicry in the whistle mode by an Atlantic bottlenosed dolphin. *Cetology* 9:1–8.

———. 1979. The whistle of the Atlantic bottlenosed dolphin *(Tursiops truncatus)*—ontogeny. In *Behavior of marine mammals,* vol. 3: *Cetaceans,* ed. H. E. Winn and B. L. Olla. New York: Plenum.

Caldwell, M. C., D. K. Caldwell, and P. Tyack. 1990. Review of the signature whistle hypothesis for the bottlenosed dolphin. In *The bottlenose dolphin,* ed. S. Leatherwood and R. R. Reeves. New York: Academic Press.

Caldwell, R. L. 1979. Cavity occupation and defensive behavior in the stomatopod *Gonodactylus festae:* Evidence for chemically mediated individual recognition. *Anim. Behav.* 27:194–201.

———. 1984. A test of individual recognition in the stomatopod *Gonodactylus festae. Anim. Behav.* 33:101–6.

———. 1986. Deceptive use of reputation by stomatopods. In *Deception: Perspectives on human and nonhuman deceit,* ed. R. W. Mitchell and N. S. Thompson. Albany: SUNY Press.

Caldwell, R. L., and H. Dingle. 1975. The ecology and evolution of agonistic behavior in stomatopods. *Naturwissenschaften* 62:214–22.

Calkins, D. G. 1978. Feeding behavior and major prey species of the sea otter, *Enhydra lutris,* in Montague Strait, Prince William Sound, Alaska. *Fishery Bull.* 76 (1): 125–31.

Camazine, S., P. K. Visscher, J. Finley, and R. S. Vetter. 1999. House-hunting by honey bee swarms: Collective decisions and individual behaviors. *Insects Sociaux* 46:348–60.

Capaldi, E. J., and D. J. Miller. 1988a. Rats classify qualitatively different reinforcers as either similar or different by enumerating them. *Bull. Psychonomic Soc.* 26:149–51.

———. 1988b. Counting in rats: Its functional significance and the independent cognitive processes that constitute it. *J. Exptl. Psychol.: Anim. Behav. Processes* 14:3–17.

Capaldi, E. J., T. M. Nawrocki, and D. R. Verry. 1983. The nature of anticipation: An inter- and intraevent process. *Anim. Learning and Behav.* 11:193–98.

Capaldi, E. A., A. D. Smith, J. L. Osborne, S. E. Fahrbach, S. M. Farris, D. R.

Reynolds, A. S. Edwards, A. Martin, G. E. Robinson, G. M. Poppy, and J. R. Riley. 2000. Ontogeny of orientation flight in the honeybee revealed by harmonic radar. *Nature* 403:537–40.

Caro, T. M. 1986a. The functions of stotting: A review of the hypotheses. *Anim. Behav.* 34:649–62.

———. 1986b. The functions of stotting in Thompson's gazelles: Some tests of the predictions. *Anim. Behav.* 34:663–84.

———.1995. *Cheetahs of the Serengeti Plains: Group living in an asocial species.* Chicago: University of Chicago Press.

Caro, T. M., and M. D. Hauser. 1992. Is there teaching in nonhuman animals? *Q. Rev. Biol.* 67:151–74.

Carruthers, P. 1989. Brute experience. *J. Philosophy* 89:258–69.

Cartmill, M., ed. In press. Animal consciousness: Historical, theoretical, and empirical perspectives. *Am. Zool.*

Cerella, J. 1979. Visual classes and natural categories in the pigeon. *J. Exptl. Psychol.: Human Perception and Performance* 5:68–77.

Chalmers, D. J. 1996. *The conscious mind:In search of a fundamental theory.* New York: Oxford University Press.

Cheney, D. L., and R. M. Seyfarth. 1982. How vervet monkeys perceive their grunts: Field playback experiments. *Anim. Behav.* 30:739–51.

———. 1988. Assessment of meaning and the detection of unreliable signals by vervet monkeys. *Anim. Behav.* 36:477–86.

———. 1990a. *How monkeys see the world: Inside the mind of another species.* Chicago: University of Chicago Press.

———. 1990b. The representation of social relations by monkeys. *Cognition* 37:167–96.

———. 1991. Truth and deception in animal communication. In *Cognitive ethology: The minds of other animals,* ed. C. A. Ristau. Hillsdale, N.J.: Lawrence Erlbaum Associates.

Chevalier-Skolnikoff, S. 1989. Spontaneous tool use and sensorimotor intelligence in *Cebus* compared with other monkeys and apes. *Behav. Brain Sci.* 12:561–627 (with commentaries).

Chisholm, A. H. 1954. The use of "tools" or "instruments." *Ibis* 96:380–83.

———. 1971. Further notes on tool-using birds. *Victorian Nat.* 88:342–43.

———. 1972. Tool-using by birds: A commentary. *Bird Watcher* 4:156–59.

Chomsky, N. 1966. *Cartesian linquistics.* New York: Harper and Row.

Christy, J. H. 1982. Burrow structure and use in the sand fiddler crab *Uca pugilator. Anim. Behav.* 30:687–94.

Churchland, P. S. 1986. *Neuropsychology: Toward a unified science of the mind-brain.* Cambridge: MIT Press.

Clark, A. 1987. From folk psychology to naive psychology. *Cog. Sci.* 11:139–54.

———. 1989. *Microcognition: Philosophy, cognitive science, and parallel distributed processing.* Cambridge: MIT Press.

Clayton, D. A. 1978. Socially facilitated behavior. *Q. Rev. Biol.* 53:373–92.

Clayton, N. S., and A. Dickinson. 1998. Episodic-like memory during cache recovery by scrub jays. *Nature* 395:272–74 (see also discussion by K. Jeffrey and J. O'Keefe, ibid. 15–16).

Clayton, N. S., and A. Jolliffe. 1995. Marsh tits *Parus palustris* use tools to store food. *Ibis* 118:554.

Cohen, D. B. 1979. *Sleep and dreaming: Origins, nature, and functions.* New York: Pergamon.

Coleman, K., J. Arendt, and D. S. Wilson. 1997. Consistent differences in individual boldness: A frequency dependent behavior among pumpkinseed sunfish. *Bull. Ecol. Soc. Amer. Suppl.* 78 (4): 70.

Coleman, K., and D. S. Wilson. 1998. Shyness and boldness in pumpkinseed sunfish: Individual differences are context-specific. *Anim. Behav.* 56:927–36.

Colgan, R. 1989. *Animal motivation.* London: Chapman and Hall.

Collias, N. 1987. The vocal repertoire of the red jungle fowl:A spectrographic classification and the code of communication. *Condor* 89:510–24.

Collias, N., and E. Collias. 1962. An experimental study of the mechanisms nest building in a weaverbird. *Auk* 79:568–95.

———. 1964. The evolution of nest-building in weaverbirds *(Ploceidae). Univ. Calif. Pub. Zool.* 73:1–239.

———. 1984. *Nest building and bird behavior.* Princeton: Princeton University Press.

Collias, N., and M. Joos. 1953. The spectrographic analysis of sound signals of the domestic fowl. *Behaviour* 5:175–88.

Connor, R. C., J. Mann, P. L. Tyack, and H. Whitehead. 1998. Social evolution in toothed whales. *Trends in Ecol. and Evol.* 13:228–32.

Couvillon, P. A., L. Arakaki, and M. E. Bitterman. 1997. Intramodal blocking in honeybees. *Anim. Learning and Behav.* 25:277–82.

Couvillon, P. A., T. G. Leiato, and M. E. Bitterman. 1991. Learning by honeybees *(Apis mellifera)* on arrival at and on departure from a feeding place. *J. Comp. Psychol.* 105:177–84.

Cowey, A., and P. Stoerig. 1995. Blindsight in monkeys. *Nature* 373:247–49.

Cowie, R. J., J. R. Krebs, and D. F. Sherry. 1981. Food storing by marsh tits. *Anim. Behav.* 29:1252–59.

Crane, J. 1975. *Fiddler crabs of the world.* Princeton: Princeton University Press.

Crawford, J. D., and X. Huang. 1999. Communication signals and sound production mechanisms of Mormyrid fish. *J. Exptl. Biol.* 202:1417–26.

Crick, F. 1994. *The astonishing hypothesis: The scientific search for the soul.* New York: Simon and Schuster.

Crick, F., and C. Koch. 1998. Consciousness and neuroscience. *Cerebral Cortex* 8:97–107.

Crist, E. 1996. Naturalists' portrayals of animal life: Engaging the verstehen approach. *Social Studies of Sci.* 26:799–838.

———. 1998. The ethological constitution of animals as natural objects: The technical writings of Konrad Lorenz and Nikolaas Tinbergen. *Biol. and Philos.* 13:61–102.

———. 1999. *Images of animals: Anthropomorphism and animal mind.* Philadelphia: Temple University Press.

Cristol, D. A., and P. V. Switzer. 1999. Avian prey-dropping II. American crows and walnuts. *Behav. Ecol.* 10:220–26.

Cristol, D. A., P. V. Switzer, K. I. Johnson, and L. S. Walke. 1997. Crows do not use automobiles as nutcrackers: Putting an anecdote to the test. *Auk* 114:296–98.

Crook, J. H. 1964. Field experiments on the nest construction and repair behaviour of certain weaverbirds. *Proc. Zool. Soc. London* 142:217–55.

———. 1980. *The evolution of human consciousness.* Oxford: Oxford University Press.

———. 1983. On attributing consciousness to animals. *Nature* 303:11–14.

———. 1987. The nature of conscious awareness. In *Mindwaves: Thoughts on intelligence, identity, and consciousness,* ed. C. Blakemore and S. Greenfield. Oxford: Basil Blackwell.

———. 1988. The experiental context of intellect. In *Machiavellian intelligence: Social expertise and the evolution of intellect in monkeys, apes, and humans,* ed. R. W. Byrne and A. Whiten. Oxford: Oxford University Press.

Croze, H. 1970. Searching image in carrion crows. *Z. Tierpsychol.* (suppl.)5:1–86.

Curio, E. 1976. *The ethology of predation.* New York: Springer.

Damasio, A. 1994. *Descartes' error: Emotion, reason, and the human brain.* New York: Avon Press.

———. 1998. Emotion in the perspective of an integrated nervous system. *Brain Res. Rev.* 26:83–86.

———. 1999. *The feeling of what happens: Body and emotion in the making of consciousness.* New York: Harcourt Brace.

Davidson, D. 1984. Communication and convention. *Synthese* 59:3–26.

Davies, N. B. 1977. Prey selection and social behaviour in wagtails. *J. Anim. Ecol.* 46:37–57.

Davis, H. 1989. Theoretical note on the moral development of rats *(Rattus norvegicus). J. Comp. Psychol.* 103:88–90.

Davis, H., K. A. MacKenzie, and S. Morrison. 1989. Numerical discrimination by rats *(Rattus norvegicus)* using body and vibrissal touch. *J. Comp. Psychol.* 103:45–53.

Davis, H., and J. Memmott. 1982. Counting behavior in animals: A critical evaluation. *Psychol. Bull.* 92:547–71.

Dawkins, M. S. 1980. *Animal suffering: The science of animal welfare.* London: Chapman and Hall.

———. 1990. From an animal's point of view: Motivation, fitness, and animal welfare. *Behav. Brain Sci.* 13:1–61 (including commentaries).

———. 1993. *Through our eyes only? The search for animal consciousness.* Oxford: Blackwell.

———. 1998. Evolution and animal welfare. *Q. Rev. Biol.* 73:305–28.

Dawkins, M. S., and T. Guilford. 1991. The corruption of honest signaling. *Anim. Behav.* 41:865–73.

Dawkins, R. 1976. *The selfish gene.* Oxford: Oxford University Press.

Dawkins, R., and J. R. Krebs. 1978. Animal signals: Information or manipulation? In *Behavioural ecology: An evolutionary approach,* ed. J. R. Krebs and N. B. Davies. Oxford: Blackwell.

Dawson, M. E., and J. J. Furedy. 1976. The role of awareness in human differential autonomic classical conditioning: The necessary-gate hypothesis. *Psychophysiology* 13:50–53.

Delius, J. D., and G. Habers. 1978. Symmetry: Can pigeons conceptualize it? *Behav. Proc.* 1:15–27.

Dennett, D. C. 1983. Intentional systems in cognitive ethology: The "Panglossian paradigm" defended. *Behav. Brain Sci.* 6:343–90 (including commentaries).

———. 1987. *The intentional stance.* Cambridge: MIT Press.

———. 1988. Precis of *The intentional stance. Behav. Brain Sci.* 11:494–546 (with commentaries).

———. 1989. Cognitive ethology: Hunting for bargains or a wild goose chase? In *Goals, no-goals, and own goals: A debate on goal-oriented and intentional behaviour,* ed. A. Montefiore and D. Noble. London: Unwin Hyman.

———. 1991. *Consciousness explained.* Boston: Little, Brown.

De Veer, M. W., and B. van den Ruud. 1999. A critical review of methodology and interpretation of mirror self-recognition research in nonhuman primates. *Anim. Behav.* 58:459–68.

Dewsbury, D. 1984. *Comparative psychology in the twentieth century.* New York: Van Nostrand/Hutchinson Ross.

Diamond, J. 1982. Evolution of bowerbirds' bowers: Animal origins of the aesthetic sense. *Nature* 297:99–102.

———. 1986a. Biology of birds of paradise and bowerbirds. *Ann. Rev. Ecol. Syst.* 17:17–37.

———. 1986b. Animal art: Variation in bower decorating style among male bowerbirds *Amblyornis inornatus. Proc. Nat. Acad. Sci.* 83:3042–46.

———. 1987. Bower building and decoration by the bowerbird *Amblyornis inornatus. Ethology* 74:177–204.

———. 1988. Experimental study of bower decoration by the bowerbird *Amblyornis inornatus,* using colored poker chips. *Amer. Naturalist* 131:631–53.

Dickinson, A. 1980. *Contemporary animal learning theory.* London: Cambridge University Press.

———. 1988. Intentionality in animal conditioning. In *Thought without language,* ed. L. Weiskrantz. New York: Oxford University Press.

Dilger, W. C. 1960. The comparative ethology of the African parrot genus *Agapornis. Z. Tierpsychol.* 17:649–85.

———. 1962. Behavior and genetics. In *Roots of behavior: Genetics, instinct, and socialization in animal behavior,* ed. E. L. Bliss. New York: Harper.

Dill, L. M. 1983. Adaptive flexibility in the foraging behavior of fishes. *Can. J. Fish. Aquat. Sci.* 40:398–408.

Dingle, H., and R. L. Caldwell. 1969. The aggressive and territorial behavior of the mantis shrimp *Gonodactyluss bredini* Manning (Crustacea: Stomatopoda). *Behaviour* 33:115–36.

Dittrich, W. 1990. Representation of faces in longtailed macaques *(Macaca fasicularis)*. *Ethology* 85:265–78.

Dol, M., S. Kasanmoentalib, S. D. Limbach, E. Rivs, and R. van den Bos, eds. 1997. *Animal consciousness and animal ethics.* Assen, The Netherlands: Van Gorcum.

Donald, M. 1991. *Origins of the modern mind: Three stages in the evolution of culture and cognition.* Cambridge: Harvard University Press.

Donchin, E. 1981. P300 and classification. In *Electrophysiological approaches to human cognitive processing,* ed. R. Galambos and S. A. Hillyard. *Neurosci. Res. Prog. Bull.* 20:157–61.

Donchin, E., and M. G. H. Coles. 1988. Is the P300 component a manifestation of context updating? *Behav. Brain Sci.* 11:343–427 (including commentaries).

Donchin, E., G. McCarthy, M. Kutas, and W. Ritter. 1983. Event-related brain potentials in the study of consciousness. In *Consciousness and self-regulation,* vol. 3, ed. R. J. Davidson, G. Schwartz, and D. Shapiro. New York: Plenum.

Double, R. 1985. The case against the case against belief. *Mind* 94:420–30.

Drieslein, R. L., and A. J. Bennett. 1979. Red fox predation on greater sandhill crane chicks. *Wilson Bull.* 91:132–33.

Driver, P. M., and D. A. Humphries. 1988. *Protean behaviour.* New York: Oxford University Press.

Duerden, J. E. 1905. On the habits and reactions of crabs bearing actinians in their claws. *Proc. Zool. Soc. London* 2:494–511.

Dugatkin, L. 1999. *Cheating monkeys and citizen bees: The nature of cooperation in animals and men.* New York: Free Press.

Dukas, R., ed. 1998. *Cognitive ecology: The evolutionary ecology of information processing and decision making.* Chicago: University of Chicago Press.

Dukas, R., and L. Real. 1993. Effects of recent experience on foraging decisions by bumblebees. *Oecologia* 94:244–46.

Dunbar, I. M. 1988. The evolution of social behavior. In *The role of behavior in evolution.* Cambridge: MIT Press.

Dupré, J. 1990. The mental lives of animals. In *Interpretation and explanation in the study of animal behavior,* ed. M. Bekoff and D. Jamieson, vol. 1. Boulder, Colo.: Westview Press.

Dyer, F. C. 1998. Cognitive ecology of navigation. On *Cognitive ecology: The evolutionary ecology of information processing and decision making,* ed. R. Dukas. Chicago: University of Chicago Press.

Edelman, G. 1989. *The remembered present: A biological theory of consciousness.* New York: Basic Books.

Edelman, G., and G. Tononi. 2000. *A universe of consciousness: How matter becomes imagination.* New York: Basic Books.

Elgar, M. A. 1989. Predator vigilance and group size in mammals and birds: A critical review of the empirical evidence. *Biol. Rev.* 64:13–33.

Englund, G., and C. Otto. 1991. Effects of ownership status, weight asymmetry, and case fit on the outcome of case contests in two populations of *Agrypnia pagetana* (Trichoptera: Phryganeidae) larvae. *Behav. Ecol. Sociobiol.* 29:113–20.

Epstein, R., R. P. Lanza, and B. F. Skinner. 1980. Symbolic communication between two pigeons *(Columbia livia domestica). Science* 207:543–45.

Esch, H. 1961. Über die Schallerzeugung beim Werbetanz der Honigbiene. *Z. vergl. Physiol.* 45:1–11.

———. 1963. Über die Auswirkung der Futterplatzqualität auf die Schallerzeugung im Werbetanz der Honigbiene. *Verb. Deutsch. Zool. Ges.* 1962:302–9.

———. 1964. Beiträge zum Problem der Entfernungsweisung in den Schawnzeltänzen der Honigbiene. *Z. vergl. Physiol.* 48:534–46.

Evans, C. S., and P. Marler. 1991. On the use of video images as social stimuli in birds: Audience effects on alarm calling. *Anim. Behav.* 41:17–26.

Evans, H. E., and M. J. West Eberhard. 1970. *The wasps.* Ann Arbor: University of Michigan Press.

Farah, M. J. 1988. Is visual imagery really visual? Overlooked evidence from neuropsychology. *Psychol. Rev.* 95:307–17.

Farthing, G. W. 1992. *The psychology of consciousness.* Englewood Cliffs, N.J.: Prentice-Hall.

Feindel, W. 1994. Cortical localization of speech: Evidence from stimulation and excision. *Discussions in Neuroscience* 10:34–45. Geneva: Elsevier Science.

Fellers, J., and G. Fellers. 1976. Tool use in a social insect and its implications for competitive interactions. *Science* 192:70–72.

Ferguson, E. S. 1977. The mind's eye: Nonverbal thought in technology. *Science* 197:827–36.

Fewell, J. H., and S. M. Bertram. 1999. Division of labor in a dynamic environment: Responses by honeybees *(Apis mellifera)* to graded changes in colony pollen stores. *Behav. Ecol. Sociobiol.* 46:171–79.

Finke, R. A. 1989. *Principles of mental imagery.* Cambridge: MIT Press.

Fishbein, W., ed. 1981. *Sleep, dreams, and memory.* New York: SP Medical and Scientific Books.

Fisher, J., and R. A. Hinde. 1949. The opening of milk-bottles by birds. *Brit. Birds* 42:347–57.

Fisher, J. A. 1987. Taking sympathy seriously: A defense of our moral psychology toward animals. *Envir. Ethics* 9:197–215.

———. 1990. The myth of anthropomorphism. In *Interpretation and explanation in the study of animal behavior,* ed. M. Bekoff and D. Jamieson. Boulder, Colo.: Westview Press.

———. 1991. Disambiguating anthropomorphism: An interdisciplinary review. In *Perspectives in ethology,* vol. 9, *Human understanding and animal awareness,* ed. P. P. G. Bateson and P. H. Klopfer. New York: Plenum.

FitzGibbon, C. D., and J. H. Fanshawe. 1988. Stotting in Thompson's gazelles: An honest signal of condition. *Behav. Ecol. Sociobiol.* 23:69–74.

Fitzpatrick, M. D. 1979. Marsh hawk drowns common gallinule. *Amer. Birds* 33:837.

Flanagan, O. 1992. *Consciousness reconsidered.* Cambridge: MIT Press.

Fletcher, D. J. C., and C. D. Michener, eds. 1987. *Kin recognition in animals.* New York: Wiley.

Fouts, D. H. 1989. Signing interactions between mother and infant chimpanzees. In *Understanding chimpanzees,* ed. P. G. Heltne and L. A. Marquardt. Cambridge: Harvard University Press.

Fouts, R. S.1997. *Next of kin: What chimpanzees have taught me about who we are.* New York: William Morrow.

Fouts, R. S., W. Chown, and L. Goodman. 1976. Translation from vocal English to American sign language in a chimpanzee *(Pan). Learning and Motivation* 7:458–75.

Fouts, R. S., and D. H. Fouts. 1989. Loulis in conversation with the cross-fostered chimpanzees. In *Teaching sign language to chimpanzees,* ed. B. T. Gardner, R. A. Gardner, and T. Van Cantfort. Albany: SUNY Press.

Fouts, R. S., D. H. Fouts, and D. Schoenfeld. 1984. Sign language conversational interactions between chimpanzees. *Sign Language Studies* 42:1–12.

Friedmann, H. 1955. The honey-guides. *U.S. Nat. Mus. Bull.* 208:1–292.

Friedmann, H., and J. Kern. 1956. The problem of cerophagy or wax-eating in the honey-guides. *Q. Rev. Biol.* 31:19–30.

Frisch, K. von. 1950. *Bees: Their vision, chemical senses, and language.* Ithaca: Cornell University Press.

———. 1967. *The dance language and orientation of bees.* Cambridge: Harvard University Press.

———. 1974. *Animal architecture.* New York: Harcourt Brace.

Frith, C. R., R. Perry, and E. Lumer. 1999. The neural correlates of conscious experience: An experimental framework. *Trends in Cog. Sci.* 3:105–14.

Galambos, R., and S. A. Hillyard, eds. 1981. Electrophysiological approaches to human cognitive processing. *Neurosci. Res. Prog. Bull.* 20:141–264.

Galef, B. G., Jr. 1988. Imitation in animals: History, definition, and interpretation of data from the psychological laboratory. In *Social learning: Psychological and biological perspectives,* ed. T. R. Zentall and B. G. Galef Jr. Hillsdale, N.J.: Lawrence Erlbaum Associates.

Gallese, V., and A. Goldman. 1998. Mirror neurons and the simulation theory of mind-reading. *Trends in Neurosci.* 2:493–501.

Gallistel, C. R. 1990. *The organization of learning.* Cambridge: MIT Press.

Gallup, G. G., Jr. 1977. Self-recognition in primates: A comparative approach to the bidirectional properties of consciousness. *Amer. Psychol.* 32:329–38.

———. 1983. Toward a comparative psychology of mind. In *Animal cognition and behavior,* ed. R. L. Mellgren. Amsterdam: North Holland.

Gardner, B. T., and R. A. Gardner. 1971. Two-way communication with an infant chimpanzee. In *Behavior of nonhuman primates,* ed. A. Schrier and F. Stollnitz. New York: Academic Press.

———. 1989b. Cross-fostered chimpanzees II: Modulation of meaning. In *Understanding chimpanzees,* ed. P. G. Heltne and L. A. Marquardt. Cambridge: Harvard University Press.

Gardner, R. A., and B. T. Gardner. 1969. Teaching sign language to a chimpanzee. *Science* 165:664–72.

———. 1984. A vocabulary test for chimpanzees *(Pan troglodytes). J. Comp. Psychol.* 98:381–404.

———. 1989a. Cross-fostered chimpanzees I: Testing vocabulary. In *Understanding chimpanzees,* ed. P. Heltne and L. A. Marquardt. Cambridge: Harvard University Press.

Gardner, R. A., B. T. Gardner, and T. E. Van Cantfort, eds. 1989. *Teaching sign language to chimpanzees.* Albany: SUNY Press.

Gayou, D. C. 1982. Tool use by green jays. *Wilson Bull.* 94:593–94.

Gelman, R., and C. R. Gallistel. 1978. *The child's understanding of number.* Cambridge: Harvard University Press.

Georgopoulis, A. P., J. T. Lurie, M. Petrides, A. B. Schwartz, and J. T. Massey. 1989. Mental rotation of the neuronal population vector. *Science* 243:234–36.

Ghazanfar, A. A., and M. D. Hauser. 1999. The neuroethology of primate vocal communication: Substrates for the evolution of speech. *Trends in Cog. Sci.* 3:377–84.

Gilliard, E. T. 1969. *Birds of paradise and and bower birds.* London: Weidenfeld and Nicolson.

Globus, G. G., G. Maxwell, and I. Savodnik, eds. 1976. *Consciousness and the brain: A scientific and philosophical inquiry.* New York: Plenum.

Glover, A. A., M. C. Onofrj, M. F. Ghilardi, and I. Bodis-Wollner. 1986. P300-like potentials in the normal monkey using classical conditioning and an auditory "oddball" paradigm. *Electroencephal. Clin. Neurophysiol.* 65:231–35.

Gochfeld, M. 1984. Antipredator behavior: Aggressive and distraction displays of shorebirds. In *Shorebirds: Breeding behavior and population,* ed. J. Burger and B. L. Olla. New York: Plenum.

Goodall, J. van Lawick. 1968. Behaviour of free-living chimpanzees of the Gombe Stream area. *Anim. Behav. Monogr.* 1:165–311.

———. 1986. *The chimpanzees of Gombe: Patterns of behavior.* Cambridge: Harvard University Press.

Goodall, J. van Lawick, and H. van Lawick. 1966. Use of tools by the Egyptian vulture *Neophron percnopterus. Nature* 212:1468–69.

Gore, J. A., and W. W. Baker. 1989. Beavers residing in caves in northern Florida. *J. Mammalogy* 70:677–78.

Gould, J. L. 1976. The dance language controversy. *Q. Rev. Biol.* 51:211–44.

———. 1979. Do honeybees know what they are doing? *Nat. Hist.* 88 (June/July): 66–75.

———. 1980. Sun compensation by bees. *Science* 207:545–47.

———. 1982. *Ethology: The mechanisms and evolution of behavior.* New York: Norton.

———. 1984. Natural history of honey bee learning. In *The biology of learning,* ed. P. Marler and H. S. Terrace. Report of the Dahlem Workshop, Berlin, Oct. 23–28, 1983. *Life Sci. Res. Rep.* 29. New York: Springer.

Gould, J. L., and C. G. Gould. 1988. *The honey bee.* New York: Scientific American Library.

———. 1994. *The animal mind.* New York: Scientific American Library, W. H. Freeman.

Gould, J. L., and W. F. Towne. 1987. Evolution of the dance language. *Amer. Naturalist* 130:317–38.

———. 1988. Honey bee learning. *Advances in Insect Physiol.* 20:55–86.

Gouzoules, H., and S. Gouzoules. 1989. Design features and developmental modification of pigtail macaque, *Macaca nemestrina,* agonistic screams. *Anim. Behav.* 37:383–401.

Gouzoules, S., H. Gouzoules, and P. Marler. 1984. Rhesus monkey *(Macaca mulatta)* screams: Representational signaling in the recruitment of agonistic aid. *Anim. Behav.* 32:182–93.

———. 1986. Vocal communication: A vehicle for the study of relationships. In *The Cayo Santiago macaques: History, behavior, and biology.* Albany: SUNY Press.

Grant, P. R. 1986. *Ecology and evolution of Darwin's finches.* Princeton: Princeton University Press.

Grater, R. K. 1936. An unusual beaver habitat. *J. Mammalogy* 17:66.

Green, E. E., R. W. Ashford, and D. H. Hartridge. 1988. Sparrowhawk killing lapwing in water. *Brit. Birds* 79:502.

Green, S. 1975. Communication by a graded vocal system in Japanese monkeys. In *Primate behavior developments in field and laboratory research,* ed. L. A. Rosenblum. New York: Academic Press.

Greene, S. L. 1983. Feature memorization in pigeon concept formation. In *Quantitative analysis of behavior: Discrimination processes,* vol. 4, ed. M. L. Commons, R. J. Herrnstein, and A. R. Wagner. Cambridge, Mass.: Ballinger.

Greenfield, P. M., and E. S. Savage-Rumbaugh. 1990. Grammatical combination in *Pan paniscus:* Processes of learning and invention in the evolution and development of language. In *"Language" and intelligence in monkeys and apes: Comparative and developmental perspectives.* New York: Cambridge University Press.

Grice, H. P. 1957. Meaning. *Philos. Rev.* 66:377–88.

Grieg, R. 1979. Death by drowning—one Cooper's hawk's approach. *Amer. Birds* 33:836.

Griffin, Donald R. 1958. *Listening in the dark.* Chap. 12. Reprint 1986, Ithaca: Cornell University Press.

———. 1976. *The question of animal awareness: Evolutionary continuity of mental experience.* New York: Rockefeller University Press.

———. 1977. Anthropomorphism. *BioScience* 27:445–46.

———. 1978. Prospects for a cognitive ethology. *Behav. Brain Sci.* 1:527–38 (with commentaries 1:555–629 and 3:615–23).

———. 1981. *The question of animal awareness.* 2d ed. New York: Rockefeller University Press.

———. 1984. *Animal thinking.* Cambridge: Harvard University Press.

———. 1985a. Animal consciousness. *Neurosci. and Biobehav. Rev.* 9:615–22.

———. 1985b. The cognitive dimensions of animal communication. *Fortschritte der Zoologie* 31:417–82.

———. 1998. From cognition to consciousness. *Anim. Cog.* 1:3–16.

———. In press. Nonhuman minds. *Philos. Topics.*

Griffiths, D., A. Dickinson, and N. Clayton. 1999. Episodic memory: What can animals remember about their past? *Trends in Cog. Sci.* 3:74–80.

Grosslight, J. H., and W. C. Zaynor. 1967. Verbal behavior in the mynah bird. In *Research in verbal behavior and some physiological implications*, ed. K. Salzinger and S. Salzinger. New York: Academic Press.

Guilford, T., and M. S. Dawkins. 1991. Receiver psychology and the evolution of animal signals. *Anim. Behav.* 42:1–14.

Gyger, M., P. Marler, and R. Pickett. 1987. Semantics of an avian alarm call system: The male domestic fowl, *Gallus domesticus*. *Behaviour* 102:15–39.

Haeseler, V. 1985. Werkzeuggebrauch bei der europäischen Grabwespe *Ammophila hungarica* Moscary 1883—*(Hymenoptera: Specidae)*. *Zool. Anz.* 215:279–86.

Hall, K. 1963. Tool-using performances as indicators of behavioral adaptability. *Current Anthrop.* 4:479–94.

Hameroff, S. R., A. W. Kaszniak, and A. C. Scott, eds. 1996. *Toward a science of consciousness: The first Tucson discussions and debates.* Cambridge: MIT Press.

———. 1998. *Toward a science of consciousness II: The second Tucson discussions and debates.* Cambridge: MIT Press.

Hamilton, C. R., and B. A. Vermeire. 1988. Complementary hemispheric specialization in monkeys. *Science* 242:1691–94.

Hanlon, R. T., and J. B. Messenger. 1996. *Cephalopod behaviour.* New York: Cambridge University Press.

Hannay, A. 1971. *Mental images: A defense.* London: George Allen and Unwin.

Hansell, M. H. 1968a. The house building behaviour of the caddis-fly larvae *Silo pallipes* Fabricius I: The structure of the house and method of house extension. *Anim. Behav.* 16:558–61.

———. 1968b. The house building behaviour of the caddis-fly larvae *Silo pallipes* Fabricius II: Description and analysis of the selection of small particles. *Anim. Behav.* 16:562–77.

———. 1968c. The house building behaviour of the caddis-fly larvae *Silo pallipes* Fabricius III: The selection of large particles. *Anim. Behav.* 16:578–84.

———. 1972. Case building behaviour of the caddis fly larva, *Lepidostoma hirtum. J. Zool. London* 167:179–92.

———. 1974. Regulation of building unit size in the house building of the caddis larvae *Lepidostoma hirtum. Anim. Behav.* 22:133–43.

———. 1984. *Animal architecture and building behaviour.* London: Longman.

———. 1987. What's so special about using tools? *New Scientist* no. 1542:54–56 (Jan. 5).

Harnad, S. 1982. Consciousness: An afterthought. *Cog. Brain Theory* 5:29–47.

Harrison, J., J. Buchwald, and K. Kaga. 1986. Cat P300 present after primary auditory cortex ablation. *Electroencephalogr. Clin. Neurophysiol.* 63:180–87.

Hartman, E. 1970. *Sleep and dreaming.* Boston: Little, Brown.

Hauser, M. D. 1993. Right hemisphere dominance for the production of facial expression in monkeys. *Science* 261:475–77.

———. 1996. *The evolution of communication.* Cambridge: MIT Press.

———. 2000. *Wild minds: What animals really think.* New York: Henry Holt.

Hauser, M. D., and K. Andersson. 1994. Left hemisphere dominance for processing vocalizations in adult, but not infant rhesus monkeys. *Proc. Nat. Acad. Sci.* 91:3946–48.

Hauser, M. D., and J. Kralik. 1997. Life beyond the mirror: A reply to Anderson and Gallup. *Anim. Behav.* 54:1568–71.

Hauser, M. D., J. Kralik, and C. Butto-Mahan. 1995. Problem solving and functional design features: Experiments on cotton-top tamarins, *Saguinus oedipus oedipus. Anim. Behav.* 57:565–82.

Hauser, M. D., J. Kralik, C. Butto-Mahan, M. Garrett, and J. Oser. 1995. Self-recognition in primates: Phylogeny and the salience of species-typical features. *Proc. Nat. Acad. Sci.* 92:10811–14.

Hayes, C. 1951. *The ape in our house.* New York: Harper.

Hayes, K., and C. Hayes. 1951. The intellectual development of a home-raised chimpanzee. *Proc. Amer. Philos. Soc.* 95:105–9.

Haynes, B. D., and E. Haynes, eds. 1966. *The grizzly bear: Portraits from life.* Norman: University of Oklahoma Press.

Haynes, P. F. R. 1976. Mental imagery. *Can. J. Philos.* 6:705–19.

Hector, D. P. 1986. Cooperative hunting and its relationship to foraging success and prey size in an avian predator. *Ethology* 73:247–57.

Hediger, H. 1947. Ist das tierliche Bewusstsein unerforschbar? *Behaviour* 1:130–37.

Heffner, H. E., and R. S. Heffner. 1984. Temporal lobe lesions and perception of species-specific vocalizations by macaques. *Science* 226:75–76.

Heinrich, B. 1984. *In a patch of fireweed.* Cambridge: Harvard University Press.

———. 1988. Raven tool use? *Condor* 90:270–71.

———. 1989. *Ravens in winter.* New York: Summit Books, Simon and Schuster.

———. 1995. An experimental investigation of insight in the common raven *Corvus corax. Auk* 112:994–1003.

———. 1999. *The mind of the raven: Investigations and adventures with wolf-birds.* New York: Cliff Street Books.

Helfman, G. S. 1986. Behavioral responses of prey fishes during predator-prey interactions. In *Predator-prey relationships. Perspectives and approaches from the study of lower vertebrates,* ed. M. E. Feder and G. V. Lauder. Chicago: University of Chicago Press.

———. 1989. Threat-sensitive predator avoidance in damselfish-trumpetfish interactions. *Behav. Ecol. Sociobiol.* 24:47–58.

Heltne, P. G., and L. A. Marquardt, eds. 1989. *Understanding chimpanzees.* Cambridge: Harvard University Press.

Hendrichs, H., F. Drechmann, and A. Stephan, eds. 1999. Minds. *Evol. and Cog.* 5 (2): 98–197.

Henry, J. D. 1986. *Red fox: The catlike canine.* Washington: Smithsonian Institution Press.

Hepper, P. 1990. *Kin recognition*. New York: Cambridge University Press.

Herman, L. M., ed. 1980. *Cetacean behavior: Mechanisms and functions*. New York: Wiley.

———. 1986. Cognition and language competencies of bottlenosed dolphins. In *Dolphin cognition and behavior: A comparative approach*, ed. R. J. Schusterman, A. J. Thomas, and F. G. Wood. Hillsdale, N.J.: Lawrence Erlbaum Associates.

———. 1987. *Receptive competencies of language-trained animals*. Advances in the Study of Behavior, no. 17. New York: Academic Press.

———. 1988. The language of animal language research: Reply to Schusterman and Gisiner. *Psychol. Rec.* 38:349–62.

Herrnstein, R. J. 1984. Objects, categories, and discriminative stimuli. In *Animal cognition*, ed. H. L. Roitblat, T. G. Bever, and H. S. Terrace. Hillsdale, N.J.: Lawrence Erlbaum. Associates.

———. 1990. Levels of stimulus control: A functional approach. *Cognition* 37:133–66.

Herrnstein, R. J., and P. A. de Villiers. 1980. Fish as a natural category for people and pigeons. In *The psychology of learning and motivation: Advances in research and theory*, vol. 14, ed. G. H. Bower. New York: Academic Press.

Herrnstein, R. J., and D. H. Loveland. 1964. Complex visual concept in the pigeon. *Science* 146:549–51.

Herrnstein, R. J., D. H. Loveland, and C. Cable. 1976. Natural concepts in pigeons. *J. Exptl. Psychol.: Anim. Behav. Processes* 2:285–302.

Herrnstein, R. J., W. Vaughan, Jr., D. B. Mumford, and S. M. Kosslyn. 1989. Teaching pigeons an abstract relational rule: Insideness. *Perception and Psychophysics* 46:56–64.

Hershberger, W. A. 1986. An approach through the looking-glass. *Anim. Learning and Behav.* 14:443–51.

Heyes, C. M. 1987. Cognisance of consciousness in the study of animal knowledge. In *Evolutionary epistemology: A multiparadigm program*, ed. W. Callebaut and R. Pinxten. Dordrecht: North Holland.

Heyes, C. M., and A. Dickinson. 1990. The intentionality of animal action. *Mind and Language* 5:87–104.

Heyes, C. M., and L. Huber, eds. 2000. *The evolution of cognition*. Cambridge: MIT Press.

Higuchi, H. 1986. Bait-fishing by the green backed heron *Ardeola striata* in Japan. *Ibis* 128:285–90.

———. 1987. Cast master. *Nat. Hist.* 96 (8): 40–43.

———. 1988a. Individual differences in bait-fishing by the green-backed heron *Ardeola striata* associated with territory quality. *Ibis* 130:39–44.

———. 1988b. Bait-fishing by green-backed herons in south Florida. *Florida Field Naturalist* 16:8–9.

Hilgard, E. R., and G. H. Bower. 1975. *Theories of learning*. 4th ed. Englewood Cliffs, N.J.: Prentice-Hall.

Hinde, R. A. 1981. Animal signals: Ethological and games-theory approaches are not incompatible. *Anim. Behav.* 29:535–42.

Hinde, R. A., and J. Fisher. 1951. Further observations on the opening of milk bottles by birds. *Brit. Birds* 44:393–96.

———. 1972. Some comments on the re-publication of two papers on the opening of milk bottles by birds. In *Function and evolution of behavior,* ed. P. H. Klopfer and J. P. Hailman. Reading, Mass.: Addison-Wesley.

Hoage, R. J., and L. Goldman, eds. 1986. *Animal intelligence: Insights into the animal mind.* Washington, D.C.: Smithsonian Institution Press.

Hobson, J. A., and R. Stickgold. 1997. The conscious state paradigm: A neurocognitive approach to waking, sleeping, and dreaming. In *The cognitive sciences,* ed. M. S. Gazzaniga. Cambridge: MIT Press.

Hodgdon, H. E. 1978. Social dynamics and behavior within an unexploited beaver *(Castor canadensis)* population. Ph.D. thesis, University of Massachusetts, Amherst.

Holland, P. C. 1979. Differential effects of omission contingencies on various components of Pavlovian appetitive conditioning. *J. Exptl. Psychol.: Anim. Beh. Processes* 5:178–93.

Hölldobler, B. 1974. Communication by tandem running in the ant *Campanotus sericeus. J. Comp. Physiol. A* 90:105–27.

———. 1977. Communication in social Hymenoptera In *How animals communicate,* ed. T. Sebeok. Bloomington: Indiana University Press.

Hölldobler, B., and E. O. Wilson, 1978. The multiple recruitment system of the African weaver ant *Oecophylla longinoda* (Latrelle) (Hymenoptera, Formicidae). *Behav. Ecol. Sociobiol.* 3:19–60.

———. 1990. *The ants.* Cambridge: Harvard University Press.

Honig, W. K., and R. K. R. Thompson. 1982. Retrospective and prospective processing in animal working memory. In *The psychology of learning and motivation: Advances in research and theory,* vol. 16, ed. G. H. Bower. New York: Academic Press.

Hopkins, C. D. 1999. Design features for electric communication. *J. Exptl. Biol.* 202:1217–28.

Horgan, I., and J. Woodward. 1985. Folk psychology is here to stay. *Philos. Rev.* 94:197–226.

Houk, J. L., and J. J. Geibel. 1974. Observations of underwater tool use by the sea otter *Enhydra lutris* Linneus. *Calif. Fish and Game* 60:207–8.

Howell, T. R. 1979. *Breeding biology of the Egyptian plover Pluvianus aegypticus.* Publications in Zoology no. 113. Berkeley: University of California Press.

Humphrey, N. K. 1980. Nature's psychologists. In *Consciousness and the physical world,* ed. B. D. Josephson and V. S. Ramachandran. New York: Pergamon.

Humphrey, N. K., and L. Weiskrantz. 1967. Vision in monkeys after removal of the striate cortex. *Nature* 215:595–97.

Hunt, G. R. 1996. Manufacture and use of hook-tools by New Caledonian crows. *Nature* 379:249–51.

Huntingford, F. A. 1982. Do inter- and intraspecific aggression vary in reaction to predation pressure in sticklebacks? *Anim. Behav.* 30:909–16.

Huntingford, F. A., and N. Giles. 1987. Individual variation in anti-predator responses in the three-spined stickleback. *Ethology* 74:205–10.

Iersel, J. J. A. van, and J. van dem Assem. 1964. Aspects of orientation in the diggerwasp *Bembix rostrata. Anim. Behav. Suppl.* 1:145–62.

Ikatura, S., and T. Matsuzawa. 1993. Acquisition of personal pronouns by a chimpanzee. In *Language and communication: Comparative perspectives,* ed. H. L. Roitblat, L. M. Herman, and P. E. Nachtigall. Hillsdale, N.J.: Lawrence Erlbaum Associates.

Immelmann, K., and C. Beer. 1989. *A dictionary of ethology.* Cambridge: Harvard University Press.

Ingold, T. 1988. The animal in the study of humanity. In *What is an animal?* ed. T. Ingold. London: Unwin Hyman.

Isack, H. A., and H.-U. Reyer. 1989. Honeyguides and honey gatherers: Interspecific communication in a symbiotic relationship. *Science* 243:1343–46.

Jacobs, L. 1989. Cache economy of the gray squirrel. *Nat. Hist.* 98 (10): 40–47.

Jacobs, L., and E. R. Liman. 1991. Grey squirrels remember the locations of buried nuts. *Anim. Behav.* 41:103–10.

James, W. 1890. *The principles of psychology.* New York: Henry Holt.

Janes, S. W. 1976. The apparent use of rocks by a raven in nest defense. *Condor* 78:409.

Jeffrey, R. C. 1985. Animal interpretation. In *Actions and events: Perspectives on the philosophy of Donald Davidson,* ed. E. Lepore and B. P. McLaughlin. Oxford: Basil Blackwell.

Jenkins, H. M., and B. R. Moore. 1973. The form of the auto-shaped response with food and water reinforcements. *J. Exp. Anal. Behav.* 20:163–81.

Jerison, H. J., and I. Jerison, eds. 1986. *Intelligence and evolutionary biology.* New York: Springer.

Johnson, D. M. 1988. "Brutes believe not." *Philosophical Psychol.* 1:279–94.

Jolly, A. 1966. Lemur social behavior and primate intelligence. *Science* 153:501–6.

Jones, T., and A. Kamil. 1973. Tool-making and tool-using in the northern bluejay. *Science* 180:1076–78.

Jouvet, M. 1979. What does a cat dream about? *Trends in Neurosci.* 2:280–82.

———. 1999. *The paradox of sleep: The story of dreaming.* Cambridge: MIT Press.

Jung, C. G. 1973. *Syncronicity: A causal connecting principle.* Princeton: Princeton University Press.

Kacelnik, A., and M. Bateson. 1996. Risky theories: The effects of variance on foraging decisions. *Amer. Zoologist* 36:402–34.

Kallander, H., and H. G. Smith. 1990. Food storing in birds: An evolutionary perspective. In *Current Ornithology,* vol. 7, ed. D. M. Powers. New York: Plenum.

Kalmeijn, A. J. 1974. The detection of electric fields from inanimate and animate sources other than electric organs. In *Handbook of sensory physiology,* ed. A. Fessard, vol. 3, pt. 3. New York: Springer.

Kamil, A. C. 1988. A synthetic approach to the study of animal intelligence. In *Nebraska symposium on motivation, 1987: Comparative perspectives in modern psychology,* ed. D. W. Leger. Lincoln: University of Nebraska Press.

———. 1998. On the proper definition of cognitive ethology. In *Animal cognition in nature: The convergence of psychology and biology in laboratory and field,* ed. R. P. Balda, I. M. Pepperberg, and A. C. Kamil. San Diego: Academic Press.

Kamil, A. C., and T. D. Sargent, eds. 1981. *Foraging behavior: Ecological, ethological, and psychological approaches.* New York: Garland STPM.

Karakashian, S. J., M. Gyger, and P. Marler. 1988. Audience effects on alarm calling in chickens *(Gallus gallus). J. Comp. Psychol.* 102:129–35.

Kellogg, W. N., and L. A. Kellogg. 1933. *The ape and the child: A study of environmental influence on early behavior.* New York: Whittlesey House. (Reprint, 1967: New York: Haffner.)

Kennedy, J. S. 1992. *The new anthropomorphism.* New York: Cambridge University Press.

Kentridge, R. W., and C. A. Heywood. 1999. The status of blindsight. *J. Consc. Studies* 6:3–11.

Kenyon, K. W. 1969. *The sea otter in the eastern Pacific Ocean.* North American Fauna, no. 68. Washington, D.C.: U.S. Bureau of Sport Fisheries and Wildlife.

Kerr, W. E. 1969. Some aspects of the evolution of social bees. *Evol. Biol.* 3:119–75.

———. 1994. Communication among *Melipona* workers (Hymenoptera: Apidae). *J. Insect Behav.* 7:123–28.

Kettlewell, H. B. D. 1955. Recognition of appropriate backgrounds by the pale and black phases of Lepidoptera. *Nature* 175:943–44.

Kettlewell, H. B. D., and D. L. T. Conn. 1977. Further background choice experiments on cryptic Lepidoptera. *J. Zool. London* 181:371–76.

Kety, S. S. 1960. A biologist examines the mind and behavior. *Science* 132:1861–70.

Kirchner, W. H., C. Dreller, and W F. Towne. 1991. Hearing in honeybees: Operant conditioning and spontaneous reactions to airborne sound. *J. Comp. Physiol. A.* 168:85–89.

Kirchner, W., M. Lindauer, and A. Michelsen. 1988. Honeybee dance communication: Acoustical indication of direction in round dances. *Naturwissenschaften* 75:629–30.

Koch, C., and G. Laurent. 1999. Complexity and the nervous system. *Science* 284:96–98.

Koehler, O. 1956a. Sprach und unbenanntes Denke. In *L'instinct dans le comportement des animaux et de l'homme.* Paris: Masson.

———. 1956b. Thinking without words. *Proc. 14th Int. Zool. Cong. Copenhagen,* 1953:75–88.

———. 1969. Tiersprache und Menschensprache. In *Kreatur Mensch: Moderne Wissenschaft auf der Suche nach dem Humanum,* 119–33 and 187–90. Munich: Heinz Moos.

Konishi, M. 1985. Birdsong: From behavior to neuron. *Ann. Rev. Neurosci.* 8:125–70.

Kosslyn, S. M. 1988. Aspects of a cognitive neuroscience of mental imagery. *Science* 240:1621–26.

———. 1994. *Image and brain: The resolution of the imagery debate.* Cambridge: MIT Press.

Kramer, B. 1990. *Electrocommunication in teleost fishes: Behavior and experiments.* New York: Springer.

———. 1997. Electric organ discharges and their relation to sex in Mormyrid fishes. *Naturwissenschaften* 84:119–21.

Krebs, J. R. 1977. Review of *The question of animal awareness. Nature* 266:792.

———. 1979. Foraging strategies and their social significance. In *Handbook of behavioral neurobiology,* vol. 3, *Social behavior and communication,* ed. P. Marler and J. G. Vandenbergh. New York: Plenum.

Krebs, J. R., and N. B. Davies. 1978. *Behavioural ecology: An evolutionary approach.* Oxford: Blackwell.

Krebs, J. R., and R. Dawkins. 1984. Animal signals: Mind reading and manipulation. In *Behavioural ecology: An evolutionary approach.* 2d ed. Sunderland, Mass.: Sinauer.

Krebs, J. R., and A. J. Inman. 1994. Learning and foraging: Individuals, groups, and populations. In *Behavioral mechanisms in evolutionary ecology,* ed. L. A. Real. Chicago: University of Chicago Press.

Krebs, J. R., M. H. MacRoberts, and J. M. Cullen. 1972. Flocking and feeding in the great tit *(Parus major)*—an experimental study. *Ibis* 114:507–30.

Kroeber, A. L. 1952. *The nature of culture.* Chicago: University of Chicago Press.

Kruuk, H. 1972. *The spotted hyena: A study of predation and social behavior.* Chicago: University Chicago Press.

Kushlan, J. A. 1973. Bill-vibrating: A prey attracting behavior of the snowy egret *Leucophayx thula. Amer. Midl. Nat.* 89:509–10.

LaMon, B., and H. P. Zeigler. 1988. Control of pecking response form in the pigeon: Topography of ingestive behaviors and conditioned keypecks with food and water reinforcers. *Anim. Learning and Behav.* 16:256–67.

Langen, T. A. 1996. Social learning of a novel foraging skill by white-throated magpie-jays (*Calocitta formosa* corvidae): A field experiment. *Ethology* 102:157–66.

Latto, R. 1986. The question of animal consciousness. *Psychol. Rec.* 36:309–14.

Lea, S. E. G. 1984. In what sense do pigeons learn concepts? In *Animal cognition,* ed. H. L. Roitblat, T. G. Bever, and H. S. Terrace. Hillsdale, N.J.: Lawrence Erlbaum Associates.

Lecours, A. R., A. Basso, S. Moraschini, and J.-L. Nespoulous. 1984. Where is the speech area and who has seen it? In *Biological perspectives on language,* ed. D. Caplan, A. R. Lecours, and A. Smith. Cambridge: MIT Press.

LeDoux, J. 1996. *The emotional brain.* New York: Simon and Schuster.

Lee, Y., and M. E. Bitterman. 1991. Learning in honeybees *(Apis mellifera)* as a function of amount of reward: Acquisition measures. *J. Comp. Psychol.* 104:152–58.

Lethmate, J. 1977. Problemlöseverhalten von Orang-utans *(Pongo pygmaeus). Fortschritte der Verhaltensforschung: Suppl. J. Comp. Ethology.* Berlin: Paul Parey.

Lewis, A. R. 1980. Patch use by gray squirrels and optimal foraging. *Ecology* 61:1371–79.

Lindahl, B. I. B. 1997. Consciousness and biological evolution. *J. Theoret. Biol.* 187:613–29.

———. In press. Consciousness, behavioural patterns and direction of biological evolution: Implications for the mind-brain problem. In *Dimensions of conscious experience,* ed. P. Pylkänen and T. Vaden. Amsterdam: John Benjamins.

Lindauer, M. 1954. Dauertänze im Bienenstock und ihre Beziehung zur Sonnenbahn. *Naturwissenschaften* 41:506–7.

———. 1955. Schwarmbienen auf Wohnungssuche. *Z. vergl. Physiol.* 37:263–324.

———. 1971. *Communication among social bees,* 2d ed. Cambridge: Harvard University Press.

Livet, P. 1992. Représentations, comportements, et psychologie expérimentale populare. In *La représentation animale: Représentations de la représentation,* ed. J. Gervet, P. Livet, and A. Tete. Nancy: Presses Universitaires de Nancy.

Lloyd, D. 1989. *Simple minds.* Cambridge: MIT Press.

Lloyd, J. E. 1996. Firefly communication and deception: "Oh, what a tangled web." In *Deception: Perspectives on human and nonhuman deceit,* ed. R. W. Mitchell and N. S. Thompson. Albany: SUNY Press.

Loeb, J. 1912. *The mechanistic conception of life.* Chicago: University of Chicago Press. Reprint (with preface by D. Fleming) 1964, Cambridge: Harvard University Press.

———. 1918. *Forced movements, tropisms, and animal conduct.* Philadelphia: Lippincott. Reprinted 1973, New York: Dover.

Lovell, H. B. 1958. Baiting of fish by a green heron. *Wilson Bull.* 70:280–81.

Lucas, J. R. 1982. The biophysics of pit construction by antlion larvae (*Myrmeleon,* Neuroptera). *Anim. Behav.* 30:651–64.

———. 1989. The structure and function of antlion pits: Slope asymmetry and predator-prey interactions. *Anim. Behav.* 38:318–30.

Lycan, W. G. 1987. *Consciousness.* Cambridge: MIT Press.

Macedonia, J. M. 1990. What is communicated in the antipredator calls of lemurs: Evidence from playback experiments with ringtailed and ruffed lemurs. *Ethology* 86:177–90.

Macedonia, J. M. and C. S. Evans. 1993. Variation among mammalian alarm call systems and the problem of meaning in animal signals. *Ethology* 93:177–97.

MacKenzie, B. D. 1977. *Behaviourism and the limits of the scientific method.* London: Routledge & Kegan Paul.

Mackintosh, N. J. 1974. *The psychology of animal learning.* New York: Academic Press.

———. 1987. Animal minds. In *Mindwaves: Thoughts on intelligence, identity, and consciousness,* ed. C. Blakemore and S. Greenfield. Oxford: Oxford University Press.

MacLean, G. L. 1968. Field studies on the sandgrouse of the Kalahari desert. *Living Bird* 7:209–35.

———. 1975. Belly-soaking in the Charadriformes. *J. Bombay Nat. Hist. Soc.* 72:74–82.

Macphail, E. M. 1998. *The evolution of consciousness.* New York: Oxford University Press.

Macphail, E. M., and S. Reilly. 1989. Rapid acquisition of a novelty versus familiarity concept by pigeons *(Columba livia). J. Exptl. Psychol.: Anim. Behav. Processes* 15:242–52.

Magurran, A. E. 1986. Individual differences in fish behaviour. In *The behaviour of teleost fishes,* ed. T. J. Pitcher. Baltimore: Johns Hopkins University Press.

Magurran, A. E., and T. J. Pitcher. 1987. Provenance, shoal size, and the sociobiology of predator-evasion behaviour in minnow shoals. *Proc. Roy. Soc.* B 229:439–65.

Mahner, M., and M. Bunge. 1997. *Foundations of biophilosophy.* Stuttgart: Springer.

Malcolm, N. 1973. Thoughtless brutes. *Proceedings and Addresses of Amer. Philos. Assn.* 46:5–20.

Malott, R. W., and J. W. Siddall. 1972. Acquisition of the people concept in pigeons. *Psychol. Reports* 31:3–13.

Marler, P., A. Dufty, and R. Pickert. 1986a. Vocal communication in the domestic chicken I: Does a sender communicate information about the quality of a food referent to a receiver? *Anim. Behav.* 34:188–94.

———. 1986b. Vocal communication in the domestic chicken II: Is a sender sensitive to the presence and nature of a receiver? *Anim. Behav.* 34:194–98.

Marler, P. and C. Evans. 1996. Bird calls: Just emotional displays or something more? *Ibis* 138:26–33.

Marler, P., S. Karakashian, and M. Gyger. 1991. Do animals have the option of withholding signals when communication is inappropriate? In *Cognitive ethology,* ed. C. A. Ristau. Hillsdale, N.J.: Lawrence Erlbaum Associates.

Marshall, A. J. 1954. *Bower birds.* London: Oxford University Press.

Marshall, A. J., R. W. Wrangham, and A. C. Arcadi. 1999. Does learning affect the structure of vocalizations in chimpanzees? *Anim. Behav.* 58:825–30.

Masataka, N. 1989. Motivational referents of contact calls in Japanese monkeys. *Ethology* 80:265–73.

Mason, W. A. 1976. Review of *The question of animal awareness. Science* 194:930–31.

Mather, J. 1994. Home choice and modification by juvenile *Octopus vulgaris* (Mollusca: Cephalpoda): Specialized intelligence and tool use? *J. Zool. London* 233:359–68.

———. 1995. Cognition in cephalopods. *Advances in the Study of Behav.* 24:317–53.

Mather, J., and R. C. Anderson. 1999. Exploration, play, and habituation in octopuses *(Octopus dofleini). J. Comp. Psychol.* 113:333–38.

Matsuzawa, T. 1985. Use of numbers by a chimpanzee. *Nature* 315:57–59.

———. 1996. Chimpanzee intelligence in nature and in captivity: Isomorphism of symbol and tool use. In *Great ape societies,* ed. W. C. McGrew, L. F. Marchant, and T. Nishida. New York: Cambridge University Press.

Matsuzawa, T., T. Asano, K. Kubita, and K. Murofushi. 1996. Acquisition and generalization of numerical labeling by a chimpanzee. In *Current perspectives in primate social dynamics: Selected proceedings of the ninth congress of the International Primatological Society,* ed. D. M. Taub and F. A. King. New York: Van Nostrand Reinhold.

Maunsell, J. H. R., and W. T. Newsome. 1987. Visual processing in the monkey extrastriate cortex. *Ann. Rev. Neurosci.* 10:363–401.

McAlpine, D. F. 1977. Notes on cave utilization by beavers. *Bull. Natl. Speleological Soc.* 39:90–91.

McCasland, J. S. 1987. Neuronal control of bird song production. *J. Neuroscience* 7:23–39.

McCleneghan, K., and J. A. Ames. 1976. A unique method of prey capture by a sea otter, *Enhydra lutris. J. Mammalogy* 57:410–12.

McComb, K., C. Packer, and A. Pusey. 1994. Roaring and numerical assessment in contests between groups of female lions, *Panthera leo. Anim. Behav.* 47:379–87.

McFarland, D. 1989. *Problems of animal behaviour.* Burnt Mill, Harlow, Essex: Longman Scientific and Technical.

McGregor, A., and S. D. Healy. 1999. Spatial accuracy in food-storing and nonstoring birds. *Anim. Behav.* 58:727–34.

McGrew, W. C. 1994. Tools compared: The material of culture. In *Chimpanzee cultures,* ed. R. W. Wrangham, W. C. McGrew, F. de Waal, and P. G. Heltne. Cambridge: Harvard University Press.

McGrew, W. C., C. E. G. Tutin, and P. J. Baldwin. 1979. Chimpanzees, tools, and termites: Cross-culture comparisons of Senegal, Tanzania, and Rio Muni. *Man* 14:185–214.

McLean, I. G., and G. Rhodes. 1991. Enemy recognition and response in birds. In *Current ornithology,* vol. 8, ed. D. M. Power. New York: Plenum.

McMahan, E. A. 1982. Bait-and-capture strategy of termite-eating assassin bug. *Insectes Sociaux* 29:346–51.

———. 1983. Bugs angle for termites. *Nat. Hist.* 92 (5): 40–47.

McMahon, B. F., and R. M. Evans. 1992. Foraging strategies of American white pelicans. *Behaviour* 120:69–89.

McQuade, D. B., E. H. Williams, and H. B. Eichenbaum. 1986. Cues used for localizing food by the gray squirrel *(Sciurus carolinensis). Ethology* 72:22–30.

Meijsing, M. 1997. Awareness, self-awareness, and perception: An essay on animal consciousness. In *Animal consciousness and animal ethics: Perspectives from The Netherlands,* ed. M. Dol, S. Kasanmoentalib, S. Lijmbach, E. Rivas, and R. van den Bos. Assen, The Netherlands: Van Gorcum.

Meinertzhagen, R. 1959. *Pirates and predators: The piratical and predatory habits of birds.* Edinburgh: Oliver and Boyd.

Mercado, E., III, S. O. Murray, R. K. Uyeyama, A. A. Pack, and L. N. Herman. 1998. Memory for recent actions in the bottlenosed dolphin *(Tursiops truncatus):* Repetition of arbitrary behaviors using an abstract rule. *Anim. Learning and Behav.* 26:210–18.

Mercado, E., III, R. K. Uyeyama, A. A. Pack, and L. M. Herman. 1999. Memory for action events in the bottlenosed dolphin. *Anim. Cog.* 2:17–25.

Meyerriecks, A. J. 1959. Foot-stirring feeding behavior in herons. *Wilson Bull.* 71:153–58.

———. 1960. *Comparative breeding behavior of four species of North American herons.* Pub. no. 2. Cambridge, Mass.: Nuttall Ornithological Club.

Michelsen, A., W. H. Kirchner, B. B. Andersen, and M. Lindauer. 1986. The tooting and quacking vibration signals of honeybee queens: A quantitative analysis. *J. Comp. Physiol. A* 158:605–11.

Michelsen, A., W. H. Kirchner, and M. Lindauer. 1986. Sound and vibrational signals in the dance language of the honeybee, *Apes mellifera. Behav. Ecol. Sociobiol.* 18:207–12.

———. 1989. Honeybees can be recruited by a mechanical model of a dancing bee. *Naturwissenschaften* 76:277–80.

Michelsen, A., W. F. Towne, W. H. Kirchner, and P. Kryger. 1987. The acoustic near field of a dancing honeybee. *J. Comp. Physiol.* 161:633–43.

Midgley, M. 1978. *Beast and man: The roots of human nature.* Ithaca, N.Y: Cornell University Press.

———. 1983. *Animals and why they matter.* Athens: University of Georgia Press.

———. 1998. One world, but a big one. In *From brains to consciousness? Essays on the new science of the mind,* ed. S. Rose. Princeton: Princeton University Press.

Miles, H. L. W. 1990. The cognitive foundations for reference in a signing orangutan. In *"Language" and intelligence in monkeys and apes,* ed. S. T. Parker and K. R. Gibson. New York: Cambridge University Press.

Miller, G. A. 1967. *The psychology of communication.* New York: Basic Books.

Millikan, G. C., and R. I. Bowman. 1967. Observations on Galapagos tool-using finches in captivity. *Living Bird* 6:23–41.

Mills, E. A. 1919. *The grizzly: Our greatest wild animal.* Boston: Houghton Mifflin.

Milner, A. D. 1998. Streams of consciousness: Visual awareness and the brain. *Trends in Cog. Sci.* 2:25–30.

Milner, A. D., and M. Goodale. 1995. *The visual brain in action.* New York: Oxford University Press.

Milner, B., L. R. Squire, and E. R. Kandel. 1998. Cognitive neuroscience and the study of memory. *Neuron* 20:445–68.

Mitchell, R. W., and N. S. Thompson. 1986. *Deception: Perspectives in human and nonhuman deceit.* Albany: SUNY Press.

Moller, A. P. 1988. False alarm calls as a means of resource usurpation in the great tit *Parus major. Ethology* 79:25–30.

Montagna, W. 1976. *Nonhuman primates in biomedical research.* Minneapolis: University of Minnesota Press.

Morgan, L. H. 1868. *The American beaver and his works.* Philadelphia: Lippincott. (Reprinted 1966 by Dover Publications, New York.)

Morgane, P. J., M. S. Jacobs, and A. Galaburda. 1986. Evolutionary morphology of the dolphin brain. In *Dolphin cognition and behavior: A comparative approach,* ed. R. J. Schusterman, J. A. Thomas, and F. G. Wood. Hillsdale, N.J.: Lawrence Erlbaum Associates.

Morrison, A. R. 1983. A window on the sleeping brain. *Scientific Amer.* 248 (4): 94–102.

Mountcastle, V. 1998. Brain science at the century's ebb. *Daedalus* 127 (2): 1–36.

Mowrer, O. H. 1950. On the psychology of "talking birds"—a contribution to language and personality theory. In *Learning theory and personality dynamics.* New York: Ronald.

———. 1960a. *Learning theory and behavior.* New York: Wiley.

———. 1960b. *Learning theory and symbolic processes.* New York: Wiley.

———, ed. 1980. *Psychology of language and learning.* New York: Plenum.

Moynihan, M., and A. F. Rodanichi. 1982. The behavior and natural history of the Caribbean reef squid *Sepioteuthis sepioidea,* with a consideration of social signals, and the defensive patterns for different and dangerous environments. Berlin: Parey (Supplement 25 to *Zeitschrift für Tierpsychologie*).

Munn, C. A. 1986a. The deceptive use of alarm calls by sentinel species in mixed-species flocks of neotropical birds. In *Deception: Perspectives on human and nonhuman deceit,* ed. R. W. Mitchell and N. S. Thompson. Albany: SUNY Press.

———. 1986b. Birds that "cry wolf." *Nature* 319:143–45.

Munn, M. B. 1988. The Amazon's gregarious giant otters. *Animal Kingdom,* September–October 1988, 34–42.

Nagel, T. 1974. What is it like to be a bat? *Philos. Rev.* 83:435–50. Reprinted in T. Nagel, 1979. *Mortal questions.* London: Cambridge University Press.

Natsoulas, T. 1978. Consciousness. *Amer. Psychologist* 33:906–14.

———. 1983. Concepts of consciousness. *J. Mind and Behav.* 4:13–59.

———. 1985. An introduction to the perceptual kind of conception of direct (reflective) consciousness. *J. Mind and Behav.* 6:333–56.

———. 1986. On the radical behaviorist conception of consciousness. *J. Mind and Behav.* 8:1–21.

———. 1988. The intentionality of retroawareness. *J. Mind and Behav.* 9:549–74.

Nelson, K. 1977. First steps in language acquisition. *J. Amer. Acad. Child Psychiat.* 16:563–83.

Netzel, H. 1977. Die Bildung des Gehauses bei *Difflugia oviformes (Rhizopoda, Testacea). Arch. Prostistenk.* 119:1–30.

Neville, H. J., and S. L. Foote. 1984. Auditory event related potentials in the squirrel monkey: Parallels to human late wave responses. *Brain Res.* 298:107–16.

Niah, J. C. 1993. The stop signal of honey bees: Reconsidering its message. *Behav. Ecol. Sociobiol.* 33:51–56.

———. 1998. The food recruitment dance of the stingless bee, *Melipona panamica. Behav. Ecol. Sociobiol.* 43:133–45.

Niah, J. C., and D. W. Roubik. 1995. A stingless bee *(Melipona panamica)* indicates food location without using a scent trail. *Behav. Ecol. Sociobiol.* 37:63–70.

Noakes, D. L. G., D. G. Lindquist, G. S. Helfman, and J. A. Ward, eds. 1983. *Predators and prey in fishes.* The Hague: Junk. Also *Envir. Biol. of Fishes* 8 (3/4): 169–328 (1983).

Norris, D. 1975. Green heron *(Butorides virescens)* uses feather lure and using apparent bait. *Amer. Birds* 29:652–54.

Norton-Griffiths, M. 1967. Some ecological aspects of the feeding behaviour of the oyster-catcher *Haematopus ostralegus* on the edible mussel *Mytilus edulis. Ibis* 109:412–24.

———. 1969. The organisation, control and development of parental feeding in the oystercatcher *(Haemotropus ostralegus). Behaviour* 34:55–114.

Nottebohm, F. 1979. Origins and mechanisms in the establishment of cerebral dominance. In *Handbook of behavioral neurobiology,* vol. 2, *Neuropsychology,* ed. M. Gazzaniga. New York: Plenum.

Oakley, D. A., ed. 1985. *Brain and mind.* London: Methuen.

Ohman, A., F. Esteves, and J. J. F. Soares. 1995. Preparedness and preattentative association learning: Electrodermal conditioning to masked stimuli. *J. Psychophysiol.* 9:99–108.

Oldroyd, B. P., T. E. Rinderer, and S. M. Buco. 1991. Honey bees dance with their super-sisters. *Anim. Behav.* 42:121–29.

Olthof, A., C. M. Eden, and W. A. Roberts. 1997. Judgments of ordinality and summation of number symbols by squirrel monkeys *(Saimiri sciureus). J. Exptl. Psychol.: Anim. Behav. Processes* 23:325–39.

Orians, G. H. 1980. *Some adaptations of marsh-nesting blackbirds.* Princeton: Princeton University Press.

Ornstein, R. I. 1972. Tool use by the New Caledonian crow *(Corvus moneduloidea). Auk* 89:674–76.

Ostwald, J., H.-U. Schnitzler, and G. Schuller. 1988. Target discrimination and target classification in echolocating bats. In *Animal sonar: Processes and performance,* ed. P. E. Nachtigal and P. W. B. Moore. New York: Plenum.

Otto, C. 1987a. Asymmetric competition for cases in *Agrypnia pagetana* (Trichoptera) larvae. *Oikos* 48:253–57.

———. 1987b. Behavioural adaptations by *Agrypnia pagetana* (Trichoptera) larvae for cases of different value. *Oikos* 50:191–96.

Owings, D. H., and E. S. Morton. 1997. The role of information in communication: An assessment/management approach. In *Perspectives in ethology,* vol. 12. New York: Plenum.

———. 1998. *Animal vocal communication: A new approach.* New York: Cambridge University Press.

Packer, C., and A. E. Pusey. 1982. Cooperation and competition within coalitions of male lions: Kin selection or game theory? *Nature* 296:740–72.

———. 1997. Divided we fall: Cooperation among lions. *Scientific Amer.* 276 (5): 52–59.

Packer, C., and L. Ruttan. 1998. The evolution of cooperative hunting. *Amer. Nat.* 132:159–94.

Page, G. 1999. *Inside the animal mind.* New York: Doubleday.

Palameta, B., and L. Lefebvre. 1985. The social transmission of a food-finding technique in pigeons: What is learned? *Anim. Behav.* 33:892–96.

Paley, W. 1851. *Natural theology: Or, Evidence of the existence and attributes of the Diety, collected from the appearances of nature.* London: Gould and Lincoln.

Panksepp, J. 1998. *Affective neuroscience.* New York: Oxford University Press.

Papaj, D. R., and A. C. Lewis, eds. *Insect learning: Ecological and evolutionary perspectives.* New York: Chapman and Hall.

Parker, S. I., and K. R. Gibson, eds. 1990. *"Language" and intelligence in monkeys and apes: Comparative and developmental perspectives.* New York: Cambridge University Press.

Parker, S. T., P. W. Mitchell, and M. L. Boccia, eds. 1994. *Self-awareness in animals and humans: Developmental perspectives.* New York: Cambridge University Press.

Partridge, L. 1978. Habitat selection. In *Behavioural ecology: An evolutionary approach,* ed. J. R. Krebs and N. B. Davies. Oxford: Blackwell.

Patenaude, F. 1983. Care of the young in a family of wild beavers, *Castor canadensis. Acta Zool. Fennica* 174:121–22.

Patenaude, F., and J. Bovet. 1983. Parturition and related behavior in wild American beavers *(Castor canadensis). Z. Saugetierkunde* 48:136–45.

———. 1984. Self-grooming and social grooming in the North American beaver, *Castor canadensis. Can. J. Zool.* 62:1872–78.

Patterson, F. C. P., and R. H. Cohn. 1994. Self-recognition and self-awareness in lowland gorillas. In *Self-awareness in animals and humans: Developmental perspectives,* ed. S. T. Parker, R. W. Mitchell, and M. Boccia. New York: Cambridge University Press.

Patterson, F. G., and E. Linden. 1981. *The education of Koko.* New York: Holt, Rinehart and Winston.

Payne, K., and R. Payne. 1985. Large scale changes over 19 years in songs of humpback whales in Bermuda. *Z. Tierpsychol.* 68:89–114.

Pearce, J. M. 1997. *Animal learning and cognition: An introduction.* 2d ed. Hove, Sussex: Psychology Press.

Pedersen, H. C., and J. B. Steen. 1985. Parental care and chick production in a fluctuating population of willow ptarmigan. *Ornis Scand.* 16:270–76.

Pennisi, E. 1999. Are our primate cousins conscious? *Science* 284:2073–76.

Pepperberg, I. M. 1981. Functional vocalization by an African Grey parrot *(Psittacus erithacus). Z. Tierpsychol.* 55:139–60.

———. 1987a. Evidence for conceptual qualitative abilities in the African grey parrot: Labeling of cardinal sets. *Ethology* 75:37–61.

———. 1987b. Acquisition of the same/different concept by an African Grey parrot *(Psittacus erithacus):* Learning with respect to categories of color, shape, and material. *Anim. Learning and Behav.* 15:423–32.

———. 1988. Comprehension of "absence" by an African Grey parrot: Learning with respect to questions of same/different. *J. Exptl. Anal. of Behav.* 50:553–64.

———. 1990. Cognition in an African Grey parrot *(Psittacus erithacus):* Further evidence of comprehension of categories and labels. *J. Comp. Psychol.* 104:41–52.

———. 1991. A communicative approach to animal cognition: A study of conceptual abilities of an African grey parrot. In *Cognitive ethology,* ed. C. A. Ristau. Hillsdale, N.J.: Lawrence Erlbaum Associates.

———. 1999. *The Alex studies: Cognitive and communicative abilities of grey parrots.* Cambridge: Harvard University Press.

Pepperberg, I. M., and M. V. Brezinsky. 1991. Acquisition of a relative class concept by an African grey parrot *(Psittacus erithacus):* Discriminations based on relative size. *J. Comp. Psychol.* 105:286–94.

Pereira, M. E., and J. M. Macedonia. 1991. Ringtailed lemur anti-predator calls denote predator class, not response urgency. *Anim. Behav.* 41:543–44.

Perrett, D. I., E. T. Rolls, and W. Caan. 1982. Visual neurones responsive to faces in monkey temporal cortex. *Exp. Brain Res.* 47:329–42.

Petersen, M. R., M. D. Beecher, S. R. Zoloth, S. Green, P. Marler, D. B. Moody, and W. C. Stebbins. 1984. Neural lateralization of vocalizations by Japanese macaques: Communicative significance is more important than acoustic structure. *Behav. Neurosci.* 96:779–90.

Petersen, M. R., M. D. Beecher, S. R. Zoloth, D. B. Moody, and W. C. Stebbins. 1978. Neural lateralization of species-specific vocalizations by Japanese macaques *(Macaca fuscata). Science* 202:324–27.

Pfungst, O. 1911. *Clever Hans (The horse of Mr. von Osten): A contribution to experimental animal and human psychology.* New York: Henry Holt. Reprinted (with preface by R. Rosenthal) New York: Holt, Rinehart & Winston, 1965.

Picton, T. W., and D. T. Stuss. 1984. Event related potentials in the study of speech and language: A critical review. In *Biological perspectives on language,* ed. D. Caplan, A. R. Lecours, and A. Smith. Cambridge: MIT Press.

Pietrewicz, A. T., and A. C. Kamil. 1981. Search images and the detection of cryptic prey: An operant approach. In *Foraging behavior: Ecological, ethological, and psychological approaches,* ed. A. C. Kamil and T. D. Sargent. New York: Garland STPM.

Pilleri, G. 1983. Ingenious tool use by the Canadian beaver *(Castor canadensis)* in captivity. In *Investigations on beavers,* vol. 1. Berne: Brain Anatomy Institute.

Pineda, J. A., S. L. Foote, H. J. Neville, and C. T. Holmes. 1988. Endogenous event-related potentials in monkeys: The role of task relevance, stimulus probability, and behavioral response. *Electroencephalog. and Clinical Neurophysiol.* 70:155–71.

Pinker, S. 1994. *The language instinct.* New York: William Morrow.

Pitcher, T. J. 1986. Functions of shoaling behaviour in teleosts. In *The behavior of teleost fishes,* ed. T. J. Pitcher. Baltimore: John Hopkins University Press.

Pitcher, T. J., D. A. Green, and A. E. Magurran. 1986. Dicing with death: Predator inspection behaviour in minnow shoals. *J. Fish Biol.* 28:439–48.

Pitcher, T. J., and A. C. House. 1987. Foraging rules for group feeders: Area copying depends upon food density in shoaling goldfish. *Ethology* 76:161–67.

Polakoff, L. M. 1998. Dancing bees and the language controversy. *Integrative Biol.*1:187–94.

Poole, J., and D. G. Lander. 1971. The pigeon's concept of pigeon. *Psychonomic Sci.* 25:157–58.

Popper, K. R. 1987. Natural selection and the emergence of mind. In *Evolutionary epistemology, rationality, and the sociology of knowledge,* ed. G. Radnitzky and W. W. Bartley III. LaSalle, Ill.: Open Court.

Porter, F. L. 1979. Social behavior in the leaf-nosed bat *Carollia perspicillata* II: Social communication. *Z. Tierpsychol.* 50:1–8.

Povinelli, D. J. 1989. Failure to find self-recognition in Asian elephants *(Elephas maximus)* in contrast to their use of mirror cues to discover hidden food. *J. Comp. Psychol.* 103:122–31.

Premack, D. 1976. *Intelligence in ape and man.* Hillsdale, N.J.: Lawrence Erlbaum Associates.

———. 1980. Representational capacity and accessibility of knowledge: The case of chimpanzees. In *Language and learning: The debate between Jean Piaget and Noam Chomsky,* ed. M. Piattelli-Palmarini. Cambridge: Harvard University Press.

———. 1983a. Animal cognition. *Ann. Rev. Psychol.* 34:351–62.

———. 1983b. The codes of man and beasts. *Behav. Brain Sci.* 6:125–67.

———. 1986. *Gavagai! or the future history of the animal language controversy.* Cambridge: MIT Press. (Reprinted with slight changes from *Cognition* 19:207–96,1985.)

Premack, D., and A. J. Premack. 1983. *The mind of an ape.* New York: Norton.

Premack, D., and G. Woodruff. 1978. Does the chimpanzee have a theory of mind? *Behavioral and Brain Sciences* 4:515–26 and commentaries.

Pruett-Jones, S., and M. Pruett-Jones. 1994. Sexual competition and courtship disruption: Why do male bowerbirds destroy each other's nests? *Anim. Behav.* 47:607–20.

Pryor, K. 1975. *Lads before the wind.* New York: Harper and Row.

Pryor, K., R. Haag, and J. O'Reilly. 1969. The creative porpoise: Training for novel behavior. *J. Exp. Anal. Behav.* 12:653–61.

Pryor, K., and I. K. Schallenberger. 1991. Social structure in spotted dolphins *(Stenella attenuata)* in the tuna purse seine fishery in the eastern tropical

Pacific. In *Dolphin societies: Discoveries and puzzles,* ed. K. Pryor and K. S. Norris. Berkeley: University of California Press.

Putnam, H. 1988. *Representation and reality.* Cambridge: MIT Press.

Pyke, G. H. 1979. Optimal foraging in bumblebees: Rule of movement between flowers within inflorescences. *Anim. Behav.* 27:1167–81.

Radner, D., and M. Radner. 1989. *Animal consciousness.* Buffalo, N.Y.: Prometheus Books.

Raichle, M. E. 1998. Behind the scenes of functional brain imaging: A historical and physiological perspective. *Proc. Natl. Acad. Sci.* 95:765–72.

Rand, A. L. 1956. Foot-stirring as a feeding habit of wood ibis and other birds. *Amer. Midl. Nat.* 55:96–100.

Rasa, O. A. 1973. Prey capture, feeding techniques, and their ontogeny in the African dwarf mongoose, *Helogale undulata rufula. Z. Tierpsychol.* 32:449–88.

———. 1976. Invalid care in the dwarf mongoose *(Helogale undulata rufula). Z. Tierpsychol.* 42:337–42.

———. 1983. A case of invalid care in wild dwarf mongooses. *Z. Tierpsychol.* 62:235–40.

Real, L. A. 1981. Uncertainty and pollinator-plant interactions: The foraging behavior of bees and wasps on artificial flowers. *Ecology* 62:20–26.

———. 1987. Objective benefit versus subjective perception in the theory of risk-sensitive foraging. *American Naturalist* 130:399–411.

Real, L. A., J. R. Ott, and E. Silverfine. 1982. On the trade-off between the mean and variance in foraging: Effects of spatial distribution and color preference. *Ecology* 63:1617–23.

Real, L. A., ed. 1994. *Behavioral mechanisms in evolutionary ecology.* Chicago: University of Chicago Press.

Reid, D. G., S. M. Herrero, and T. E. Code. 1988. River otters as agents of water loss from beaver ponds. *J. Mammalogy* 69:100–7.

Reid, J. B. 1982. Tool-use by a rook *(Corvus frugilegus)* and its causation. *Anim. Behav.* 30:1212–16.

Reiss, D., B. McCowan, and L. Marino. 1997. Communicative and other cognitive characteristics of bottlenose dolphins. *Trends in Cog. Sci.* 1:140–45.

Reynolds, J. D. 1985. Sandhill crane use of nest markers as cues for predation. *Wilson Bull.* 97:106–8.

Rice, C. E. 1967a. Human echo perception. *Science* 155:656–64.

———. 1967b. The human sonar system. In *Animal sonar systems,* ed. R. G. Busnel. Jouy-en-Josas, France: Laboratoire de Physiologie Acoustique.

Richard, P. B. 1960a. Un parc à castors dans la région de Paris. *Mammalia* 24:545–55.

———. 1960b. Essai préliminaire sur l'adaptation à des problèmes simples chez le castor *(Castor fiber). J. de Psychologie normale et pathologique* 57 (4): 421–30.

———. 1967. Le déterminisme de la construction des barrages chez le castor du Rhône. *La Terre et la Vie* 114 (4): 338–470.

———. 1980. *Les castors.* Poitiers: Balland.

———. 1983. Mechanisms and adaptation in the constructive behaviour of the beaver *(C. fiber L.). Acta Zoologica Fennica* 174:150–58.

Richards, D. G., J. P. Wolz, and L. M. Herman. 1984. Vocal mimicry of computer-generated sounds and vocal labeling of objects by the bottlenosed dolphin *Tursiops truncatus. J. Comp. Psychol.* 90:10–28.

Richards, R. J. 1987. *Darwin and the emergence of evolutionary theories of mind and behavior.* Chicago: University of Chicago Press.

Richner, H., and P. Heeb. 1995. Is the information center hypothesis a flop? *Advances in the Study of Behav.* 24:1–45.

Richner, H., and C. Marclay. 1991. Evolution of avian roosting behaviour: A test of the information centre hypothesis and of a critical assumption. *Anim. Behav.* 41:433–38.

Riedman, M. L. 1990. *Sea otters.* Monterey, Calif.: Monterey Bay Aquarium.

Riedman, M. L., and J. A. Estes. 1990. The sea otter *(Enhydra lutris):* Behavior, ecology and natural history. Biological Report no. 90(14). Washington, D.C.: U.S. Fish and Wildlife Service.

Ristau, C. A. 1983a. Intentionalist plovers or just dumb birds? *Behav. and Brain Sci.* 6:373–75.

———. 1983b. Language, cognition, and awareness in animals? In *The role of animals in bio-medical research,* ed. J. A. Sechzer. *Annals N.Y. Acad. Sci.* 406:170–86.

———. 1983c. Symbols and indication in apes and other species? *J. Exp. Psychol.: General* 112:498–507.

———, ed. 1991. *Cognitive ethology: The minds of other animals.* Hillsdale, N.J.: Lawrence Erlbaum Associates.

Ristau, C. A., and D. Robbins. 1982. Language in the great apes: A critical review. *Advances in the Study of Behav* 12:142–255.

Ritchie, B. G., and D. M. Fragaszy. 1988. Capuchin monkey *(Cebus apella)* grooms her infant's wounds with tools. *Amer. J. Primatol.* 16:345–48.

Rizzolatti, G., and M. A. Arbib. 1998. Language within our grasp. *Trends in Neurosci.* 21:188–94.

Roberts, M. 1997. *The man who listens to horses.* New York: Random House.

Roberts, W. A. 1998. *Principles of animal cognition.* New York: McGraw-Hill.

Roberts, W. A., and D. S. Mazmanian. 1988. Concept learning at different levels of abstraction by pigeons, monkeys, and people. *J. Exp. Psychol.: Anim. Behav. Processes* 14:247–60.

Roitblat, H. L. 1987. *Introduction to comparative cognition.* New York: Freeman.

Roitblat, H. L., T. G. Bever, and H. S. Terrace, eds. 1984. *Animal cognition.* Hillsdale, N.J.: Lawrence Erlbaum Associates.

Roitblat, H. L., L. M. Herman, and P. E. Nachtigall, eds. 1993. *Language and communication: Comparative perspectives.* Hillsdale, N.J.: Lawrence Erlbaum Associates.

Roitblat, H. L., and J.-A. Meyer, eds. 1995. *Comparative approaches to cognitive science.* Cambridge: MIT Press.

Rollin, B. E. 1989. *The unheeded cry: Animal consciousness, animal pain, and science.* Oxford: Oxford University Press.

———. 1990. How the animals lost their minds: Animal mentation and scientific ideology. In *Interpretation and explanation in the study of animal behavior,* ed. M. Bekoff and D. Jamieson. Boulder, Colo.: Westview Press.

Rose, S., ed. 1998. *From brains to consciousness? Essays on the new sciences of the mind.* Princeton: Princeton University Press.

Rosen, B. R., R. L. Buckner, and A. M. Dale. 1998. Event-related functional MRI: Past, present, and future. *Proc. Nat. Acad. Sci.* 95:773–80.

Rosin, R. 1978. The honey-bee "language" controversy. *J. Theoret. Biol.* 72:589–602.

———. 1980. The honey bee "dance language" hypothesis and the foundation of biology and behavior. *J. Theoret. Biol.* 87:457–81.

———. 1984. Further analysis of the honey bee "dance language" controversy I: Presumed proofs for the "dance language" hypothesis by Soviet scientists. *J. Theoret. Biol.* 107:417–42.

———. 1988. Do honey bees still have a "dance language"? *Amer. Bee J.* 128:267–68.

Ross, D. 1971. Protection of hermit crabs *(Dardanus* spp.) from octopus by commensal sea anemones *(Calliactus* spp.) *Nature* 230:401–2.

Routley, R. 1981. Alleged problems in attributing beliefs and intentionality to animals. *Inquiry* 24:385–417.

Rowley, I. 1962. "Rodent-run" distraction display by a passerine: The superb blue wren *Malurus cyaneus* (L) *Behaviour* 19:170–76.

Russell, E. S. 1935. Valence and attention in animal behaviour. *Biotheoretica* 1:91–99.

Russow, L.-M. 1987. Stich on the foundations of cognitive psychology. *Synthese* 70:401–13.

Ryden. H. 1989. *Lily pond.* New York: William Morrow.

Sahraie, A., L. Weisskrantz, J. L. Barbur, A. Simmons, S. C. R. Williams, and M. J. Brammer. 1997. Pattern of neuronal activity associated with conscious and unconscious processing of visual signals. *Proc. Natl. Acad. Sci.* 94:9406–11.

Sandeman, D. C., J. Tautz, and M. Lindauer. 1996. Transmission of vibration across honeycombs and its detection by bee leg receptors. *J. Exptl. Biol.* 199:2585–94.

Sanford, D. H. 1986. Review of *From folk psychology to cognitive science: The case against belief. Philos. and Phenomenological Research* 47:149–54.

Sargent, T. D. 1981. Antipredator adaptations of underwing moths. In *Foraging behavior: Ecological, ethological and psychological approaches,* ed. A. C. Kamil and T. D. Sargent. New York: Garland.

Savage-Rumbaugh, E. S. 1986. *Ape language: From conditioned response to symbol.* New York: Columbia University Press.

Savage-Rumbaugh, S., and R. Lewin. 1994. *Kanzi: The ape at the brink of the human mind.* New York: Wiley.

Savage-Rumbaugh, S., M. A. Romski, W. D. Hopkins, and R. A. Sevcik. 1989. Symbol acquisition and use by *Pan troglodytes, Pan paniscus, Homo sapiens.* In *Understanding chimpanzees,* ed. P. G. Heltne and L. A. Marquardt. Cambridge: Harvard University Press.

Savage-Rumbaugh, S., S. G. Shanker, and T. J. Taylor. 1998. *Apes, language, and the human mind.* New York: Oxford University Press.

Savory, T. H. 1959. *Instinctive living: A study of invertebrate behaviour.* London: Pergamon.

Sayigh, L. S., P. L. Tyack, R. S. Wells, and M. D. Scott. 1990. Signature whistles of free-ranging bottlenosed dolphins *Tursiops truncatus:* Stability of mother-offspring comparisons. *Behav. Ecol. Sociobiol.* 26:247–60.

Schacter, D. L. 1996. *Searching for memory: The brain, the mind, and the past.* New York: Basic Books.

Schacter, D. L., C. Y.-P. Chiu, and K. N. Ochsner. 1993. Implicit memory: A selective review. *Ann. Rev. Neurosci.* 16:159–82.

Schaller, G. B. 1972. *The Serengeti lion: A study of predator-prey relations.* Chicago: University of Chicago Press.

Scheel, D., and C. Packer. 1991. Group hunting behaviour of lions: A search for cooperation. *Anim. Behav.* 41:697–709.

Schmidt-Koenig, K. 1979. *Avian orientation and navigation.* New York: Academic Press.

Schnitzler, H.-U., D. Menne, R. Kober, and K. Heblich. 1983. The acoustical image of fluttering insects in echolocating bats. In *Neuroethology and behavioral physiology: Roots and growing points,* ed. F. Huber and H. Markl. New York: Springer.

Schram, F. R. 1986. *Crustacea.* Oxford: Oxford University Press.

Schultz, D. 1975. *A history of modern psychology.* 2d ed. New York: Academic Press.

Schusterman, R. J., and R. Gisiner. 1988a. Artificial language comprehension in dolphins and sea lions: The essential cognitive skills. *Psychol. Rec.* 38:311–48.

———. 1988b. Animal language research: Marine mammals re-enter the controversy. In *Intelligence and evolutionary biology,* ed. H. J. Jerison and I. Jerison. New York: Springer.

———. 1989. Please parse the sentence: Animal cognition in the procrustean bed of linguistics. *Psychol. Rec.* 39:3–18.

Schusterman, R. J., and K. Krieger. 1984. California sea lions are capable of semantic comprehension. *Psychol. Rec.* 34:3–23.

Schwartz, B., and H. Lacey. 1982. *Behaviorism, science, and human nature.* New York: Norton.

Searle, J. R. 1984a. *Intentionality: An essay in the philosophy of mind.* New York: Cambridge University Press.

———. 1984b. Intentionality and its place in nature. *Synthese* 61:3–16.

———. 1990a. Consciousness, explanatory inversion, and cognitive science. *Behav. Brain Sci.* 13:585–642 (including commentaries).
———. 1990b. Is the brain's mind a computer program? *Scientific Amer.* 262 (1): 326–31.
———. 1992. *The rediscovery of the mind.* Cambridge: MIT Press.
———. 1998. How to study consciousness scientifically. *Brain Research Rev.* 26:379–87.
Sebeok, T. A., and R. Rosenthal. 1981. The Clever Hans phenomenon: Communication with horses, whales, apes, and people. *Ann. N.Y. Acad. Sci.* 364:1–311.
Seeley, T. D. 1977. Measurement of nest cavity volume by the honey bee *(Apis mellifera). Behav. Ecol. Sociobiol.* 2:201–27.
———. 1985. *Honeybee ecology.* Princeton: Princeton University Press.
———. 1986. Social foraging by honey bees: How colonies allocate foragers among patches of flowers. *Behav. Ecol. Sociobiol.* 19:343–54.
———. 1989a. Social foraging in honey bees: How nectar foragers assess their colony's nutritional status. *Behav. Ecol. Sociobiol.* 24:181–99.
———. 1989b. The honey bee colony as a superorganism. *Amer. Scientist* 77:546–53.
———. 1992. The tremble dance of the honeybee: Messages and meanings. *Beh. Ecol. and Sociobiol.* 31:375–83.
———. 1995. *The wisdom of the hive: The social physiology of honey bee colonies.* Cambridge: Harvard University Press.
Seeley, T. D., and S. C. Buhrman. 1999. Group decision making in swarms of honey bees. *Behav. Ecol. Sociobiol.* 45:19–31.
———. Nest-site selection in honey bees: How well do swarms implement the "best-of-N" decision rule? Typescript.
Seeley, T. D., and W. H. Towne. 1992. Tactics of dance choice in honey bees: Do foragers compare dances? *Behav. Ecol. Sociobiol.* 30:59–69.
Seibt, U. 1982. Zahlbegriff und Zählverhalten bei Tieren: Neue Versuche und Deutungen. *Z. Tierpsychol.* 60:325–41.
Seidenberg, M. S., and L. A. Petitto. 1979. Signing behavior in apes: A critical review. *Cognition* 7:177–215.
Seyfarth, R. M. 1984. What the vocalizations of monkeys mean to humans and what they mean to the monkeys themselves. In *The meaning of primate signals,* ed. R. Harre and V. Reynolds. Cambridge: Cambridge University Press.
———. 1987. Vocal communication and its relation to language. In *Primate societies,* ed. B. B. Smuts et al. Chicago: University of Chicago Press.
Seyfarth, R. M., and D. L. Cheney. 1994. The evolution of social cognition in primates. In *Behavioral mechanisms in evolutionary ecology,* ed. L. A. Real. Chicago: University of Chicago Press.
Seyfarth, R. M., and D. L. Cheney. In press. Social self-awareness in monkeys. *Am. Zool.*
Seyfarth, R. M., D. L. Cheney, and P. Marler. 1980. Vervet monkey alarm calls: Evidence for predator classification and semantic communication. *Anim. Behav.* 28:1070–94.

Shanks, D. R., and M. F. St. John. 1994. Characteristics of dissociable human learning systems. *Beh. and Brain Sci.* 17:367–447 (including commentaries).
Sheets-Johnstone, M. 1998. Consciousness: A natural history. *J. Consciousness Studies* 5:260–94.
Sheinberg, D. L., and N. K. Logothetis. 1997. The role of temporal cortical areas in perceptual organization. *Proc. Natl. Acad. Sci.* 94:3408–13.
Sherry, D. F. 1984. Food storage by black-capped chickadees: Memory for the location and contents of caches. *Anim. Behav.* 32:451–64.
Sherry, D. F., and B. G. Galef. 1984. Cultural transmission without imitation: Milk bottle opening by birds. *Anim. Behav.* 32:937–38.
———. 1990. Social learning without imitation: More about milk bottle opening by birds. *Anim. Behav.* 40:987–89.
Sherry, D. F., J. R. Krebs, and R. J. Cowie. 1981. Memory for the location of stored food in marsh tits. *Anim. Behav.* 29:1260–66.
Shettleworth, S. J. 1972. Constraints on learning. In *Advances in the study of behavior,* vol. 4, ed D. S. Lehrman, R. A. Hinde, and E. Shaw. New York: Academic Press.
———. 1998. *Cognition, evolution, and behavior.* New York: Oxford University Press.
Shettleworth, S. J., and J. R. Krebs. 1982. How marsh tits find their hoards: The roles of site preference and spatial memory. *J. Exptl. Psychol.: Anim. Behav. Processes* 18:219–35.
Shinoda, A., and M. E. Bitterman. 1987. Analysis of the overlearning-extinction effect in honeybees. *Anim. Learning and Behav.* 15:93–96.
Short, L. L., and J. F. M. Horne. 1985. Behavioral notes on the nest-parasitic Afrotropical honeyguides (Aves: Indicatoidae). *Amer. Museum Novitates* 2825:1–46.
Shuster, G., and P. W. Sherman. 1998. Tool use by naked mole-rats. *Anim. Cog.* 1:71–74.
Siegel, R. K., and W. K. Honig. 1970. Pigeon concept formation: Successive and simultaneous acquisition. *J. Exptl. Anal. Behav.* 13:385–90.
Siewert, C. P. 1998. *The significance of consciousness.* Princeton: Princeton University Press.
Silverman, P. S. 1983. Attributing mind to animals: The role of intuition. *J. Social Biol. Struct.* 6:223–47.
Simmons, K. E. L. 1955. The nature of the predator-reactions of waders towards humans: With special reference to the role of the aggressive-, escape-, and brooding-drives. *Behaviour* 8:130–73.
Sisson, R. E. 1974. Aha! It really works! *Natl. Geog.* 144:142–47.
Skinner, B. F. 1960. Pigeons in a pelican. *Amer. Psychol.* 15:28–37.
———. 1974. *About behaviorism.* New York: Random House.
Skutch, A. F. 1954. The parental strategems of birds. *Ibis* 96:544–64 and 97:118–42.
———. 1976. *Parent birds and their young.* Austin: University of Texas Press.
———. 1996. *The minds of birds.* College Station: Texas A&M University Press.

Slobodchikoff, C. N., J. Kiriazis, C. Fischer, and E. Creef. 1991. Semantic information distinguishing individual predators in the alarm calls of Gunnison's prairie dogs. *Anim. Behav.* 42:713–19.

Smith, A. F. 1986. An "informational" perspective on manipulation. In *Deception: Perspectives on human and nonhuman deceit,* ed. R. W. Mitchell and N. S. Thompson. Albany: SUNY Press.

Smith, W. J. 1991. Animal communication and the study of cognition. In *Cognitive ethology: The minds of other animals,* ed. C. A. Ristau. Hillsdale, N.J.: Lawrence Erlbaum Associates.

Smitha, B., J. Thakar, and M. Watve. 2000. Theory of mind and appreciation of geometry by the small green bee eater *(Merops orientalis)*. *Current Sci.* 76:574–77.

Smolker, R. A., J. Mann, and B. Smuts. 1993. Use of signature whistles during separations and reunions by wild bottlenosed dolphin mothers and infants. *Behav. Ecol. Sociobiol.* 33:393–402.

Snowdon, C. T. 1991. Review of *Cognitive ethology. Science* 251:813–14.

Sober, E. 1983. Mentalism and behaviorism in comparative psychology. In *Comparing behavior: Studying man studying animals.* Hillsdale, N.J.: Lawrence Erlbaum Associates.

Sommer, W. J., J. Matt, and H. Leuthold. 1990. Consciousness of attention and expectancy as reflected in event-related potentials and reaction times. *J. Exptl. Psychol.: Language, Memory, and Cog.* 16:902–15.

Sonerud, G. A. 1988. To distract display or not: Grouse hens and foxes. *Oikos* 51:233–37.

Spelke, E. S. 1994. Initial knowledge: Six suggestions. *Cognition* 50:431–45.

Srinivasan, M., and S.-W. Zhang. 1998. Probing perception in a miniature brain: Pattern recognition and maze navigation in honeybees. *Zoology: Analysis of Complex Systems* 101:246–59.

Stander, P. 1992. Cooperative hunting in lions: The role of the individual. *Behav. Ecol. Sociobiol.* 29:445–54.

Stanford, C. B. 1998. *Chimpanzees and red colobus: The ecology of predator and prey.* Cambridge: Harvard University Press.

Stent, G. S. 1987. The mind-body problem. *Science* 236:990–92.

Stevens, T. S. 1888. Notes on an intelligent parrot. *J. Trenton Nat. Hist. Soc.* no. 3:347–56.

Stich, S. P. 1983. *From folk psychology to cognitive science: The case against belief.* Cambridge: MIT Press.

Stoddard, P. K. 1999. Predation enhances complexity in the evolution of electric fish signals. *Nature* 400:254–56.

Stoyva, J., and J. Kamiya. 1968. Electrophysiological studies of dreaming as the prototype of a new strategy in the study of consciousness. *Psychol. Rev.* 75:192–205.

Straub, R. O., and H. S. Terrace. 1981. Generalization of serial learning in the pigeon. *Anim. Learn. Behav.* 9:454–68.

Struhsaker, T. T. 1967. *The red colobus monkey.* Chicago: University of Chicago Press.

Stuss, D. T., T. W. Picton, and A. M. Cerri. 1986. Searching for the name of pictures: An event-related potential study. *Psychophysiol.* 23:215–23.

Suarez, S. D., and G. G. Gallup Jr. 1981. Self-recognition in chimpanzees and orangutans, but not gorillas. *J. Hum. Evol.* 10:175–88.

Sugiyama, Y., and J. Koman. 1979. Tool-using and -making behavior in wild chimpanzees at Bossou, Guinea. *Primates.* 20:513–24.

Sullivan, M. G. 1979. Blue grouse hen–black bear confrontation. *Can. Field-Nat.* 93:200.

Supa, M., M. Cotzin, and K. M. Dallenbach. 1944. "Facial vision": The perception of obstacles by the blind. *Amer. J. Psychol.* 57:133–83.

Switzer, P. V., and D. A. Cristol. 1999. Avian prey-dropping behavior I. The effects of prey characteristics and prey loss. *Behav. Ecol.* 10:213–19.

Tautz, J. 1996. Honeybee waggle dance: Recruitment success depends on the dance floor. *J. Exptl. Biol.* 199:1375–81.

———. 1998. Communication in social bees: New approaches. *Abstracts of the 26th Göttingen Neurobiology Conference.* March 26–29, 1998:302–6.

Tayler, C. K., and G. S. Saayman. 1973. Imitative behavior of Indian Ocean bottlenosed dolphins *(Tursiops aduncus)* in captivity. *Behaviour* 44:286–98.

Taylor, J. G. 1999. *The race for consciousness.* Cambridge: MIT Press.

Terrace, H. S. 1979. *Nim.* New York: Knopf.

———. 1984. Animal cognition. In *Animal cognition,* ed. H. L. Roitblat, T. G. Bever, and H. S. Terrace. Hillsdale, N.J.: Lawrence Erlbaum Associates.

———. 1987. Thoughts without words. In *Mindwaves: Thoughts on intelligence, identity, and consciousness,* ed. C. Blakemore and S. Greenfield. Oxford: Basil Blackwell.

Terrace, H. S., L. A. Petitto, and T. G. Bever. 1979. Can an ape create a sentence? *Science* 206:891–902.

Thatcher, R. W., and E. R. John. 1977. *Foundation of cognitive processes.* Hillsdale, N.J.: Lawrence Erlbaum Associates.

Thomas, R. K., and B. Lorden. 1993. Numerical competence in animals: A conservative view. In *The development of numerical competence: Animal and human models,* ed. Boysen, S. T., and E. J. Capaldi. Hillsdale, N.J.: Lawrence Erlbaum Associates.

Thorpe, W. H. 1963. *Learning and instinct in animals,* 2d ed. Cambridge: Harvard University Press.

———. 1972. Duetting and antiphonal song in birds. *Behaviour,* Supp. 18.

Thorpe, W. H., and M. E. W. North. 1965. Origin and significance of the power of vocal imitation: With special reference to the antiphonal singing of birds. *Nature* 208:219–22.

Thouless, C. R., J. H. Fabshawe, and B. C. R. Bertram. 1989. Egyptian vultures *Neophron percnopterus* and ostrich *Struthio camelus* eggs: The origins of stone-throwing behaviour. *Ibis* 131:9–15.

Tinbergen, J. M. 1981. Foraging decisions in starlings *(Sturnus vulgaris). Ardea* 69:1–67.

Tinklepaugh, O. L. 1928. An experimental study of representative factors in monkeys. *J. Comp. Psychol.* 8:197–236.

Todt, D. 1975. Social learning of vocal patterns and models of their application in the grey parrot. *Z. Tierpsychol.* 39:178–88.

Tolman, E. C. 1932. *Purposive behavior in animals and men.* New York: Appleton-Century.

———. 1937. The acquisition of string-pulling by rats—conditioned reflex or sign gestalt? *Psychol. Rev.* 44:195–211.

———. 1959. Principles of purposive behavior. In *Psychology: A study of a science,* ed. S. Koch. *Study I. Conceptual and systematic.* Vol. 2, *General systematic formulations: Learning and special processes.* New York: McGraw-Hill.

Tomasello, M., and J. Call. 1997. *Primate cognition.* New York: Oxford University Press.

Tong, F., K. Nakayama, J. Thomas Vaughan, and N. Kanwisher. 1998. Binocular rivalry and visual awareness in human extrastriate cortex. *Neuron* 21:753–59.

Tononi, G., and G. M. Edelman. 1998. Consciousness and complexity. *Science* 282:1846–51.

Tononi, G., R. Sinivasan, D. P. Russell, and G. N. Edelman. 1998. Investigating neural correlates of conscious perception by frequency-tagged neuromagnetic responses. *Proc. Nat. Acad. Sci.* 95:3198–3203.

Topoff, H. 1977. The pit and the antlion. *Nat. Hist.* 86 (4): 65–71.

Towne, W. F. 1985. Acoustic and visual cues during the waggle dance of four honeybee species. *Behav. Ecol. Sociobiol.* 16:185–87.

Towne, W. F., and J. L. Gould. 1988. The spatial precision of the honey bees—dance communication. *J. Insect Behav.* 1:129–55.

Towne, W. F., and W. H. Kirchner. 1989. Hearing in honey bees: Detection of air-particle oscillation. *Science* 244:686–88.

Tyack, P. L. 1986. Whistle repertoires of two bottlenosed dolphins, *Tursiops truncatus:* Mimicry of signature whistles? *Behav. Ecol. Sociobiol.* 18:251–57.

———. 1993. Animal language research needs a broader comparative and evolutionary framework. In *Language and communication: Comparative perspectives,* ed. H. L. Roitblat, L. M. Herman, and P. E. Nachtigall. Hillsdale, N.J.: Lawrence Erlbaum Associates.

Tyack, P. L., and L. S. Sayigh. 1997. Vocal learning in cetaceans. In *Social influences on vocal development,* ed. C. Snowdon and M. Hausberger. New York: Cambridge University Press.

Umiker-Sebeok, J., and T. A. Seboek. 1981. Clever Hans and smart simians: The self-fulfilling prophecy and kindred methodological pitfalls. *Anthropos, Internationale Z. Völker- und Sprachenkunde.* 76:89–165.

Van der Steen, W. J. 1997. Fighting against umbrellas: An essay on consciousness. In *Animal consciousness and animal ethics: Perspectives from The Netherlands,* ed. M. Doll, S. Kasanmoentalib, S. Lijmbach, E. Rivas, and R. van den Bos. Assen, The Netherlands: Van Gorcum.

Vander Wall, S. B. 1990. *Food hoarding in animals.* Chicago: University of Chicago Press.

Vander Wall, S. B., and R. P. Balda. 1981. Ecology and evolution of food storing behavior in conifer-seed-caching corvids. *Z. Tierpsychol.* 56:217–42.

Van Eerden, M. R., and B. Voslamber. 1995. Mass fishing by cormorants *Phalacrocorax carbo sinensis* at Lake Ijsselmeer, The Netherlands: A recent and successful adaptation to a turbid environment. *Ardea* 83:199–212.

Vauclair, J. 1992. Psychologie cognitive et représentations animales. In *La Représentation animale: Représentation de représentation,* ed. J. Gervet, P. Livet, and A. Tête. Nancy: Presses Universitaires de Nancy.

———. 1996. *Animal cognition: An introduction to modern comparative psychology.* Cambridge: Harvard University Press.

———. 1997. Mental states in animals: Cognitive ethology. *Trends in Cog. Sci.* 1:35–39.

Vaughan, W., Jr., and S. L. Greene. 1984. Pigeon visual memory capacity. *J. Exptl. Psychol.: Anim. Behav. Processes* 10:256–71.

Vellenga, R. E. 1970. Behaviour of the male satin bower-bird at the bower. *Aust. Bird Bander* 8:3–11.

———. 1980. Distribution of bowers of the Satin Bowerbird at Leura, NSW, with notes on parental care, development and independence of young. *Emu* 80:97–102.

Velmans, M. 1991. Is human information processing conscious? *Behav. Brain Sci.* 14:651–726 (including commentaries).

Verleger, R. 1988. Event-related potentials and cognition: A critique of the context updating hypothesis and an alternative interpretation of P3. *Behav. Brain Sci.* 11:343–427 (including commentaries).

Vince, M. A. 1961. String pulling in birds III: The successful response in greenfinches and canaries. *Behaviour* 17:103–29.

Visscher, P. K., and S. Camazine. 1999a. The mystery of swarming bees: From individual behaviors to collective decisions. In *Information processing in social insects,* ed. C. Detrain, J. L. Deneubourg, and J. M. Pasteels. Basel: Birkhäuser.

———. 1999b. Collective decisions and cognition in bees. *Nature* 397:400.

Visscher, P. K., J. Shepardson, L. McCart, and S. Camazine. 1999. Vibration signal modulates the behavior of house-hunting honey bees *(Apis mellifera). Ethology* 105:759–69.

Vogel, G. 1999. Chimpanzees in the wild show stirrings of culture. *Science* 284:2070–73.

Voslamber, B., M. Platteeuw, and M. R. Van Eerden. 1995. Solitary foraging in sand pits by breeding cormorants, *Phalacrocorax carbo sinensis:* Does specialized knowledge about fishing sites and fish behaviour pay off? *Ardea* 83:213–22.

Vries, H. de, and J. B. Biesmeijer. 1998. Modelling collective foraging by means of individual behaviour rules in honey-bees. *Behav. Ecol. Sociobiol.* 44:109–24.

Waal, F. de. 1982. *Chimpanzee politics.* London: Jonathan Cape.

———. 1986. Deception in the natural communication of chimpanzees. In

Deception: Perspectives on human and nonhuman deceit, ed. R. W. Mitchell and N. S. Thompson. Albany: SUNY Press.

———. 1989. *Peacemaking among primates.* Cambridge: Harvard University Press.

———. 1990. Do rhesus mothers suggest friends to their offspring? *Primates* 31:597–600.

———. 1996. *Good natured: The origins of right and wrong in humans and other animals.* Cambridge: Harvard University Press.

Walker, S. 1983. *Animal thought.* London: Routledge and Kegan Paul.

Wallace, J. B., and F. F. Sherberger. 1975. The larval dwelling and feeding structure of *Macronema transversum* (Walker) *(Trichoptera: Hydropsychidae). Anim. Behav.* 23:592–96.

Walraff, H. G. 1996. Seven theses on pigeon homing deduced from empirical findings. *J. Exptl. Biol.* 199:105–11.

Wallman, J. 1992. *Aping language.* New York: Oxford University Press.

Walsh, J. F., J. Grunewald, and B. Grunewald. 1985. Green-backed herons *(Butorides striatus)* possibly using a lure and using apparent bait. *J. Ornithol.* 126:439–42.

Walters, J. R. 1990. Anti-predator behavior of lapwings: Field evidence of discriminative abilities. *Wilson Bull.* 102:49–70.

Walther, E. R. 1969. Flight behaviour and avoidance of predators in Thompson's gazelle *(Gazella thompsoni* Guenther 1884). *Behaviour* 34:184–221.

Ward, P., and A. Zahavi. 1973. The importance of certain assemblages of birds as "information centres" for food finding. *Ibis* 115:517–34.

Warham, J. 1962. Field notes on Australian bower-birds and cat-birds. *Emu* 62:1–30.

Washburn, M. F. 1917. *The animal mind: A textbook of comparative psychology.* 2d ed. New York: Macmillan.

Wasserman, E. A. 1981. Comparative psychology returns: A review of Hulse, Fowler, and Honig's *Cognitive processes in animal behavior. J. Exptl. Anal. Behav.* 35:248–57.

———. 1982. Further remarks on the role of cognition in the comparative analysis of behavior. *J. Exptl. Anal. Behav.* 38:211–16.

———. 1983. Is cognitive psychology behavioral? *Psychol. Rec.* 33:6–11.

———. 1984. Animal intelligence: Understanding the minds of animals through their behavioral "ambassadors." In *Animal cognition,* ed. H. L. Roitblat, T. G. Bever, and H. S. Terrace. Hillsdale, N.J.: Lawrence Erlbaum Associates.

———. 1985. Comments on *Animal Thinking. Amer. Scientist* 73:6.

———. 1995. The conceptual abilities of pigeons. *Amer. Scientist* 83:246–55.

———. 1997. The science of animal cognition: Past, present, and future. *J. Exptl. Psychol: Anim. Behav. Processes* 23:123–35.

Wasserman, E. A., R. E. Kiedinger, and R. S. Bhatt. 1988. Conceptual behavior in pigeons: Categories, subcategories, and pseudocategories. *J. Exptl. Psychol.: Anim. Behav. Processes* 14:235–46.

Watson, J. B. 1929. *Psychology from the standpoint of a behaviorist.* Philadelphia: Lippincott.

Weber, N. A. 1972. Gardening ants: The attines. *Memoirs Amer. Philos. Soc.* 92:1–146.

Weidenmüller, A., and T. D. Seeley. In press. Imprecision in waggle dances for nearby food sources: Error or adaptation? *Behav. Ecol. Sociobiol.* 46:190–99.

Weiner, J. 1994. *The beak of the finch.* New York: Knopf.

Weiskrantz, L. 1997. *Consciousness lost and found: A neuropsychological exploration.* New York: Oxford University Press.

———, ed. 1985. *Animal intelligence.* New York: Oxford University Press.

———. 1988. *Thought without language.* New York: Oxford University Press.

Wemelsfelder, F. 1997. Investigating the animal's point of view: An inquiry into a subject-based method of measurement in the field of animal welfare. In *Animal consciousness and animal ethics: Perspectives from the Netherlands,* ed. M. Doll, S. Kasanmoentalib, S. Lijmbach, E. Rivas, and R. van den Bos. Assen, The Netherlands: Van Gorcum.

Wenner, A. M. 1959. The relationship of sound production during the waggle dance of the honey bee to the distance of the food source. (Abstract.) *Bull. Entomol. Soc. Amer.* 5:142.

———. 1962. Sound production during the waggle dance of the honey bee. *Anim. Behav.* 10:79–95.

———. 1989. Concept-centered versus organism-centered biology. *Amer. Zoologist* 29:1177–97.

Wenner, A. M., and P. H. Wells. 1990. *Anatomy of a controversy: The question of "language" among bees.* New York: Columbia University Press.

———. 2000. Stinging criticism. *Amer. Scientist* 88:3–4. See also reply by J. Niah, 4–5.

Wenner, A. M., P. H. Wells, and F. J. Rohlf. 1967. An analysis of the waggle dance and recruitment in honey bees. *Physiol. Zool.* 40:317–44.

West, M. M., and A. P. King. 1988. Female visual displays affect the development of male song in the cowbird. *Nature* 334:244–46.

Westby, G. W. M. 1988. The ecology, discharge diversity, and predatory behaviour of gymnotiform electric fish in the coastal streams of French Guiana. *Behav. Ecol. Sociobiol.* 22:341–54.

Westergaard, G. C., and D. M. Fragaszy. 1987. The manufacture and use of tools by capuchin monkeys *(Cebus apella). J. Comp. Psychol.* 101:159–68.

Wheeler, W. M. 1930. *Demons of the dust.* New York: Norton.

Whiteley, C. H. 1973. *Mind in action: An essay in philosophical psychology.* London: Oxford University Press.

Whiten, A. 1998. Imitation of the sequential structure of actions by chimpanzees *(Pan troglodytes). J. Comp. Psychol.* 112:270–81.

———, ed. 1991. *Natural theories of mind: Evolution, development, and simulation of everyday mindreading.* Oxford: Basil Blackwell.

Whiten, A., and R. W. Byrne. 1988. Tactical deception in primates. *Behav. Brain Sci.* 11:233–73 (including commentaries).

———, eds. 1997. *Machiavellian intelligence II: Extensions and evaluations.* New York: Cambridge University Press.

Whiten, A., J. Goodall, W. C. McGrew, T. Nishida, V. Reynolds, Y. Sugiyama, E. G. Tutin, R. W. Wrangham, and C. Boesch. 1999. Culture in chimpanzees. *Nature* 399:682–85.

Wiggins, G. B. 1977. *Larvae of North American caddisflies.* Toronto: University of Toronto Press.

Wilcox, R. S., and R. R. Jackson. 1998. Cognitive abilities of Araneophagic jumping spiders. In *Animal cognition in nature: The convergence of psychology and biology in laboratory and field,* ed. R. P. Balda, I. M. Pepperberg, and A. C. Kamil. San Diego: Academic Press.

Wilder, M. B., G. R. Farley, and A. Starr. 1981. Endogenous late positive component of the evoked potential in cats corresponding to the P300 in humans. *Science* 211:605–6.

Wiley, R. H. 1983. The evolution of communication: Information and manipulation. In *Animal behaviour,* ed. T. R. Halliday and P. J. B. Slater, vol. 2: *Communication.* New York: Freeman.

Wilkie, D. M., R. J. Wilson, and S. Kardal. 1989. Pigeons discriminate pictures of a geographical location. *Anim. Learning and Behav.* 17:163–71.

Williams, D. D., A. F. Tavares, and E. Bryant. 1987. Respiratory device or camouflage? A case for the caddisfly. *Oikos* 50:42–52.

Williams, G. C. 1997. *The pony fish's glow, and other cues to plan and purpose in nature.* New York: Basic Books.

Wilson, D. S., K. Coleman, A. B. Clark, and L. Biederman. 1993. The shy-bold continuum in pumpkinseed sunfish *(Lepomis gibbosus):* An ecological study of a psychological trait. *J. Comp. Psychol.* 107:25–60.

Wilsson, L. 1971. Observations and experiments on the ethology of the European beaver *(Castor fiber L.):* A study in the development of phylogenetically adapted behaviour in a highly specialized mammal. *Viltrevy: Swedish Wildlife* 8 (3): 115–266.

Wiltschko, R. 1996. The function of olfactory input in pigeon orientation: Does it provide navigational information or play another role? *J. Exptl. Biol.* 199:113–19.

Winkler, H. 1982. Das Jagdverhalten des Glockenreihers *Egretta ardesica. J. f. Ornithol.* 123:307–14.

Winson, J. 1985. *Brain and psyche: The biology of the unconscious.* Garden City, N.Y.: Doubleday.

Wittenberger, J. F. 1981. *Animal social behavior.* Boston: Duxbury Press.

Woolfenden, J. 1985. *The California sea otter: Saved or doomed?* Pacific Grove, Calif.: Boxwood Press.

Wrangham, R. W., W. C. McGrew, F. de Waal, and P. G. Heltne, eds. 1994. *Chimpanzee cultures.* Cambridge: Harvard University Press.

Wright, A. A., R. G. Cook, J. J. Rivera, S. F. Sands, and J. D. Delius. 1988. Concept learning by pigeons: Matching-to-sample with trial unique video picture stimulation. *Anim. Learning and Behav.* 16:436–44.

Wright, R. V. S. 1972. Imitative learning of a flaked stone technology—the case of an orangutan. *Mankind* 8:296–308.

Wright, W. H. 1909. *The grizzly bear.* New York: Scribner's.

Würsig, B. 1983. The question of dolphin awareness approached through studies in nature. *Cetus* 5:4–7.

Yerkes, R. M. 1925. *Almost human.* New York: Century.

Yerkes, R. M., and A. W. Yerkes. 1929. *The great apes: A study of anthropoid life.* New Haven: Yale University Press.

Yoerg, S. I. 1991. Ecological frames of mind: The role of cognition in behavioral ecology. *Q. Rev. Biol.* 66:287–301.

Yoerg, S. I., and A. C. Kamil. 1991. Prospects for a more cognitive ethology. In *Cognitive ethology: The minds of other animals,* ed. C. A. Ristau. Hillsdale, N.J.: Lawrence Erlbaum Associates.

Zach, R. 1978. Selection and dropping of whelks by northwestern crows. *Behaviour* 67:134–47.

Zahavi, A., and A. Zahavi. 1997. *The handicap principle: A missing piece of Darwin's puzzle.* New York: Oxford University Press.

Zanen, P. O., and R. T. Cardé. 1999. Directional control by male gypsy moths of upwind flight along a pheromone plume in three wind speeds. *J. Comp. Physiol. A* 184:21–36.

Zentall, T. R., D. E. Hogan, C. A. Edwards, and E. Hearst. 1980. Oddity learning in the pigeon as a function of the number of incorrect alternatives. *J. Exptl. Psychol.: Anim. Behav. Proc.* 6:278–99.

Zoloth, S., and S. Green. 1979. Monkey vocalizations and human speech: Parallels in perception? *Brain Behav. Evol.* 16:430–42.

Zoloth, S., R. Petersen, M. D. Beecher, S. Green, P. Marler, D. B. Moody, and W. Stebbins. 1979. Species-specific perceptual processing of vocal sounds by monkeys. *Science* 204:870–73.

Zuberbühler, K., D. L. Cheney, and R. M. Seyfarth. 1999. Conceptual semantics in a nonhuman primate. *J. Comp. Psychol.* 113:33–42.

Zuberbühler, K., R. Noe, and R. M. Seyfarth. 1997. Diana monkeys' long-distance calls: Messages for conspecifics and predators. *Anim. Behav.* 53:589–604.

INDEX